Physical
Science

Physical Science

Second Edition

Bill W. Tillery
Arizona State University

Wm. C. Brown Publishers

Dubuque, Iowa•Melbourne, Australia•Oxford, England

Book Team

Editor *Craig S. Marty*
Developmental Editor *Robert Fenchel*
Production Editor *Kay J. Brimeyer*
Art Editor *Carla Goldhammer*
Photo Editor *Shirley Charley/Lori Gockel*
Permissions Editor *Mavis M. Oeth*

Wm. C. Brown Publishers
A Division of Wm. C. Brown Communications, Inc.

Vice President and General Manager *Beverly Kolz*
National Sales Manager *Vincent R. Di Blasi*
Assistant Vice President, Editor-in-Chief *Edward G. Jaffe*
Director of Marketing *John W. Calhoun*
Marketing Manager *Elizabeth Robbins*
Advertising Manager *Amy Schmitz*
Director of Production *Colleen A. Yonda*
Manager of Visuals and Design *Faye M. Schilling*

Art Manager *Janice Roerig*
Publishing Services Manager *Karen J. Slaght*
Permissions/Records Manager *Connie Allendorf*

Wm. C. Brown Communications, Inc.

Chairman Emeritus *Wm. C. Brown*
Chairman and Chief Executive Officer *Mark C. Falb*
President and Chief Operating Officer *G. Franklin Lewis*
Corporate Vice President, President of WCB Manufacturing *Roger Meyer*

Cover and interior design by David C. Lansdon

Cover photo © Kindra Clineff

Section cover photo credits:
Physics: © James H. Karales/Peter Arnold, Inc.
Chemistry: © Steven Fuller/Peter Arnold, Inc.
Astronomy: © Kent Wood/Peter Arnold, Inc.
Earth Science: © Scott Blackman/Tom Stack & Associates

Printed in the United States of America by Wm. C. Brown Communications, Inc.,
2460 Kerper Boulevard, Dubuque, IA 52001

10 9 8 7 6 5 4 3 2 1

Contents

This is the complete table of contents for *Physical Science, 2/e.* If you are using one of the customized textbooks from our *Foundations of Physical Science Series,* please note that only a portion of this table of contents applies to your text.

Contents

Contents

Contents

Preface

Physical Science is a straightforward, easy-to-read, but substantial introduction to the fundamental behavior of matter and energy. It is intended to serve the needs of nonscience majors who are required to complete one or more physical science courses. It introduces basic concepts and key ideas while providing opportunities for students to learn reasoning skills and a new way of thinking about their environment. No prior work in science is assumed. The language, as well as the mathematics, is as simple as can be practical for a college-level science course.

The second edition was designed to build on the strengths of the first edition by eliminating errors, fine-tuning various details, combining topics where appropriate, and, expanding topical coverage where otherwise indicated for greater flexibility. The physics section was reduced from nine chapters to seven, while the earth science section was expanded from five chapters to eight.

Organization

The book is divided into four sections: physics, chemistry, astronomy, and the earth sciences. With laboratory studies, *Physical Science* contains enough material for a two-semester course. The chapters and sections are flexible: the instructor can determine topic sequence and depth of coverage. *Physical Science* can also serve as a text in a one-semester astronomy and earth science course, or in other combinations.

Special Treatment

Physical Science is based on two fundamental assumptions arrived at as the result of years of experience and observation from teaching the course: (a) that students taking the course often have very limited background and/or aptitude in the natural sci-

ences; and (b) that this type of student will better grasp the ideas and principles of physical science if they are discussed with minimal use of technical terminology and detail. In addition, it is critical that the student be able to see relevant applications of the material to everyday life. Special interest areas such as environmental concerns are not isolated in an arbitrary section or chapter; they are discussed where they occur naturally throughout the text.

Each chapter presents historical background where appropriate, uses everyday examples in developing concepts, and follows a logical flow of presentation. The historical chronology, of special interest to the humanistically inclined nonscience major, serves to humanize the science being presented. The use of everyday examples appeals to the nonscience major, typically accustomed to reading narration, not scientific technical writing, and also tends to bring relevancy to the material being presented. The logical flow of presentation is helpful to students not accustomed to thinking about relationships between what is being read and previous knowledge learned, a useful skill in understanding the physical sciences. Worked examples help students integrate concepts and understand the use of relationships called equations. They also serve as a problem solving model; consequently, special attention is given to *complete* unit work and to the clear, fully expressed use of mathematics. Where appropriate, chapters contain one or more activities that use everyday materials rather than specialized laboratory equipment. These activities are intended to bring the science concepts closer to the world of the student. The activities are supplemental and can be done as optional student activities or as demonstrations.

Pedagogical Devices

Physical Science also contains a number of innovative learning aids. Each chapter begins with an *introductory overview* and a brief *outline* that help students to organize their thoughts for the coming chapter materials. Each chapter ends with a brief *summary* that organizes the main concepts presented, a *summary of equations* (where appropriate) written both with words and with symbols, a list of page-referenced *key terms,* a set of *multiple-choice questions* with nearby answers for immediate correction or reinforcement of major understandings, a set of *thought questions* for discussion or essay answers, and, *two sets of problem exercises* with complete solutions for one set provided in the appendix. The two sets are nearly parallel in early chapters, but they become progressively less so in successive chapters. The set with the solutions provided is intended to be a model to help students through assigned problems in the other set. In trial classroom testing, this approach proved to be a tremendous improvement over the traditional "odd problem answers." The "odd answer only" approach provided students little help in learning problem solving skills, unless it was how to work a problem backward.

Finally, each chapter of *Physical Science* also includes a boxed feature that discusses topics of special human or environmental concern (the use of seat belts, acid rain, and air pollution, for example), topics concerning interesting technological applications (passive solar homes, solar cells, and catalytic converters, for example), or topics on the cutting edge of scientific research (quarks, superstrings, and deep-ocean exploration, for example). All boxed features are informative materials that are supplementary in nature.

Supplementary Materials

Physical Science is accompanied by a variety of supplementary materials, including an instructor's manual for the text, a laboratory manual, an instructor's manual for the laboratory manual, a student study guide, WCB testpak—a computer disk containing multiple choice test items, slides, and overhead transparencies. The laboratory manual, written and classroom tested by the author, presents a selection of traditional laboratory exercises specifically written for the interest and abilities of nonscience majors. When the laboratory manual is used with *Physical Science,* students will have an opportunity to master basic scientific principles and concepts, learn new problem solving and thinking skills, and understand the nature of scientific inquiry from the perspective of hands-on experiences.

The instructor's manual, also written by the text author, provides a chapter outline, an introduction/summary of each chapter, suggestions for discussion and demonstrations, additional multiple choice questions (with answers), and answers and solutions to all end-of-chapter questions and exercises not provided in the text.

The student study guide provides a solid foundation for nonscience students by stressing conceptual understanding as well as techniques for successful problem solving. All the examples and illustrations are new and different from the examples and illustrations of the text, which tends to maintain interest as it adds a new dimension to student understanding of course concepts and skills.

The author has attempted to present an interesting, helpful program that will be useful to both students and instructors. Comments and suggestions about how to do a better job of reaching this goal are welcome. Any comments about the text or other parts of the program should be addressed to:

Bill W. Tillery
Department of Physics and Astronomy
Arizona State University
Tempe, AZ 85287-1504

To the Reader

*T*his text includes a variety of aids to the reader that should make your study of physical science more effective and enjoyable. These aids are included to help you clearly understand the concepts and principles that serve as the foundation of the physical sciences.

Overview

Chapter One provides an overview or orientation to what the study of physical science, in general, and this text in particular, are all about. It discusses the fundamental methods and techniques used by scientists to study and understand the world around us. It also explains the problem solving approach used throughout the text so that you can more effectively apply what you have learned.

Section Introductions

Each section begins with an introduction designed to give you a quick preview of the topics that are covered, of how the topics are related to build your knowledge of them, and of the key things to learn from reading them.

Chapter Outlines

The chapter outline includes all the major topic headings and subheadings within the body of the chapter. It gives you a quick glimpse of the chapter's contents and helps you locate sections dealing with particular topics.

Introductory Overview

Each chapter begins with an introductory overview that is set off in blue type to call attention to it. The overview previews the chapter's contents and what you can expect to learn from reading the chapter. After reading the introduction, browse through the chapter, paying particular attention to topic headings and illustrations so that you get a feel for the kinds of ideas included within the chapter.

Bold-Faced Terms/Italicized

As you read each chapter you will notice that various words appear darker than the rest of the text, and others appear in italics. The darkened words, or bold-faced terms, signify key terms that you will need to understand and remember to fully comprehend the material in which they appear. The italicized words are meant to emphasize their importance in understanding explanations of ideas and concepts discussed.

Examples

Each topic discussed in the chapter contains one or more concrete examples of problems and solutions to problems as they apply to the topic at hand. Through careful study of these examples you can better appreciate the many uses of problem solving in the physical sciences.

Activities

As you look through each chapter you will find one or more activities features that are highlighted with a blue bar. These activities are simple investigative exercises that you can perform at home or in the classroom to demonstrate important concepts and reinforce your understanding of them.

Boxed Readings

One or more boxed readings or features are included on topics of special interest to the general population as well as the science community. These features serve to underscore the relevance of physical science in confronting the many issues we face in our day-to-day lives.

End-of-Chapter Features

At the end of each chapter you will find the following materials: (a) a Summary that highlights the key elements of the chapter; (b) a Summary of Equations to reinforce your retention of them; (c) a Listing of Key Terms that are page-referenced; (d) a multiple choice quiz entitled Applying the Concepts to test your comprehension of the material covered; (e) Questions for Thought designed to challenge you to demonstrate your understanding of the topics; and (f) a section entitled Exercises. There are two groups of exercises, Group A and Group B. The Group A exercises have complete solutions worked out in Appendix D. The Group B exercises are similar to those in Group A but do not contain answers in the text. By working through the Group A exercises and checking your answers in Appendix D you will gain confidence in tackling the exercises in Group B, and thus, reinforce your problem solving skills.

End-of-Text Materials

At the back of the text you will find appendices that will give you additional background details, charts, and answers to chapter exercises. There is also a glossary of all key terms, an index organized alphabetically by subject matter, and special tables printed on the inside covers for reference use.

Acknowledgments

Many constructive suggestions, new ideas, and invaluable advice were provided by reviewers through several stages of manuscript development. Special thanks and appreciation to those who reviewed all or part of the manuscript for both editions of the text. They include: Lawrence H. Adams, Polk Community; Charles L. Bissell, Northwestern State University of Louisiana; W. H. Breazeale, Jr., Francis Marion College; William Brown, Montgomery College; Carla G. Davis, Danville Area Community College; Floretta Haggard, Rogers State College; Robert G. Hamerly, University of Northern Colorado; Eric Harms, Brevard Community College; L.

D. Hendrick, Francis Marion College; Abe Korn, New York City Tech College; William Luebke, Modesto Junior College; Douglas L. Magnus, St. Cloud State University; L. Whit Marks, Central State University; Jesse C. Moore, Kansas Newman College; Michael D. Murphy, Northwest Alabama Community College; David L. Vosburg, Arkansas State University; Steven Carey, Mobile College; Stan Celestian, Glendale Community College; Robert Larson, St. Louis Community College; Dennis Englin, The Master's College; Virginia Rawlins, University of North Texas. Lastly, I wish to acknowledge the very special contributions of my wife, Patricia Northrup Tillery, whose assistance and support throughout the revision were invaluable.

Publisher's Note to the Instructor

Physical Science, 2/e, has been developed to fit the special needs of your course. There are several binding options that have been developed to ensure that you, the instructor, ask your students to purchase only the material you choose to teach.

Physical Science, 2/e, by Bill W. Tillery is available as a complete, paperbound text or as custom separates of each section of the text. You can purchase these sections separately or ask us to package them together to make a custom teaching package. To purchase the custom separates, please order the appropriate custom separate from the table provided, or contact your local Wm. C. Brown Representative.

Binding Option	Description	ISBN Number
Physical Science, 2/e	The full length text, paperbound at a reduced price, when compared with casebound texts.	2–13818–01
Physics: Foundations of Physical Science	Section One of *Physical Science,* 2/e	2–17218–01
Chemistry: Foundations of Physical Science	Section Two of *Physical Science,* 2/e	2–17216–01
Astronomy: Foundations of Physical Science	Section Three of *Physical Science,* 2/e	2–17219–01
Earth Science: Foundations of Physical Science	Section Four of *Physical Science,* 2/e	2–17217–01
Physics & Chemistry Combined	Section One and Two of *Physical Science,* 2/e	2–17220–01
Choose Your Own Version	Any combination of the above books can be packaged together to fit your course needs.	Contact your local sales representative.

Physical Science

Two

Chemistry

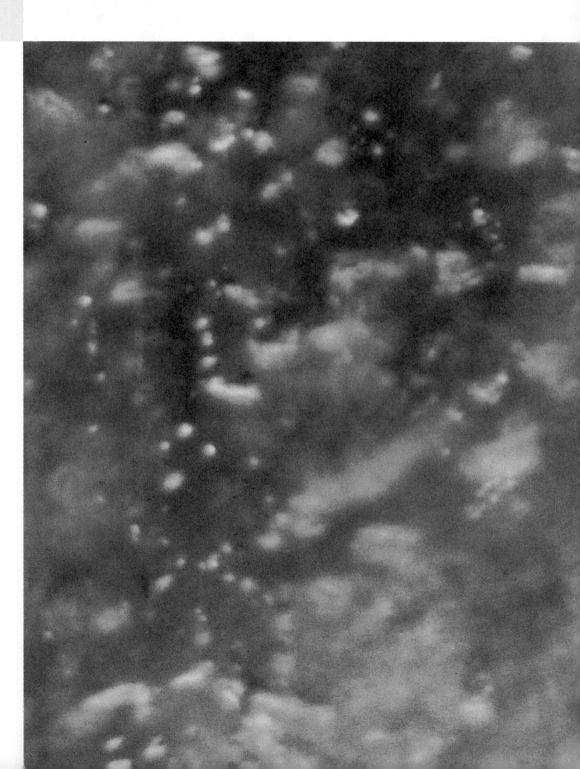

This section is an introduction to the structure of matter and the changes it undergoes. This is the field of science known as chemistry, a broad area of study concerned with the form and properties of matter on a large scale and the atoms, molecules, and electrons of matter on a much smaller scale. You might wonder about the boundary between physics (the behavior of matter) and chemistry (how matter is put together and how it changes). Scientific fields of study, and the boundaries between the fields, are purely arbitrary and human-made definitions made for convenience of study or made out of habit. Today the fields overlap, and it is often difficult to tell where one field ends and another begins. This is especially true with some of the more complex and vast modern-day fields such as biochemistry, geophysics, physical chemistry, and physical geochemistry. Space travel and the modern-day problems of a weakening ozone shield, increasing air pollution, acid rain, and the greenhouse effect all require the combined efforts of scientists from many fields of study.

The study of chemistry gives you an op-

A familiar item from your everyday surroundings, photographed at a very close range. Can you reach agreement with other students about the identity of the item?

portunity to learn thinking and reasoning skills as you study the forms, the structure, and the changes that matter undergoes. The photograph, for example, is a very close-range picture of a common item from the world around you. To identify what the item is requires a knowledge of some structures and properties of the overwhelming variety of matter in your surroundings. With this knowledge you are able to reason, thinking out what the item might or might not be. Without further information, however, you might not be able to identify the grain of table salt in the close-range photograph.

In this section you will examine matter at a much closer range than shown in the photograph. You will consider the smallest, discreet unit of matter called a molecule, and in some substances an atom. You will learn about (1) the structure of an atom and how people learned about the structure, (2) patterns that exist between matter with similar atomic structures, (3) how and why certain kinds of matter interact, forming new chemical substances, (4) ways of expressing the reactions of matter and what can be learned from these expressions, (5) water solutions of acids, bases, and salts and how they are used in your everyday life, (6) petroleum, petroleum products, and the chemicals of life, and (7) reactions that can take place involving the innermost part of an atom, the nucleus, and why these reactions involve tremendous amounts of energy. These new concepts are presented from a viewpoint of understanding your environment, removing the apparent mystery and helping you to understand a little bit more of your science-oriented culture.

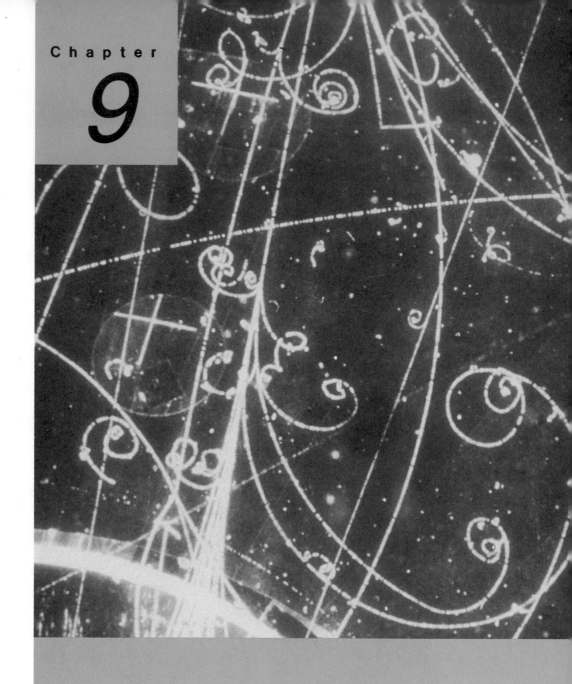

Chapter

9

Outline

Atomic Structure

You can see the tracks of hundreds of atomic particles in this bubble chamber. Current research indicates that protons and neutrons are made of even smaller particles called quarks.

Many materials used today are relatively new, created in the last few decades. These new materials are the result of modern chemical research, produced and manufactured through controlled chemical reactions. The new materials include synthetic fibers, from nylon to polyesters, and plastics, from polyethylene to Teflon. They also include water-based paints and super adhesives used in construction. The manufactured materials are lighter, stronger, and have special properties not found in natural materials. Today, such synthetic materials are used extensively in buildings, clothing, automobiles, and airplanes. The packaging, preserving, and marketing of many convenience foods are also made possible by the products of chemical research as are manufactured vitamins and drugs that help keep you healthy. From synthetic fibers to synthetic drugs, there are millions of products today that are the direct result of chemical research.

The countless numbers of new products resulting from chemical research demonstrate understandings about matter and how it is put together. These understandings start with the most basic unit of matter, the atom. Perhaps you have wondered how incredibly tiny atoms were discovered and how they can be studied. Atoms are so tiny that they are invisible to any optical device. Even more incredible is the study of the innermost parts of these invisible atoms and the development of knowledge of how they are put together. You will soon know the answer to questions about how atoms were discovered and studied (figure 9.1). This chapter contains the essence of the fascinating story of how the atomic concept was discovered and developed.

The development of the modern atomic model illustrates how modern scientific understanding comes from many different fields of study. For example, you will learn how studies of electricity led to the discovery that atoms have subatomic parts called electrons. The discovery of radioactivity led to the discovery of more parts, a central nucleus that contains protons and neutrons. Information from the absorption and emission of light was used to construct a model of how these parts are put together, a model resembling a miniature solar system with electrons circling the nucleus. The "solar system" model had initial, but limited, success and was inconsistent with other understandings about matter and energy. Modifications of this model were attempted, but none solved the problems. Then the discovery of wave properties of matter led to an entirely new model of the atom.

The atomic model will be put to use in later chapters to explain the countless varieties of matter and the changes that matter undergoes. In addition, you will learn how these changes can be manipulated to make new materials, from drugs to ceramics. In short, you will learn how understanding the atom and all the changes it undergoes not only touches your life directly but shapes and affects all parts of civilization. Basic to all of this is understanding the atom.

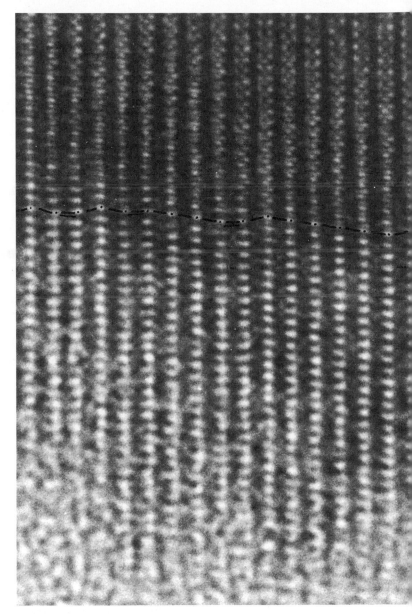

Figure 9.1

This electron microscope high-resolution image shows magnification of the thin edge of a piece of mica. The white dots are "empty tunnels" between layers of silicon-oxygen tetrahedrons, and the black dots are potassium atoms that bond the tetrahedrons together. Note the 10 Angstrom width, which is 0.000001 mm.

First Definition of the Atom

The atom concept is very old, dating back to ancient Greek philosophers some 2,500 years ago. The ancient Greeks thought, reasoned, and speculated about the basis for their surroundings. Consider, as an example, that you have observed water in puddles, ponds, lakes, rivers, and perhaps an ocean. You recognize rain, snow, sleet, dew, and ice to be water. You have also observed that plants and animals contain water, and that water can be produced by heating wood and other substances. If you think about all the forms, places, and things that water is a part of, you might begin to get the idea that water is a basic, fundamental substance that makes up other materials. Over time, this kind of reasoning led the ancient Greeks to the concept that everything is made up of four basic, fundamentally different substances, or "elements." These elements were earth, air, fire, and water. From logical considerations of what they observed, the ancient Greeks considered all substances to be composed of these four elements in varying proportions.

The ancient Greeks also reasoned about the way that pure substances are put together. A glass of water, for example, appears to be the same throughout. Is it the same? Two plausible, but conflicting, ideas were possible as an intellectual exercise.

The water could have a continuous structure, that is, it could be completely homogeneous throughout. The other idea was that the water only appears to be continuous but is actually *discontinuous*. This means that if you continue to divide the water into smaller and smaller volumes, you would eventually reach a limit to this dividing, a particle that could not be further subdivided. This model was developed by Leucippus and Democritus in the fourth century B.C. Democritus called the indivisible particle an *atom,* from a Greek word meaning "uncuttable."

Democritus speculated that the atoms of a substance were separated by *empty space* and that the atoms of the four elements had different shapes. Since this is an intellectual exercise you can imagine anything you wish—as long as it is logically consistent with what is observed. For example, you might imagine that atoms of water might be round and smooth since water pours and flows. Atoms of earth might have cubic shapes that would prevent them from pouring and flowing. Democritus speculated that one substance changed to another by separation or combination of atoms.

When Aristotle organized and recorded all the ancient Greek knowledge that was available during this time, it was about one hundred years after Democritus had proposed the existence of atoms. Aristotle adopted the continuous, four element model because of perceived logical problems with the empty space between the atoms—he did not believe a vacuum could exist in nature. The idea of four elements—earth, air, fire, and water—with a continuous structure became the accepted model for the next thousand years. If anyone during this period did support the idea of atoms, and a few did, they were definitely in the minority.

Atomic Structure Discovered

In the 1600s, Robert Boyle's work with gases provided evidence to reject Aristotle's idea that a vacuum could not exist. In 1661, Boyle published *The Skeptical Chemist,* in which he rejected Aristotle's four element theory, too. Boyle defined an element as a "simple substance" that could not be broken down into anything simpler. Today, an **element** is defined as a pure substance that cannot be broken down to anything simpler by chemical or physical means. Water is a pure substance, but it can be broken down into oxygen and hydrogen, so water is not an element. Oxygen and hydrogen are pure substances that cannot be broken down into anything simpler, so they are elements. Oxygen, silicon, iron, gold, and aluminum are common elements, and over one hundred elements are known today. Elements will be considered in the next chapter. The development of a model of the atom is the topic of interest in this chapter, a model that will explain how elements are different.

It was information about how elements combine that led John Dalton in the early 1800s to bring back the ancient Greek idea of hard, indivisible atoms. In general, he had noted, as others had, that certain elements always combined with other elements in fixed ratios. For example, 1 gram of oxygen always combined with 13 grams of lead to produce 14 grams of a yellow compound called lead oxide (figure 9.3). Dalton reasoned that this must mean that the oxygen and lead were made up of individual particles called *atoms,* not a form of matter that was

A

B

Figure 9.2

Reasoning the existence of atoms from the way elements combine in fixed weight ratios. (*a*) If matter were a continuous, infinitely divisible material there would be no reason for one amount to go with another amount. (*b*) If matter is made up of discontinuous, discrete units (atoms), then the units would combine in a fixed weight ratio that equals the weight of the combination.

completely homogeneous in structure. If elements were continuous, there would be no reason for the oxygen and lead to combine in fixed ratios. If elements were composed of atoms, however, then whole atoms would combine to make the compound. By way of analogy, on a macroscopic scale you could consider both peanut butter and jelly to be homogeneous, continuous substances. If you combine peanut butter and jelly there is no reason for a particular amount of peanut butter to go with a particular amount of jelly. Crackers and uniform slices of cheese, however, could both be considered on a macroscopic scale to be individual units. If you combine a slice of cheese and a cracker, the combination would have a total weight with part contributed by a cracker and part contributed by the cheese slice. A fixed ratio of the weight of a slice of cheese to the weight of an individual cracker would result in the same total weight each time. Using this kind of reasoning, Dalton theorized that elements must be composed of atoms (figure 9.2).

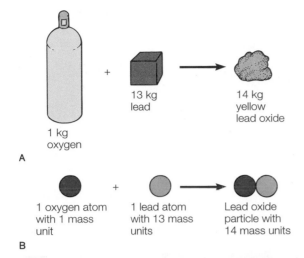

A

B

Figure 9.3

(a) Oxygen and lead combine to form yellow lead oxide in a ratio of 1:13 for a total of 14. (b) If 1 atom of oxygen combines with 1 atom of lead, the fixed ratio in which oxygen and lead combine must mean that 1 atom of lead is 13 times more massive than 1 atom of oxygen.

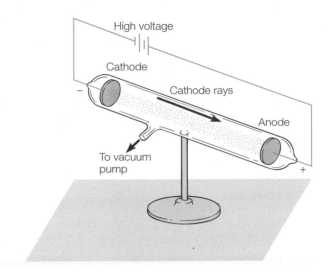

Figure 9.4

A vacuum tube with metal plates attached to a high-voltage source produces a greenish beam called cathode rays. These rays move from the cathode (negative charge) to the anode (positive charge).

During the 1800s Dalton's concept of hard, indivisible atoms was familiar to most scientists. Yet, the existence of atoms was not generally accepted by all scientists. There was skepticism about something that could not be observed directly. Strangely, full acceptance of the atom came in the early 1900s with the discovery that the atom was not indivisible after all. The atom has parts that give it an internal structure. The first part to be discovered was the *electron,* a part that was discovered through studies of electricity.

Discovery of the Electron

Scientists of the 1800s were interested in understanding the nature of electricity, but it was impossible to see anything but the effects of a current as it ran through a wire. To observe a current directly, they tried to produce a current by itself, away from matter, by pumping the air from a tube and then running a current through the empty space. By 1885, a good air pump was invented, and this pump could evacuate the air from a glass tube until the pressure was 1/10,000 of normal air pressure. When metal plates inside such a tube were connected to the negative and positive terminals of a high-voltage source (figure 9.4), a greenish beam was observed that seemed to move from the cathode (negative terminal) through the empty tube and collect at the anode (positive terminal). Since this mysterious beam seemed to come out of the cathode it was said to be made of **cathode rays.**

Just what the cathode rays were became a source of controversy. Some scientists suggested that cathode rays were atoms of the cathode material. But the properties of cathode rays were found to be the same when the cathode was made of any material. Perhaps the rays were a form of light (figure 9.5)? But the rays were observed to be deflected by a magnet, not a behavior observed with light. What *were* the cathode rays?

The mystery of the cathode rays was finally solved by the English physicist J. J. Thomson, who provided the answer in 1897. Thomson had placed a positively charged metal plate on

Figure 9.5

What appears to be visible light in this vacuum tube is produced by cathode ray particles striking a detecting screen. You know it is not light, however, since the beam can be pulled or pushed away by a magnet and since it is attracted to a positively charged metal plate. These are not the properties of light, so cathode rays must be something other than light.

one side of the beam and a negatively charged metal plate on the other side (figure 9.6). The beam was deflected toward the positive plate and away from the negative plate. Since it was known that unlike charges attract and like charges repel, this meant that the beam must be composed of *negatively charged particles.*

The cathode ray was also deflected when caused to pass between the poles of a magnet. Thomson knew that moving charges are deflected by a magnetic field. He found more information by adjusting the electric charge on the plates above and below the beam, then measuring the deflection. The same procedure was used with a measured magnetic field. A greater charge on a particle would result in a greater deflection by an

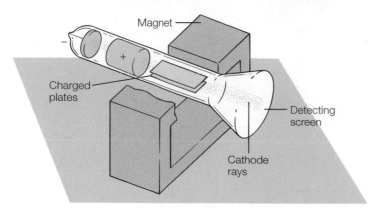

Figure 9.6

A cathode ray passed between two charged plates is deflected toward the positively charged plate. The ray is also deflected by a magnetic field. By measuring the deflection by both, J. J. Thomson was able to calculate their ratio of charge to mass. He was able to measure the deflection because the detecting screen was coated with zinc sulfide, a substance that produces a visible light when struck by a charged particle.

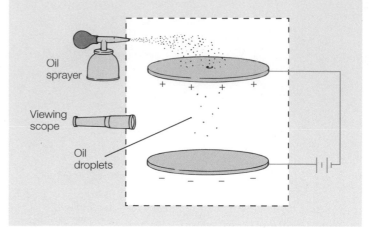

Figure 9.7

Millikan measured the charge of an electron by balancing the pull of gravity on oil droplets with an upward electrical force. Knowing the charge to mass ratio that Thomson had calculated, Millikan was able to calculate the charge on each droplet. He found that all the droplets had a charge of 1.60×10^{-19} coulomb or multiples of that charge. The conclusion was that this had to be the charge of an electron.

electric field, and a greater moving mass would be more difficult to deflect than a lesser moving mass. By balancing the deflections made by the magnet with the deflections made by the electric field, Thomson could thus determine the *ratio of the charge to mass* for an individual particle. Today, the charge to mass ratio is considered to be 1.7584×10^{11} coulomb/kilogram. The significant part of Thomson's experiments was finding that the charge to mass ratio was the *same* no matter what gas was in the tube and of what materials the electrodes were made. Thomson was convinced that he had discovered a fundamental particle, the stuff of which atoms are made.

Thomson did not propose any special name for the particle. Some time earlier, the term **electron** had been proposed as a name for the unit of charge gained or lost when atoms became ions. Not long after Thomson reported his findings in 1897, the existence of the fundamental particle was generally accepted and everyone started calling the particles electrons.

A method for measuring the charge and mass of the electron was worked out by an American physicist, Robert A. Millikan, around 1906. Millikan used an apparatus such as the one illustrated in figure 9.7 to measure the charge indirectly. Small droplets of mineral oil sprayed into the apparatus could be observed with a magnifier, and measurements could be made as the droplets drifted downwards. With a vertical electric field turned on, the droplets would drift upwards at a rate depending on the electric charge on each droplet. Measuring the rise and fall of the droplets as the electric field is turned on and off enabled Millikan to deduce the charge from the speed of fall and rise.

Millikan found that none of the droplets had a charge less than one particular value (1.60×10^{-19} coulomb) and that larger charges on various droplets were always multiples of this unit of charge. Since all of the droplets carried the single unit of charge or multiples of the single unit, the unit of charge was assumed to be the charge of a single electron.

Knowing the charge of a single electron and knowing the charge to mass ratio that Thomson had measured now made it possible to calculate the mass of a single electron. The mass of an electron was thus determined to be about 9.11×10^{-31} kg, or about 1/1,840 of the mass of the lightest atom, hydrogen.

Thomson had discovered the negatively charged electron, and Millikan had measured the charge and mass of the electron. But atoms themselves are electrically neutral. If an electron is part of an atom, there must be something else that is positively charged, canceling the negative charge of the electron. The next step in the sequence of understanding atomic structure would be to find what is neutralizing the negative charge and to figure out how all the parts are put together.

About 1900, Thomson proposed a model for what was known about the atom at the time. He suggested that an atom could be a blob of positively charged matter in which electrons were stuck like "raisins in plum pudding." If the mass of a hydrogen atom is due to the electrons embedded in a positively charged matrix, then 1,840 electrons would be needed together with sufficient positive matter to make the atom electrically neutral.

About this same time, 1896, radioactivity was discovered by Antonie Becquerel, which was soon described in terms of alpha, beta, and gamma rays. The details of radioactivity are considered in chapter 15. For now, all you need to know is what was known in 1907, that alpha particles are very fast, massive, and positively charged particles that are spontaneously given off from radioactive elements. The experimental application of alpha particles would lead to the discovery of the nucleus, then other parts of the atom.

The Nucleus

The nature of radioactivity and matter were the research interests of an English physicist, Ernest Rutherford. In 1907, Rutherford was studying the scattering of alpha particles directed toward a thin sheet of metal. As shown in figure 9.8, alpha particles from a radioactive source were allowed to move through

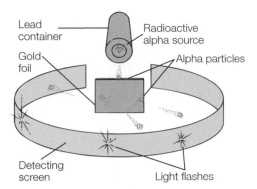

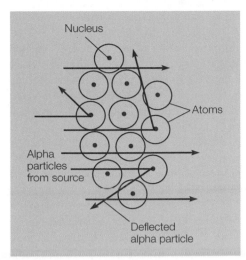

Figure 9.8

Rutherford and his co-workers studied alpha particle scattering from a thin metal foil. The alpha particles struck the detecting screen, producing a flash of visible light. Measurements of the angles between the flashes, the metal foil, and the source of the alpha particles showed that the particles were scattered in all directions, including straight back toward the source.

a small opening of a lead container, so only a narrow beam of the massive, fast-moving particles would penetrate a very thin sheet of gold. The alpha particles were then detected by plates covered with zinc sulfide, which produced a small flash of light when struck by the positively charged alpha particle.

Activities

The luminous dial of a watch or clock contains a mixture of zinc sulfide and a radioactive substance. Obtain a watch or clock with a luminous dial and a magnifying glass. Wait in a completely dark room about ten minutes until your eyes adjust to the darkness. Observe the glowing parts of the dial with the magnifying glass. Is the glow continuous or is it made up of tiny flashes of light? What would flashes of light mean?

Rutherford and his co-workers found that most of the alpha particles went straight through the foil, and just as expected, none were scattered by more than a few degrees. Then he suggested that one of his young co-workers check to see if any alpha particles were scattered through very large angles. He really did not expect any great amount of scattering since alpha particles were very fast, massive particles with a great deal of energy and electrons were not very massive in comparison. When a young co-worker reported that alpha particles were deflected at very large angles and some were even reflected backwards, Rutherford was astounded. Rutherford could account for the large deflections and backward scattering only by assuming that the massive, positively charged particles were repelled by a massive positive charge concentrated in a small region of the atom (figure 9.9). He concluded that an atom must have a tiny, massive, and positively charged **nucleus** surrounded by electrons. Since the electrons had an opposite charge than the nucleus, the electrons must be moving or else they would be attracted to it.

From measurements of the scattering, Rutherford estimated the radius of the nucleus to be approximately 10^{-13} cm. Other researchers had estimated the radius of the atom to be

Figure 9.9

Rutherford's nuclear model of the atom explained the alpha scattering experiment as positive alpha particles experiencing a repulsive force from the positive nucleus. Measurements of the various angles of scattering gave Rutherford a means of estimating the size of the nucleus.

Figure 9.10

From measurements of alpha particle scattering, Rutherford estimated the radius of an atom to be 100,000 times greater than the radius of the nucleus. This ratio is comparable to the thickness of a dime to the length of a football field.

on the order of 10^{-8} cm, so the electrons must be moving around the nucleus at a distance 100,000 times the radius of the nucleus. To visualize this spatial relationship, think of the thickness of a dime, which is about 1 mm thick. A distance 100,000 times the thickness of a dime is 100,000 mm, or 100 m. Thus, if the radius of a nucleus were the thickness of a dime, the electrons would be moving around the nucleus at a distance of about 100 m away, or about the length of a football field. If the radius of a nucleus were about the same size as the thickness of a dime, the atom would be about two football fields wide (figure 9.10). As you can see, the volume of an atom is mostly empty space.

Atomic Structure

Rutherford announced his conclusions and evidence for the existence of the atomic nucleus in 1911. This was a revolutionary development that would soon lead to more developments and more evidence about atomic structure. In 1917, Rutherford was able to break up the nucleus of a nitrogen atom by using alpha particles and identified the discrete unit of positive charge he called a **proton.** Rutherford also speculated about the existence of a neutral particle in the nucleus, a **neutron.** The neutron was eventually identified in 1932 by James Chadwick.

Today, the number of *protons* in the nucleus of an atom is called the **atomic number.** An element is made up of atoms that all have the same number of protons in their nucleus, so all atoms of an element have the same atomic number. Hydrogen has an atomic number of 1, so any atom that has one proton in its nucleus is an atom of the element hydrogen. Today, there are 109 different kinds of elements with a different number of protons. The *neutrons* of the nucleus, along with the protons, contribute to the mass of an atom. Atomic mass (and atomic weight) will be considered in the next chapter.

So the atom has a tiny, massive nucleus containing positively charged protons and neutral neutrons. Negatively charged electrons, equal in number to the protons, are moving around at a distance of about 100,000 times the radius of the nucleus. How are these electrons moving? It might occur to you, as it did to Rutherford and others, that an atom might be similar to a miniature solar system. This idea would picture the nucleus in the role of the sun, electrons in the role of moving planets in their orbits, and electrical attractions between the nucleus and electrons in the role of gravitational attraction. There are, however, significant problems with this idea. If electrons were moving in circular orbits, they would continually change their direction of travel and would therefore be accelerating. According to the Maxwell model of electromagnetic radiation, an accelerating electric charge emits electromagnetic radiation such as light. If an electron gave off light, it would lose energy. The energy loss would mean that the electron could not maintain its orbit, and it would be pulled into the oppositely charged nucleus. The atom would collapse as electrons spiraled into the nucleus. Since atoms do not collapse like this, there is a significant problem with the "solar system" model of the atom.

The Bohr Model

Niels Bohr was a young Danish student who visited Rutherford's laboratory in 1912 and became very interested in questions about the solar system model of the atom. He wondered what determined the size of the electron orbits and the energies of the electrons. He wanted to know why orbiting electrons did not give off electromagnetic radiation. Seeking answers to questions such as these led Bohr to incorporate the *quantum concept* of Planck and Einstein with Rutherford's model to describe the electrons in the outer part of the atom. This quantum concept will be briefly reviewed before proceeding with the development of Bohr's model of the hydrogen atom.

The Quantum Concept

In the year 1900, Max Planck introduced the idea that matter emits and absorbs energy in discrete units that he called **quanta.**

Planck had been trying to match data from experiments with data that could be predicted from the theory of electromagnetic radiation. In order to match the experimental findings with the theory, he had to assume that specific, discrete amounts of energy were associated with different frequencies of radiation. In 1905, Albert Einstein extended the quantum concept to light, stating that light consists of discrete units of energy that are now called **photons.** The energy of a photon is directly proportional to the frequency of vibration, and the higher the frequency of light the greater the energy of the individual photons. In addition, the interaction of a photon with matter is an "all-or-none" affair, that is, matter absorbs an entire photon or none of it. The relationship between frequency (f) and energy (E) is

$$E = hf \qquad \text{equation 9.1}$$

where h is the proportionality constant known as *Planck's constant* (6.63×10^{-34} J·sec). This relationship means that higher-frequency light, such as ultraviolet, has more energy than lower-frequency light, such as red light.

Example 9.1

What is the energy of a photon of red light with a frequency of 4.60×10^{14} Hz?

Solution

$f = 4.60 \times 10^{14}$ Hz

$h = 6.63 \times 10^{-34}$ J·sec

$E = ?$

$$E = hf$$

$$= (6.63 \times 10^{-34} \text{ J·sec})\left(4.60 \times 10^{14} \frac{1}{\text{sec}}\right)$$

$$= (6.63 \times 10^{-34})(4.60 \times 10^{14}) \text{ J·} \cancel{\text{sec}} \times \frac{1}{\cancel{\text{sec}}}$$

$$= \boxed{3.05 \times 10^{-19} \text{ J}}$$

Example 9.2

What is the energy of a photon of violet light with a frequency of 7.30×10^{14} Hz? (Answer: 4.84×10^{-19} J)

Atomic Spectra

Planck was concerned with hot solids that emit electromagnetic radiation. The nature of this radiation, called blackbody radiation, depends on the temperature of the source. When this light is passed through a prism it is dispersed into a *continuous spectrum,* with one color gradually blending into the next as in a rainbow. Today, it is understood that a continuous spectrum comes from solids, liquids, and dense gases because the atoms interact and all frequencies within a temperature-determined range are emitted. Light from an incandescent gas, on the other hand, is dispersed into a **line spectrum,** narrow lines of colors with no light between the lines (figure 9.11). The atoms in the incandescent gas are able to emit certain characteristic frequencies, and each frequency is a line of color that represents

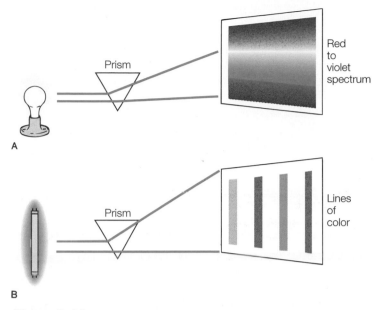

A

B

Figure 9.11

(a) Light from incandescent solids, liquids, or dense gases produces a continuous spectrum as atoms interact to emit all frequencies of visible light. (b) Light from an incandescent gas produces a line spectrum as atoms emit certain frequencies that are characteristic of each element.

a definite value of energy. The line spectra are specific for a substance, and increased or decreased temperature changes only the intensity of the lines of colors. Thus hydrogen always produces the same colors of lines in the same position. Helium has its own specific set of lines, as do other substances. Line spectra are a kind of "fingerprint" that can be used to identify a gas. A line spectrum might also extend beyond visible light into ultraviolet, infrared, and other electromagnetic regions.

In 1885, a Swiss mathematics teacher named J. J. Balmer was studying the regularity of spacing of the hydrogen line spectra. Balmer was able to develop an equation that fit all the visible lines. By assigning a value of n of 3, 4, 5, and 6 to the four lines he found the wavelengths fit the equation

$$\frac{1}{\lambda} = R\left(\frac{1}{2^2} - \frac{1}{n^2}\right)$$ **equation 9.2**

when R is a constant of 1.097×10^7 1/m.

Balmer's findings were:

Violet line	(n = 6)	$\lambda = 4.1 \times 10^{-7}$ m
Violet line	(n = 5)	$\lambda = 4.3 \times 10^{-7}$ m
Blue-green line	(n = 4)	$\lambda = 4.8 \times 10^{-7}$ m
Red line	(n = 3)	$\lambda = 6.6 \times 10^{-7}$ m

These four lines became known as the **Balmer series.** Other series were found later, outside the visible part of the spectrum (figure 9.12). The equations of the other series were different only in the value of n and the number in the other denominator.

Such regularity of observable spectral lines must reflect some unseen regularity in the atom. At this time it was known that hydrogen had only one electron. How could one electron produce series of spectral lines with such regularity?

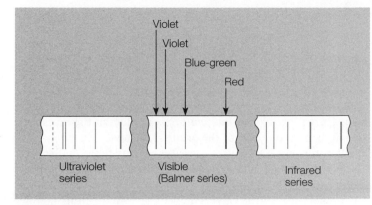

Figure 9.12

Atomic hydrogen produces a series of characteristic line spectra in the ultraviolet, visible, and infrared parts of the total spectrum. The visible light spectra always consist of two violet lines, a blue-green line, and a bright red line.

Example 9.3

Calculate the wavelength of the violet line (n = 6) in the hydrogen line spectra according to Balmer's equation.

Solution

n = 6

R = 1.097×10^7 1/m

λ = ?

$$\frac{1}{\lambda} = R\left(\frac{1}{2^2} - \frac{1}{n^2}\right)$$

$$= 1.097 \times 10^7 \frac{1}{m}\left(\frac{1}{2^2} - \frac{1}{6^2}\right)$$

$$= 1.097 \times 10^7 \left(\frac{1}{4} - \frac{1}{36}\right)\frac{1}{m}$$

$$= 1.097 \times 10^7 (0.222)\frac{1}{m}$$

$$\frac{1}{\lambda} = 2.44 \times 10^6 \frac{1}{m}$$

$$\lambda = \boxed{4.11 \times 10^{-7} \text{ m}}$$

Bohr's Theory

An acceptable model of the hydrogen atom would have to explain the characteristic line spectra and their regularity described by Balmer. In fact, a successful model should be able to *predict* the occurrence of each color line as well as account for its origin. By 1913, Bohr was able to do this by applying the quantum concept to a "solar system" model of the atom. He began by considering the single hydrogen electron to be a single "planet" revolving in a circular orbit around the nucleus. He assumed that the electron could occupy more than one circular orbit and that it would have a different state of energy, which depended on the radius of the orbit it was in at any particular time. Each orbit was assigned a number *n,* which could be any whole number 1, 2, 3, and so on out from the nucleus. These were known as the orbit **quantum numbers.** The following describes how orbit quantum numbers were used with definitions

Atomic Structure

of (1) allowed orbits, (2) radiationless orbits, and (3) quantum jumps to describe the **Bohr model** of the atom.

Allowed Orbits

An electron can revolve around an atom only in specific allowed orbits. Bohr considered the electron to be a particle with a known mass in motion around the nucleus. Rotational motion is measured by *angular momentum,* a product of the mass (m), velocity (v), and radius of the orbit (r), or mvr. Conservation of angular momentum requires a greater velocity for a smaller orbit and less velocity for a larger orbit. It was Bohr's assumption that the allowed orbits are those for which the angular momentum of the electron (mvr) equaled the orbit quantum number ($n = 1, 2, 3, \ldots$) times Planck's constant (h) divided by 2π, or

$$mvr = n\frac{h}{2\pi} \qquad \text{equation 9.3}$$

(Note that $2\pi r$ describes a circumference and that mv describes momentum.) Bohr used this relationship to determine the allowed orbits because differences between the energy levels it describes fit exactly with the differences between the frequencies (and thus energies) of the line spectra described by Balmer. Bohr did not have a reason for this assumption but used it simply because it worked.

Bohr also assumed that the force of electrical attraction between the electron (q_1) and proton (q_2) must be equal to the centripetal force if the electron moves in a circular orbit. This means that the force according to Coulomb's law of electrical attraction must be equal to the centripetal force described by Newton's second law of motion, or

$$\frac{kq_1q_2}{r^2} = \frac{mv^2}{r} \qquad \text{equation 9.4}$$

This describes a relationship that is very similar to the relationship between the gravitational forces between the earth and moon and the centripetal force that keeps the moon in its orbit around the earth (see chapter 3). The earth and moon are attracted by the force of gravity, however, not electrical forces.

Equations 9.3 and 9.4 can be solved for r, which will yield the distances, or radii, of the allowed orbits. When r is obtained for the first quantum number ($n = 1$), the radius of the closest orbit is found to be 0.529×10^{-10} m, which is known as the calculated *Bohr radius.* The next allowed orbit has a radius of 2.12×10^{-10} m, the next a radius of 4.46×10^{-10} m, and so on. According to the Bohr model, electrons can exist *only* in one of these allowed orbits and nowhere else.

Radiationless Orbits

An electron in an allowed orbit does not emit radiant energy as long as it remains in the orbit. It had been understood since the development of Maxwell's theory of electromagnetic radiation that an accelerating electron should emit an electromagnetic wave, such as light, which would move off into space from the electron. Bohr recognized that electrons moving in a circular orbit are accelerating, since they are changing direction continuously. Yet, light was not observed to be emitted from hydrogen atoms in their normal state. Bohr decided that the situation must be different for orbiting electrons, and that electrons

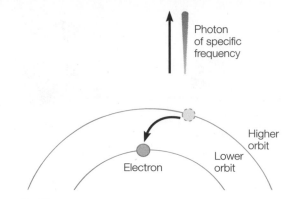

Figure 9.13

Each time an electron makes a "quantum leap," moving from a higher energy orbit to a lower energy orbit, it emits a photon of a specific frequency and energy value.

could stay in their allowed orbits and *not* give off light. He postulated this rule as a way to make his theory consistent with other scientific theories.

Quantum Jumps

An electron gains or loses energy only by moving from one allowed orbit to another (figure 9.13). The reference level for the potential energy of an electron is considered to be zero when the electron is *removed* from an atom. The electron, therefore, has a lower and lower potential energy at closer and closer distances to the nucleus and has a negative value when it is in some allowed orbit. By way of analogy, you could consider ground level as a reference level where the potential energy of some object equals zero. But suppose there are two basement levels below the ground. An object on either basement level would have a gravitational potential energy less than zero, and work would have to be done on each object to bring each back to the zero level. Thus, each object would have a negative potential energy. The object on the lowest level would have the largest negative value of energy, since more work would have to be done on it to bring it back to the zero level. Therefore, the object on the lowest level would have the *least* potential energy, and this would be expressed as the *largest negative value.*

Just as the objects on different basement levels, the electron has a definite negative potential energy in each of the allowed orbits. Bohr calculated the energy of an electron in the orbit closest to the nucleus to be -2.17×10^{-18} J, which is called the energy of the lowest state. The energy of electrons is expressed in units of the **electron volt** (eV). An electron volt is defined as the energy of an electron moving through a potential of one volt. Since this energy is charge times voltage (from $V = w/q$), 1.00 eV is equivalent to 1.60×10^{-19} J. Therefore the energy of an electron in the innermost orbit is its energy in joules divided by 1.60×10^{-19} J/eV, or -13.6 eV.

Bohr found that the energy of each of the allowed orbits could be found from the simple relationship of

$$E_n = \frac{E_1}{n^2} \qquad \text{equation 9.5}$$

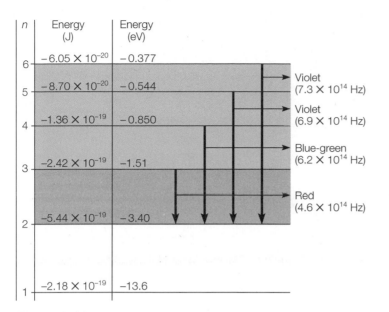

Figure 9.14

An energy level diagram for a hydrogen atom, not drawn to scale. The energy levels (n) are listed on the left side, followed by the energies of each level in J and eV. The color and frequency of the visible light photons emitted are listed on the right side, with the arrow showing the orbit moved from and to.

Figure 9.15

These fluorescent lights emit light as electrons of mercury atoms inside the tubes gain energy from the electric current. As soon as they can, the electrons drop back to a lower energy orbit, emitting photons with ultraviolet frequencies. Ultraviolet radiation strikes the fluorescent chemical coating inside the tube, stimulating the emission of visible light.

where E_1 is the energy of the innermost orbit (-13.6 eV), and n is the quantum numbers for the orbits, or 1, 2, 3, and so on. Thus, the energy for the second orbit ($n = 2$) is $E_2 = -13.6$ eV$/4 = -3.40$ eV. The energy for the third orbit out ($n = 3$) is $E_3 = -13.6$ eV$/9 = -1.51$ eV, and so forth (figure 9.14). Thus, the energy of each orbit is *quantized,* occurring only as definite values.

In the Bohr model, the energy of the electron is determined by which allowable orbit it occupies. The only way that an electron can change its energy is to jump from one allowed orbit to another in quantum "jumps." An electron must *acquire* energy to jump from a lower orbit to a higher one. Likewise an electron *gives up* energy when jumping from a higher orbit to a lower one. Such jumps must be all at once, not part way and not gradual. By way of analogy, this is very much like the gravitational potential energy that you have on the steps of a staircase. You have the lowest potential on the bottom step and the greatest amount on the top step. Your potential energy is quantized because you can increase or decrease it by going up or down a number of steps, but you cannot stop between the steps.

An electron acquires energy from high temperatures or from electrical discharges to jump to a higher orbit. An electron jumping from a higher to a lower orbit gives up energy in the form of light. A single photon is emitted when a downward jump occurs, and the *energy of the photon is exactly equal to the difference in the energy level* of the two orbits. If E_1 represents the lower energy level (closest to the nucleus) and E_h represents a higher energy level (farthest from the nucleus), the energy of the emitted photon is

$$hf = E_h - E_1 \qquad \textbf{equation 9.6}$$

where h is Planck's constant and f is the frequency of the emitted light (figure 9.15).

The energy level diagram in figure 9.14 shows the energy states for the orbits of a hydrogen atom. The lowest energy state, $n = 1$, is known as the **ground state** (or normal state). The higher states, $n = 2$, $n = 3$, and so on, are known as the **excited states.** The electron in a hydrogen atom would, under normal conditions, be located in the ground state ($n = 1$). But high temperatures or electric discharge can give the electron sufficient energy to jump to one of the excited states. Once in an excited state, the electron immediately jumps back to a lower state as shown by the arrows in the figure. The length of the arrow represents the frequency of the photon that the electron emits in the process. A hydrogen atom can give off only one photon at a time, and the many lines of a hydrogen line spectrum come from many atoms giving off many photons at the same time.

As you can see, the energy level diagram in figure 9.14 shows how the change of known energy levels from known orbits results in the exact energies of the color lines in the Balmer series. Bohr's theory did offer an explanation for the lines in the hydrogen spectrum with a remarkable degree of accuracy. However, the model did not have much success with larger atoms. Larger atoms had spectra lines that could not be explained by the Bohr model with its single quantum number. A German physicist, A. Sommerfield, tried to modify Bohr's model by adding elliptical orbits in addition to Bohr's circular orbits. It soon became apparent that the "patched up" model, too, was not adequate. Bohr had made the rule that there were radiationless orbits without an explanation, and he did not have an explanation for the quantized orbits. There was something fundamentally incomplete about the model.

Example 9.4

An electron in a hydrogen atom jumps from the excited energy level $n = 4$ to $n = 2$. What is the frequency of the emitted photon?

Solution

The frequency of an emitted photon can be calculated from equation 9.6, $hf = E_h - E_1$. The values for the two energy levels can be obtained from figure 9.14. (Note: E and E_1 must be in joules. If the values are in electron volts, they can be converted to joules by multiplying by the ratio of joules per electron volt, or $(eV)(1.6 \times 10^{-19} \text{ J/eV}) = \text{joules.})$

$E_h = -1.36 \times 10^{-19}$ J

$E_1 = -5.44 \times 10^{-19}$ J

$h = 6.63 \times 10^{-34}$ J·sec

$f = ?$

$$hf = E_h - E_1 \therefore f = \frac{E_h - E_1}{h}$$

$$f = \frac{(-1.36 \times 10^{-19} \text{ J}) - (-5.44 \times 10^{-19} \text{ J})}{6.63 \times 10^{-34} \text{ J·sec}}$$

$$= \frac{4.08 \times 10^{-19}}{6.63 \times 10^{-34}} \frac{\cancel{J}}{\cancel{J}\cdot\text{sec}}$$

$$= 6.15 \times 10^{14} \frac{1}{\text{sec}}$$

$$= \boxed{6.15 \times 10^{14} \text{ Hz}}$$

This is approximately the blue-green line in the hydrogen line spectrum.

Quantum Mechanics

The Bohr model of the atom successfully accounted for the line spectrum of hydrogen and provided an understandable mechanism for the emission of photons by atoms. However, the model did not predict the spectra of any atom larger than hydrogen, and there were other limitations. A new, better theory was needed. The roots of a new theory would again come from experiments with light. Experiments with light had established that sometimes light behaves like a stream of particles and other times it behaves like a wave (see chapter 8). Eventually scientists began to accept that light has both wave properties and particle properties, which is now referred to as the *wave-particle duality of light*. This dual nature of light was recognized in 1905 when Einstein applied Planck's quantum concept to the energy of a photon with the relationship found in equation 9.1, $E = hf$, where E is the energy of a photon particle and f is the frequency of the associated wave, and h is Planck's constant.

Matter Waves

In 1923 Louis de Broglie, a French physicist, reasoned that symmetry is usually found in nature, so if a particle of light has a dual nature then particles such as electrons should too. De Broglie reasoned further that if this is true, an electron in its circular path around the nucleus would have to have a partic-

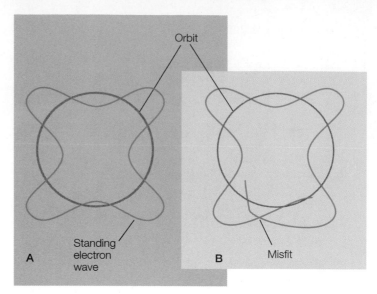

Figure 9.16

(a) Schematic of de Broglie wave, where the standing wave pattern will just fit in the circumference of an orbit. This is an allowed orbit. (b) This orbit does not have a circumference that will match a whole number of wavelengths; it is not an allowed orbit.

ular wavelength that would fit into the circumference of the orbit (figure 9.16). The circumference of the orbit must be a whole number of wavelengths long or

$$\text{circumference} = (\text{number})(\text{wavelength})$$

or

$$2\pi r = n\lambda \qquad \text{equation 9.7}$$

$$\text{where } n = 1, 2, 3, \ldots$$

De Broglie derived a relationship from equations concerning light and energy, which was

$$\lambda = \frac{h}{mv} \qquad \text{equation 9.8}$$

where λ is the wavelength, m is mass, v is velocity, and h is again Planck's constant. This equation means that any moving particle has a wavelength that is associated with its mass and velocity. In other words, de Broglie was proposing a wave-particle duality of matter, the existence of **matter waves.** According to equation 9.8, *any* moving object should exhibit wave properties. However, an ordinary-sized object would have wavelengths so small that they could not be observed. This is different for electrons because they have such a tiny mass.

Example 9.5

What is the de Broglie wavelength associated with a 0.150 kg baseball with a velocity of 50.0 m/sec?

Solution

m = 0.150 kg

v = 50.0 m/sec

h = 6.63×10^{-34} J·sec

$\lambda = ?$

$$\lambda = \frac{h}{mv}$$

$$= \frac{6.63 \times 10^{-34} \text{ J} \cdot \text{sec}}{(0.150 \text{ kg})\left(50.0 \dfrac{m}{\text{sec}}\right)}$$

$$= \frac{6.63 \times 10^{-34}}{(0.150)(50.0)} \frac{\text{J} \cdot \text{sec}}{\text{kg} \times \dfrac{m}{\text{sec}}}$$

$$= \frac{6.63 \times 10^{-34}}{7.50} \frac{\dfrac{\text{kg} \cdot \text{m}^2}{\text{sec}^2} \cdot \text{sec}}{\dfrac{\text{kg} \cdot \text{m}}{\text{sec}}}$$

$$= \boxed{8.84 \times 10^{-35} \text{ m}}$$

What is the de Broglie wavelength associated with an electron with a velocity of 6.00×10^6 m/sec?

Solution

$m = 9.11 \times 10^{-31}$ kg

$v = 6.00 \times 10^6$ m/sec

$h = 6.63 \times 10^{-34}$ J·sec

$\lambda = ?$

$$\lambda = \frac{h}{mv}$$

$$= \frac{6.63 \times 10^{-34} \text{ J} \cdot \text{sec}}{(9.11 \times 10^{-31} \text{ kg})\left(6.00 \times 10^6 \dfrac{m}{\text{sec}}\right)}$$

$$= \frac{6.63 \times 10^{-34}}{5.47 \times 10^{-24}} \frac{\text{J} \cdot \text{sec}}{\text{kg} \times \dfrac{m}{\text{sec}}}$$

$$= 1.21 \times 10^{-10} \frac{\dfrac{\text{kg} \cdot \text{m}^2}{\text{sec}^2} \cdot \text{sec}}{\dfrac{\text{kg} \cdot \text{m}}{\text{sec}}}$$

$$= \boxed{1.21 \times 10^{-10} \text{ m}}$$

The baseball wavelength of 8.84×10^{-35} m is much too small to be detected or measured. The electron wavelength of 1.21×10^{-10} m, on the other hand, is comparable to the distances between atoms in a crystal, so a beam of electrons through a crystal should produce diffraction.

The idea of matter waves was soon tested after de Broglie published his theory. Experiments with a beam of light passing through a very small opening or by the edge of a sharp-edged obstacle were described in chapter 8. In these experiments the beam of light produces diffraction and interference patterns. This was part of the evidence for the wave nature of light, since such results could only be explained by waves, not particles. When similar experiments were performed with a beam of electrons, *identical* wave property behaviors were observed. This and many related experiments showed without doubt that electrons have both wave properties and particle properties. And, as was the case with light waves, measurements of the electron interference patterns provided a means to measure the wavelength of electron waves.

Recall that waves confined on a fixed string establish resonate modes of vibration called *standing waves* (see chapter 6). Only certain fundamental frequencies and harmonics can exist on a string, and the combination of the fundamental and overtones gives the stringed instrument its particular quality. The same result of resonate modes of vibrations is observed in *any* situation where waves are confined to a fixed space. Characteristic standing wave patterns depend on the wavelength and wave velocity for waves formed on strings, in enclosed columns of air, or for any kind of wave in a confined space. Electrons are confined to the space near a nucleus, and electrons have wave properties, so an electron in an atom must be a confined wave. Does an electron form a characteristic wave pattern? This was the question being asked in about 1925 when Heisenberg, Schrödinger, Dirac, and others applied the wave nature of the electron to develop a new model of the atom based on the mechanics of electron waves. The new theory is now called **wave mechanics,** or **quantum mechanics.**

Activities

Obtain a long spiral spring such as a "slinky." Connect one end to the other, forming a circle. Suspend the circular spring with rubber bands so that the spring makes a horizontal circle. Wiggle one part of the spring at various frequencies until standing waves are formed around the circle. From your observations, explain why n is a whole number in the equation $2\pi r = n\lambda$ (equation 9.7).

Wave Mechanics

Erwin Schrödinger, an Austrian physicist, treated the atom as a three-dimensional system of waves to derive what is now called the *Schrödinger equation.* Instead of the simple circular planetary orbits of the Bohr model, solving the Schrödinger equation results in a description of three-dimensional shapes of the patterns that develop when electron waves are confined by a nucleus. Schrödinger first considered the hydrogen atom, calculating the states of vibration that would be possible for an electron wave confined by a nucleus. He found that the frequency of these vibrations, when multiplied by Planck's constant, matched exactly, to the last decimal point, the observed energies of the quantum states of the hydrogen atom ($E = hf$). The conclusion is that the wave nature of the electron is the important property to consider for a successful model of the atom.

The quantum mechanics theory of the atom proved to be very successful as it confirmed all the known experimental facts and predicted new discoveries. The theory does have some of the same quantum ideas as the Bohr model; for example, an electron emits a photon when jumping from a higher state to a lower one. The Bohr model, however, considered the particle nature of an electron moving in a circular orbit with a definitely assigned position at a given time. Quantum mechanics considers the wave nature, with the electron as a confined wave with well-defined shapes and frequencies. A wave is not localized like a particle and is spread out in space. The quantum mechanics model is, therefore, a series of orbitlike smears, or fuzzy statistical representations, of where the electron might be found.

The Quantum Mechanics Model

The quantum mechanics model is a highly mathematical treatment of the mechanics of matter waves. In addition, the wave properties are considered as three-dimensional problems, and three quantum numbers are needed to describe the fuzzy electron cloud. The mathematical detail will not be presented here. The following is a qualitative description of the main ideas in the quantum mechanics model. It will describe the results of the mathematics and will provide a mental visualization of what it all means.

First, understand that the quantum mechanical theory is not an extension or refinement of the Bohr model. The Bohr model considered electrons as particles in circular orbits that could be only certain distances from the nucleus. The quantum mechanical model, on the other hand, considers the electron as a wave and considers the energy of its harmonics, or modes, of standing waves. In the Bohr model the location of an electron was certain—in an orbit. In the quantum mechanical model the electron is a spread-out wave.

Quantum mechanics describes the energy state of an electron wave with four *quantum numbers,* in terms of its (1) distance from the nucleus, (2) energy sublevel, (3) orientation in space, and (4) direction of spin.

The **principal quantum number,** called *n,* describes the *main energy level* of an electron in terms of its most probable distance from the nucleus. The lowest energy state possible is closest to the nucleus and is assigned the principal quantum number of 1 ($n = 1$). Higher states are assigned progressively higher positive whole numbers of $n = 2$, $n = 3$, $n = 4$, and so on. Electrons with higher principal quantum numbers have higher energies and are located farther from the nucleus.

The **angular momentum quantum number** defines energy sublevels within the main energy levels. Each sublevel is identified with a letter. The first four of these letters, in order of increasing energy, are s, p, d, and f. The choice of these letters goes back to spectral studies when the spectral lines were described as *s*harp, *p*rincipal, *d*iffuse, and *f*ine. The letter s represents the lowest sublevel and the letter f represents the highest sublevel. A principal quantum number and a letter indicating the angular momentum quantum number are combined to identify the main energy state and energy sublevel of an electron. For an electron in the lowest main energy level, n = 1, and in the lowest sublevel, s, the number and letter are 1s (read as "one-s"). Thus 1s indicates an electron that is as close to the nucleus as possible in the lowest energy sublevel possible.

As stated in the Bohr model, the location of an electron was certain, that is, in an orbit. In the quantum mechanical model the electron is spread out and knowledge of its location is very uncertain. The **Heisenberg uncertainty principle** states that you cannot measure both the momentum and the exact po-

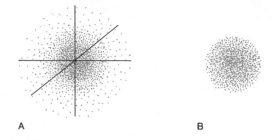

Figure 9.17

(*a*) An electron distribution sketch representing probability regions where an electron is most likely to be found. (*b*) A boundary surface, or contour, that encloses about 90 percent of the electron distribution shown in (*a*). This three-dimensional space around the nucleus, where there is the greatest probability of finding an electron, is called an orbital.

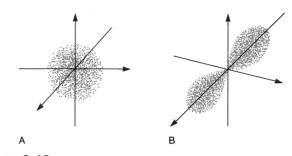
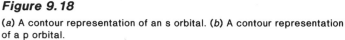

Figure 9.18

(*a*) A contour representation of an s orbital. (*b*) A contour representation of a p orbital.

sition of an electron at the same time. The location of the electron can only be described in terms of *probabilities* where it might be at a given instant. The probability of location is described by a fuzzy region of space called an **orbital.** An orbital defines the space where an electron is likely to be found. Orbitals have characteristic three-dimensional shapes and sizes and are identified with electrons of characteristic energy levels (figure 9.17).

An orbital shape represents where an electron could probably be located at any particular instant. This "probability cloud" could likewise have any particular orientation in space, and the direction of this orientation is uncertain. On the other hand, an external magnetic field applied to an atom produces different energy levels that are related to the orientation of the orbital to the magnetic field. The orientation of an orbital in space is described by the **magnetic quantum number.** This number is related to the energies of orbitals as they are oriented in space relative to an external magnetic field, a kind of energy sub-sublevel. In general, the lowest energy sublevel (s) has only one orbital orientation. The next higher energy sublevel (p) can have three orbital orientations (figure 9.18). The d sublevel can have five orbital orientations, and the highest sublevel of f can have a total of seven different orientations (table 9.1).

High-resolution studies of the hydrogen line spectra revealed details that were not known when the Bohr model of the atom was first developed. These studies showed, for example, that what was previously believed to be a single red line was

Table 9.1

Quantum numbers and electron distribution to $n = 4$

Main Energy Level	Energy Sublevels	Maximum Number of Electrons	Maximum Number of Electrons per Main Energy Level
$n = 1$	s	2	2
$n = 2$	s	2	
	p	6	8
$n = 3$	s	2	
	p	6	
	d	10	18
$n = 4$	s	2	
	p	6	
	d	10	
	f	14	32

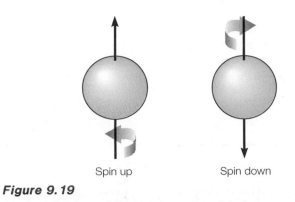

Spin up   Spin down

Figure 9.19

Experimental evidence supports the concept that electrons can be considered to spin one way or the other as they move about an orbital under an external magnetic field.

actually two lines that were very close together. The only way to explain the splitting was to consider the electron to be spinning on its axis like a top. Such a spinning movement would cause the electron to produce a magnetic field. The energy of the electron would depend on which way the electron magnetic field was aligned with an external magnetic field. Thus, an electron spinning one way (say clockwise) would have a different energy than one spinning the other way (say counterclockwise). These two spin orientations are described by the **spin quantum number** (figure 9.19).

Electron spin is an important property of electrons that helps determine the electronic structure of an atom. As it turns out, two electrons spinning in opposite directions produce unlike magnetic fields that are attractive, balancing some of the normal repulsion from two like charges. Two electrons of opposite spin, called an **electron pair,** can thus occupy the same orbital. This was summarized in 1924 by Wolfgang Pauli, a German physicist. His summary, now known as the **Pauli exclusion principle,** states that *no two electrons in an atom can have the same four quantum numbers.* This provides the key for understanding the electron structure of atoms.

Electron Configuration

Recall that the energy of electrons is measured by considering an unattached electron to have an energy value of zero. It takes energy to remove an electron from an atom, so a bound electron must have *less* energy than a free one. The arrangement of electrons in orbitals is called the **electron configuration.** When in the ground state, electrons always adopt the lowest possible energies consistent with the Pauli exclusion principle, arranging themselves in the lowest energy orbitals as close to the nucleus as possible. The lowest possible energy level is $n = 1$ and the lowest sublevel is s. *The number of electrons that can occupy this orbital is limited.* According to the Pauli exclusion principle, no two electrons in an atom can have all four of their quantum numbers exactly the same. If one electron is in the first energy level ($n = 1$) in the orbital of the lowest energy (s), a

second electron can occupy this same $n = 1$, s orbital only if it has a different spin orientation. Thus, the exclusion principle states there can only be two electrons in the $n = 1$, s orbital and, as it works out, there can only be *a maximum of two electrons in any given orbital.* An atom of helium has two electrons and both can occupy the $n = 1$, s orbital because they have opposite spins. In general, electrons occupy the orbitals in an order starting with the lowest energy. Before you can describe the electron arrangement, you need to know how many electrons are present in an atom.

An atom is electrically neutral, so the number of protons (positive charge) must equal the number of electrons (negative charge). The atomic number therefore identifies the number of electrons as well as the number of protons:

atomic number = number of protons = number of electrons

Now that you have a means of finding the number of electrons, consider the various energy levels to see how the electron configuration is determined. There are four things to consider: (1) the main energy level, (2) the energy sublevel, (3) the number of orbital orientations, and (4) the electron spin. Recall that the lowest energy level is $n = 1$, and successive numbers identify progressively higher energy levels. Recall also that the energy sublevels, in order of increasing energy, are s, p, d, and f. This electron configuration is written in shorthand with 1s standing for the lowest energy sublevel of the first energy level. A superscript gives the number of electrons present in a sublevel. Thus, the electron configuration for a helium atom, which has two electrons, is written as

$$1s^2$$

This combination of symbols has the following meaning:

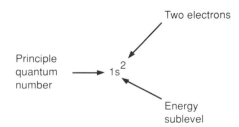

Two electrons

Principle quantum number $\longrightarrow$ 1s^2

Energy sublevel

Atomic Structure

Table 9.2

Electron configuration for the first twenty elements

Atomic Number	Element	Electron Configuration
1	Hydrogen	$1s^1$
2	Helium	$1s^2$
3	Lithium	$1s^2 2s^1$
4	Beryllium	$1s^2 2s^2$
5	Boron	$1s^2 2s^2 2p^1$
6	Carbon	$1s^2 2s^2 2p^2$
7	Nitrogen	$1s^2 2s^2 2p^3$
8	Oxygen	$1s^2 2s^2 2p^4$
9	Fluorine	$1s^2 2s^2 2p^5$
10	Neon	$1s^2 2s^2 2p^6$
11	Sodium	$1s^2 2s^2 2p^6 3s^1$
12	Magnesium	$1s^2 2s^2 2p^6 3s^2$
13	Aluminum	$1s^2 2s^2 2p^6 3s^2 3p^1$
14	Silicon	$1s^2 2s^2 2p^6 3s^2 3p^2$
15	Phosphorus	$1s^2 2s^2 2p^6 3s^2 3p^3$
16	Sulfur	$1s^2 2s^2 2p^6 3s^2 3p^4$
17	Chlorine	$1s^2 2s^2 2p^6 3s^2 3p^5$
18	Argon	$1s^2 2s^2 2p^6 3s^2 3p^6$
19	Potassium	$1s^2 2s^2 2p^6 3s^2 3p^6 4s^1$
20	Calcium	$1s^2 2s^2 2p^6 3s^2 3p^6 4s^2$

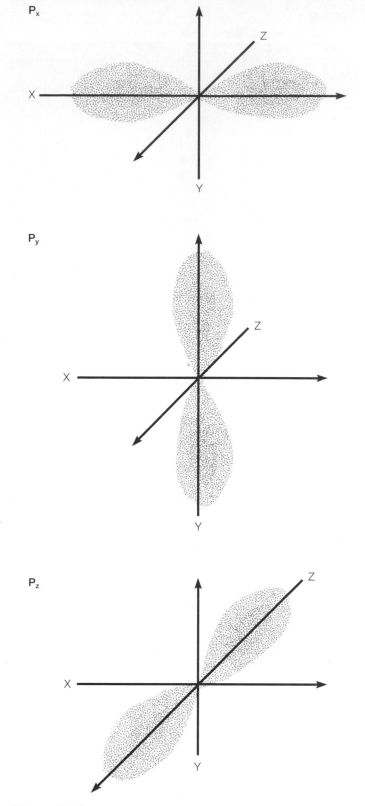

The symbols mean an atom with two electrons in the s sublevel of the first main energy level.

Table 9.2 gives the electron configurations for the first twenty elements. The configurations of the p energy sublevel have been condensed in this table. There are three possible orientations of the p orbital, each with two electrons (figure 9.20). This is shown as p^6, which is a condensation of the three possible p orientations. Note that the number of electrons in all the orbitals add up to the atomic number. Note also that as you proceed from a lower atomic number to a higher one, the higher element has the same configuration as the element before it with the addition of one more electron. In general, it is then possible to begin with the simplest atom, hydrogen, and add one electron at a time to the order of energy sublevels and obtain the electron configuration for all the elements. The exclusion principle limits the number of electrons in any orbital, and allowances will need to be made for the more complex behavior of atoms with many electrons.

The energies of the orbital are not fixed as you progress through the atomic numbers, and there are several factors that influence their energies. The first orbitals are filled in a straightforward 1s, 2s, 2p, 3s, then 3p order. Then the order becomes contrary to what you might expect. One useful way of

Figure 9.20

There are three possible orientations of the p orbital and these are called p_x, p_y, and p_z. Each orbital can hold two electrons, so a total of six electrons are possible in the three orientations, thus the notation p^6.

The Quark

Some understanding about how matter is put together came with the discovery of the electron, proton, and neutron—three elementary particles that make up an atom. In the early 1900s a particle outside the atom, the *photon,* was verified by experimental evidence. Two other particles were verified in the 1930s, the *neutrino* ("little neutral one") and the *positron* (a positively charged electron). By the mid-1930s a total of six elementary particles were known. Since that time, high-energy accelerator experiments made it possible to collide particles with great violence, probing the inner parts of atoms and how they are put together. A multitude of elementary particles are now known to exist. This feature is about these elementary particles.

There are now thought to be twelve elementary particles that make up matter. The elementary particles can be divided into three main groups: (1) *leptons,* which exist independently, (2) *quarks,* which exist together, making up a third group, and (3) *hadrons.*

Leptons are a group of fundamental particles that include the familiar electron, the muon (an overweight relative of the electron), and three types of neutrinos. In radioactive decay, an electron (beta particle) is emitted with an electron neutrino. For each lepton there is a corresponding antiparticle, or antilepton, with the same mass but opposite electric charge.

Hadrons are a group of composite particles with an internal structure, so they are not elementary particles. There are two subgroups of hadrons: (1) the *mesons* (meaning "intermediate mass," between electrons and protons) and (2) the *baryons* (meaning "greater mass"). Hundreds of short-lived hadrons have been identified that exist briefly after high-energy collisions. Among the more stable are the baryons named protons and neutrons.

Hadrons are composed of different combinations of fundamental particles called quarks. Five kinds of quarks, fancifully called *flavors,* have been identified and a sixth is believed to exist. The existing quarks are called up, down, sideways (or strange), charm, and bottom (or beauty). The sixth flavor yet to be discovered will be named top (or truth). Each flavor carries a fractional charge that is either $-1/3$ or $+2/3$. Antiquarks have equal but opposite charges. In order to explain how identical quarks could combine as observed, each flavor was assigned three quantum states that are called *color.* Each flavor can carry a charge of red, green, or blue. Antiquarks carry a corresponding anticolor, for example, a red quark has an antiquark of the color cyan, a green quark has an antiquark of the color magenta, and a blue quark has an antiquark of the color yellow. The idea of quark color was designed to follow the allowable combination of quarks and antiquarks according to the exclusion principle. Hadrons do not have a color charge, so the sum of the quark colors making up the hadron must result in a white hadron. Baryons are made up of three quarks, so a combination of a red, a green, and a blue quark would be acceptable since this would result in a white baryon. Mesons are made up of a quark and an antiquark so a combination of a blue quark and a yellow antiquark would be acceptable since this would result in a white meson.

The story of subnuclear elementary particles is by no means complete. There is no explanation at present for why quarks and leptons exist. Are there more fundamental particles? Answers to this and more questions await further research. Eventually the studies of fundamental particles will reveal the fundamental laws of nature, explaining why matter is put together the way it is. The next chapter in this story will likely be written at the Superconducting Supercollider.

---

figuring out the order in which orbitals are filled is illustrated in figure 9.21. Each row of this matrix represents a principal energy level with possible energy sublevels increasing from left to right. The order of filling is indicated by the diagonal arrows. There are exceptions to the order of filling as shown by the matrix, but it works for most of the elements.

Figure 9.21 ▶

A matrix showing the order in which the orbitals are filled. Start at the top left, then move from the head of each arrow to the tail of the one immediately below it. This sequence moves from the lowest energy level to the next higher level for each orbital.

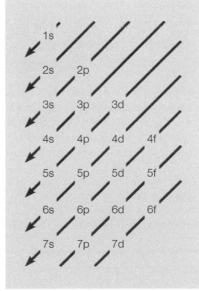

Example 9.6

Certain strontium (atomic number 38) compounds are used to add the pure red color to flares and fireworks. Write the electron configuration for strontium.

Solution

First, note that an atomic number of 38 means a total of thirty-eight electrons. Second, refer to the order-of-filling matrix in figure 9.21. Remember that only two electrons can occupy an orbital, but there are three orientations of the p orbital for a total of six electrons. There are likewise five possible orientations of the d orbital for a total of ten electrons. Starting at the lowest energy level, two electrons go in 1s making $1s^2$, then two go in 2s making $2s^2$. That is a total of four electrons so far. Next $2p^6$ and $3s^2$ use eight more electrons for a total of twelve so far. The $3p^6$, $4s^2$, $3d^{10}$, and $4p^6$ use up twenty-four more electrons for a total of thirty-six. The remaining two go into the next orbital $5s^2$ and the complete answer is

Strontium: $1s^2\ 2s^2\ 2p^6\ 3s^2\ 3p^6\ 4s^2\ 3d^{10}\ 4p^6\ 5s^2$

Summary

Attempts at understanding matter date all the way back to ancient Greek philosophers, who viewed matter as being composed of *elements,* or simpler substances. Two models were developed that considered matter to be (1) *continuous,* or infinitely divisible, or (2) *discontinuous,* made up of particles called *atoms.*

During the 1600s Robert Boyle provided experimental evidence to reject the ancient Greek ideas of continuous matter made up of earth, air, fire, and water. Boyle reasoned there were *elements,* which could not be broken down to anything simpler, and compounds, which were made up of combinations of elements.

In the early 1800s Dalton published an *atomic theory,* reasoning that matter was composed of hard, indivisible atoms that were joined together or dissociated during chemical change.

When a good air pump was invented in 1885, *cathode rays* were observed to move from the negative terminal in an evacuated glass tube. The nature of cathode rays was a mystery. The mystery was solved in 1887 when Thomson discovered they were negatively charged particles now known as *electrons.* Thomson had discovered the first elementary particle of which atoms are made and measured their charge to mass ratio.

Rutherford developed a "solar system" model based on experiments of alpha particles scattered from a thin sheet of metal. This model had a small, massive, and positively charged *nucleus* surrounded by moving electrons. These electrons were calculated to be at a distance 100,000 times the radius of the nucleus, so the volume of an atom is mostly empty space. Later, Rutherford proposed that the nucleus contained two elementary particles, *protons* with a positive charge and *neutrons* with no charge. The *atomic number* is the number of protons in an atom.

Bohr developed a model of the hydrogen atom to explain the characteristic *line spectra* emitted by hydrogen. His model specified that (1) electrons can move only in allowed orbits, (2) electrons do not emit radiant energy when they remain in an orbit, and (3) electrons move from one allowed orbit to another when they gain or lose energy. When an electron jumps from a higher orbit to a lower one it gives up energy in the form of a single photon. The energy of the photon corresponds to the difference in energy between the two levels. The Bohr model worked well for hydrogen but not other atoms.

De Broglie proposed that moving particles of matter (electrons) should have wave properties like moving particles of light (photons). His derived equation, $\lambda = h/mv$, showed that these *matter waves* were only measurable for very small particles such as electrons. De Broglie's proposal was tested experimentally, and the experiments confirmed that electrons do have wave properties.

Schrödinger and others used the wave nature of the electron to develop a new model of the atom called *wave mechanics,* or *quantum mechanics.* This model was found to confirm exactly all the experimental data as well as predict new data. The quantum mechanical model describes the energy state of the electron in terms of quantum numbers based on the wave nature of the electron. The quantum numbers defined the *probability* of the location of an electron in terms of a fuzzy region of space called *orbitals.*

Summary of Equations

9.1

$$\text{energy} = (\text{Planck's constant})\ (\text{frequency})$$

$$E = hf$$

where $h = 6.63 \times 10^{-34}\ \text{J}\cdot\text{sec}$

9.2

$$\frac{1}{\text{wavelength}} = \text{constant}\left(\frac{1}{2^2} - \frac{1}{\text{number}^2}\right)$$

$$\frac{1}{\lambda} = R\left(\frac{1}{2^2} - \frac{1}{n^2}\right)$$

where $R = 1.097 \times 10^7\ 1/\text{m}$

9.3

$$\text{angular momentum} = \left(\begin{array}{c}\text{orbit}\\\text{quantum}\\\text{number}\end{array}\right)\left(\frac{\text{Planck's constant}}{2\pi}\right)$$

$$mvr = n\frac{h}{2\pi}$$

where $h = 6.63 \times 10^{-34}\ \text{J}\cdot\text{sec}$ and $n = 1, 2, 3, \ldots$ for an orbit

9.4

$$\text{electrical force} = \text{centripetal force}$$

$$\text{Coulomb's law} = \text{Newton's second law for circular motion}$$

$$\frac{kq_1q_2}{r^2} = \frac{mv^2}{r}$$

9.5

$$\text{energy state of orbit number} = \frac{\text{energy state of innermost orbit}}{\text{number squared}}$$

$$E_n = \frac{E_1}{n^2}$$

where $E_1 = -13.6\ \text{eV}$ and $n = 1, 2, 3, \ldots$

9.6

$$\begin{array}{c}\text{energy}\\\text{of}\\\text{photon}\end{array} = \left(\begin{array}{c}\text{energy state}\\\text{of}\\\text{higher orbit}\end{array}\right) - \left(\begin{array}{c}\text{energy state}\\\text{of}\\\text{lower orbit}\end{array}\right)$$

$$hf = E_h - E_l$$

where $h = 6.63 \times 10^{-34}\ \text{J}\cdot\text{sec}$, E_h and E_l must be in joules

9.7

circumference of orbit = (whole number)(wavelength)

$$2\pi r = n\lambda$$

where $n = 1, 2, 3, \ldots$

9.8

$$\text{wavelength} = \frac{\text{Planck's constant}}{(\text{mass})(\text{velocity})}$$

$$\lambda = \frac{h}{mv}$$

where $h = 6.63 \times 10^{-34}$ J·sec

Key Terms

angular momentum quantum number (p. 192)
atomic number (p. 186)
Balmer series (p. 187)
Bohr model (p. 188)
cathode rays (p. 183)
electron (p. 184)
electron configuration (p. 193)
electron pair (p. 193)
electron volt (p. 188)
element (p. 182)
excited states (p. 189)
ground state (p. 189)
Heisenberg uncertainty principle (p. 192)
line spectrum (p. 186)
magnetic quantum number (p. 192)

matter waves (p. 190)
neutron (p. 186)
nucleus (p. 185)
orbital (p. 192)
Pauli exclusion principle (p. 193)
photons (p. 186)
principal quantum number (p. 192)
proton (p. 186)
quanta (p. 186)
quantum mechanics (p. 191)
quantum numbers (p. 187)
spin quantum number (p. 193)
wave mechanics (p. 191)

Applying the Concepts

1. According to the modern definition, which of the following is an element?
   a. water
   b. iron
   c. air
   d. All of the above.

2. John Dalton reasoned that atoms exist from the evidence that
   a. elements could not be broken down into anything simpler.
   b. water pours and flows when in the liquid state.
   c. elements always combined in certain fixed ratios.
   d. peanut butter and jelly could be combined in any ratio.

3. The electron was discovered through experiments with
   a. radioactivity.
   b. light.
   c. matter waves.
   d. electricity.

4. Thomson was convinced that he had discovered a subatomic particle, the electron, from the evidence that
   a. the charge to mass ratio was the same for all materials.
   b. cathode rays could move through a vacuum.
   c. electrons were attracted toward a negatively charged plate.
   d. the charge was always 1.60×10^{-19} coulomb.

5. The existence of a tiny, massive, and positively charged nucleus was deduced from the observation that
   a. fast, massive, and positively charged alpha particles moved straight through metal foil.
   b. alpha particles were deflected by a magnetic field.
   c. some alpha particles were deflected by metal foil.
   d. none of the above were correct.

6. According to Rutherford's calculations, the volume of an atom is mostly
   a. occupied by protons and neutrons.
   b. filled with electrons.
   c. occupied by tightly bound protons, electrons, and neutrons.
   d. empty space.

7. Atomic number is the number of
   a. protons.
   b. protons plus neutrons.
   c. protons plus electrons.
   d. protons, neutrons, and electrons in an atom.

8. All atoms of an element have the same
   a. atomic number.
   b. number of electrons.
   c. number of protons.
   d. All of the above are correct.

9. The main problem with a "solar system" model of the atoms is that
   a. electrons move in circular, not elliptical orbits.
   b. the electrons should lose energy since they are accelerating.
   c. opposite charges should attract one another.
   d. the mass ratio of the nucleus to the electrons is wrong.

10. The energy of a photon
    a. varies inversely with the frequency.
    b. is directly proportional to the frequency.
    c. varies directly with the velocity, not the frequency.
    d. is inversely proportional to the velocity.

11. The frequency of a particular color of light is equal to
    a. Eh.
    b. h/E.
    c. E/h.
    d. Eh/2.

12. A photon of which of the following has the most energy?
    a. red light
    b. orange light
    c. green light
    d. blue light

13. The lines of color in a line spectrum from a given element
    a. change colors with changes in the temperature.
    b. are always the same with a regular spacing pattern.
    c. are randomly spaced, having no particular pattern.
    d. have the same colors with a spacing pattern that varies with the temperature.

14. Hydrogen, with its one electron, produces a line spectrum in the visible light range with
    a. one color line.
    b. two color lines.
    c. three color lines.
    d. four color lines.

15. Using the laws of motion for moving particles and the laws of electrical attraction, Bohr calculated that electrons could
    a. move only in orbits of certain allowed radii.
    b. move, as the planets, in orbits of any distance from the nucleus.
    c. move in orbits that had distances that matched the separation of line color distances in the line spectrum.
    d. move in orbits of variable distances that are directly proportional to the velocity of the electrons.

16. According to the Bohr model, an electron gains or loses energy only by
    a. moving faster or slower in an allowed orbit.
    b. jumping from one allowed orbit to another.
    c. being completely removed from an atom.
    d. jumping from one atom to another atom.

17. When an electron in a hydrogen atom jumps from an orbit farther from the nucleus to an orbit closer to the nucleus, it
    a. emits a single photon with an energy equal to the energy difference of the two orbits.
    b. emits four photons, one for each of the color lines observed in the line spectrum of hydrogen.
    c. emits a number of photons dependent on the number of orbit levels jumped over.
    d. None of the above are correct.

18. The Bohr model of the atom
    a. explained the color lines in the hydrogen spectrum.
    b. could not explain the line spectrum of atoms larger than hydrogen.
    c. had some made-up rules without explanations.
    d. All of the above are correct.

19. The proposal that matter, like light, has wave properties in addition to particle properties was
    a. verified by diffraction experiments with a beam of electrons.
    b. never tested since it was known to be impossible.
    c. tested mathematically, but not by actual experiments.
    d. verified by physical measurement on a moving baseball.

20. The quantum mechanics model of the atom is based on
    a. the quanta, or measured amounts of energy of a moving particle.
    b. the energy of a standing electron wave that can fit into an orbit.
    c. calculations of the energy of the three-dimensional shape of a circular orbit of an electron particle.
    d. Newton's laws of motion, but scaled down to the size of electron particles.

21. The Bohr model of the atom described the energy state of electrons with one quantum number. The quantum mechanics model uses how many quantum numbers to describe the energy state of an electron?
    a. one
    b. two
    c. four
    d. ten

22. An electron in the lowest main energy level and the second sublevel is described by the symbols
    a. 1s.
    b. 2s.
    c. 1p.
    d. 2p.

23. The probability of the space where an electron is likely to be found is described by a(an)
    a. circular orbit.
    b. elliptical orbit.
    c. orbital.
    d. geocentric orbit.

24. Two electrons can occupy the same orbital because they have different
    a. principal quantum numbers.
    b. angular momentum quantum numbers.
    c. magnetic quantum numbers.
    d. spin quantum numbers.

Answers

1. b   2. c   3. d   4. a   5. c   6. d   7. a   8. d   9. b
10. b   11. c   12. d   13. b   14. d   15. a   16. b   17. a
18. d   19. a   20. b   21. c   22. c   23. c   24. d

Questions for Thought

1. What reason did Dalton have for bringing back the ancient Greek idea of matter being composed of hard, indivisible atoms?

2. What was the experimental evidence that Thomson had discovered the existence of a subatomic particle when working with cathode rays?

3. Describe the experimental evidence that led Rutherford to the concept of a nucleus in an atom.

4. What is the main problem with a solar system model of the atom?

5. Compare the size of an atom to the size of its nucleus.

6. What does atomic number mean? How is the atomic number related to what an atom is an element of? How is atomic number related to the number of electrons in an atom?

7. An atom has 11 protons in the nucleus. What is the atomic number? What is the name of this element? What is the electron configuration of this atom?

8. Describe the three main points in the Bohr model of the atom.

9. Why do the energies of electrons in an atom have negative values? (Hint: It is *not* because of the charge of the electron)

10. Which has the lowest energy, an electron in the first energy level ($n = 1$) or an electron in the third energy level ($n = 3$)? Explain.

11. What is similar about the Bohr model of the atom and the quantum mechanical model? What are the fundamental differences?

12. What is the difference between a hydrogen atom in the ground state and one in the excited state?

Exercises

Group A—Solutions Provided in Appendix

1. A neutron with a mass of 1.68×10^{-27} kg moves from a nuclear reactor with a velocity of 3.22×10^3 m/sec. What is the de Broglie wavelength of the neutron?

2. Calculate the energy (a) in eV and (b) in joules for the sixth energy level ($n = 6$) of a hydrogen atom.

3. How much energy is needed to move an electron in a hydrogen atom from $n = 2$ to $n = 6$? Give the answer (a) in joules and (b) in eV. (See figure 9.14 for needed values)

4. What frequency of light is emitted when an electron in a hydrogen atom jumps from $n = 6$ to $n = 2$?

5. How much energy is needed to completely remove the electron from a hydrogen atom in the ground state?

6. Thomson determined the charge to mass ratio of the electron to be -1.76×10^{11} coulomb/kilogram. Millikan determined the charge on the electron to be -1.60×10^{-19} coulomb. According to these findings, what is the mass of an electron?

7. Assume that an electron wave making a standing wave in a hydrogen atom has a wavelength of 1.67×10^{-10} m. Considering the mass of an electron to be 9.11×10^{-31} kg, use the de Broglie equation to calculate the velocity of an electron in this orbit.

8. Using any reference you wish, write the complete electron configurations for (a) boron, (b) aluminum, (c) potassium.

9. Explain how you know that you have the correct *total* number of electrons in your answer for 8a, 8b, and 8c.

10. Refer to figure 9.21 *only* and write the complete electron configurations for (a) argon, (b) zinc, (c) bromine.

Group B—Solutions Not Given

1. An electron with a mass of 9.11×10^{-31} kg has a velocity of 4.3×10^6 m/sec in the innermost orbit of a hydrogen atom. What is the de Broglie wavelength of the electron?

2. Calculate the energy (a) in eV and (b) in joules of the third energy level ($n = 3$) of a hydrogen atom.

3. How much energy is needed to move an electron in a hydrogen atom from the ground state ($n = 1$) to $n = 3$? Give the answer (a) in joules, (b) in eV.

4. What frequency of light is emitted when an electron in a hydrogen atom jumps from $n = 2$ to the ground state ($n = 1$)?

5. How much energy is needed to completely remove an electron from $n = 2$ in a hydrogen atom?

6. If the charge to mass ratio of a proton is 9.58×10^7 coulomb/kilogram and the charge is 1.60×10^{-19} coulomb, what is the mass of the proton?

7. An electron wave making a standing wave in a hydrogen atom has a wavelength of 8.33×10^{-11} m. If the mass of the electron is 9.11×10^{-31} kg, what is the velocity of the electron according to the de Broglie equation?

8. Using any reference you wish, write the complete electron configurations for (a) nitrogen, (b) phosphorus, (c) chlorine.

9. Explain how you know that you have the correct *total* number of electrons in your answer for 8a, 8b, and 8c.

10. Referring to figure 9.21 *only,* write the complete electron configuration for (a) neon, (b) sulfur, (c) calcium.

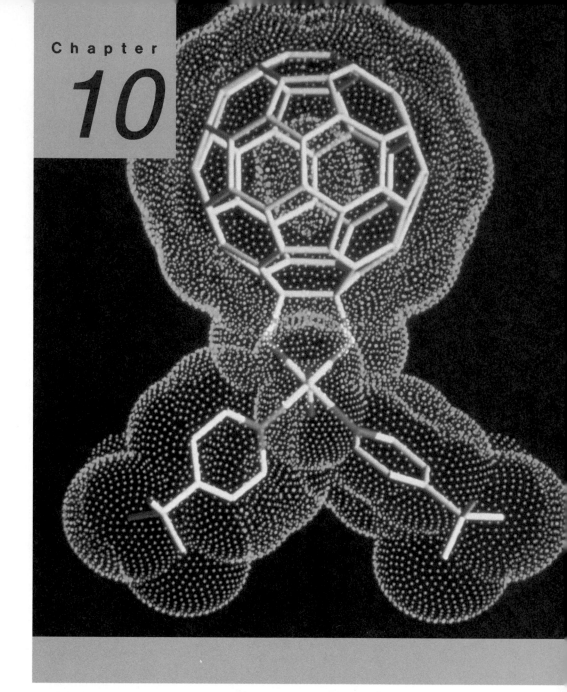

Elements and the
Periodic Table

One molecule of Buckminsterfullerene
(buckeyball) is made of 60 carbon atoms.

In chapter 9 you learned how the concept of the atom was developed, evolving into the modern model of a fuzzy cloud of matter with an internal structure. We considered this internal structure and how the electrons make up the fuzzy cloud in atoms of different elements. But there was no discussion about the meaning or implications of the different structures or how they could be used to understand the properties of matter. That is the goal of this chapter, to understand matter based on the electron structure of atoms.

The behavior of matter seems bewildering when you consider all of its different kinds, forms, and shapes and all the changes it undergoes. Things seem bewildering and confusing because you cannot make connections between behaviors; that is, there are no apparent patterns. Often in such situations the act of grouping or classifying things by similar properties is helpful. Classifying helps you to find patterns of similarities and identities of groups. Once you have a pattern you will want to find a reason for its existence. You are now on your way to an understanding that not only accounts for the patterns but also provides a means of organizing all the information as well.

This chapter presents an example of gaining understanding through grouping, which is also known as classifying (figure 10.1). We will consider several different ways that matter can be classified, for example, into classes of mixtures and pure substances. Pure substances can be further subdivided into groups of elements and compounds. Some of the more interesting elements will be described, along with how they were named and their symbols. Matter can be classified other ways, including changes in matter. All changes are either physical or chemical, and chemical changes are the key to grouping according to the electron structure of atoms. After an introduction of a few new properties of atoms, the periodic table will be discussed in terms of its systematic classification of elements. This periodic classification can be used to predict how elements react with one another, making the study of matter much easier.

Figure 10.1

Classification is the arranging into groups or categories according to some criteria. The act of classifying creates a pattern that helps you to recognize and understand the behavior of fish, chemicals, or any matter in your surroundings. These fish, for example, are classified as salmon because they ascend rivers from the sea for breeding.

Classifying Matter

The universe is made up of just two basics, matter and energy. **Matter** is usually defined as anything that occupies space and has mass. It is the substance of any particle or object, from the parts that make up atoms to the bulk of a giant star. Between these two extremes of size there is an overwhelming and complex variety of observable sizes, shapes, and kinds of matter on the earth. One way to make thinking about such a variety a little less complex is to *classify.* Classifying is the act of mentally making groups based on similar properties. Classifying helps to organize your thinking and to find patterns that might otherwise go unnoticed. The following considers several different ways of classifying matter by considering different properties.

Metals and Nonmetals

Humans learned thousands of years ago that certain earthlike materials placed on a bed of glowing coals produced a new substance, a metal. The smelting of copper and tin dates back to about 3500 B.C., the beginning of the bronze age. Iron smelting dates back to about 1500 B.C., and steel was first made about 1200 B.C. These metals all have very different properties than the earthlike materials they were extracted from. Today a **metal** is recognized as a kind of matter having the following physical properties:

1. metallic *luster,* the way a shiny metal reflects light to have the look of gold, copper, and other metals (figure 10.2);
2. high heat and electrical *conductivity;*
3. *malleability,* which means that you can roll it or pound it into a thin sheet; and
4. *ductility,* which means that you can pull it into a wire.

All metals are solids at room temperature except mercury, which is a liquid.

A **nonmetal** is a kind of matter that does not have a metallic luster, is a poor conductor of heat and electricity, and when

Elements and the Periodic Table

Figure 10.2

Most matter can be classified as metals or nonmetals according to physical properties. Aluminum, for example, is a lightweight kind of matter that can be melted and rolled into a thin sheet or pulled into a wire. Here you see aluminum pop cans that have been compressed into 1,600 lb bales for recycling, destined to again be formed into a new pop can, aluminum foil, or perhaps an aluminum wire.

solid, is a brittle material that cannot be pounded or pulled into new shapes. Nonmetals occur as solids, liquids, or gases at room temperature.

Most unprotected metals do not last long when exposed to air. Iron, for example, rusts to a reddish, earthlike material with nonmetallic properties. In time, most metals seem to return to the same kind of nonmetallic, earthlike materials from which they were extracted. Eventually humans began to wonder about the transformation of nonmetallic, earthlike materials to metals and why they returned to nonmetallic forms. Today, such questions are considered in **chemistry,** the science concerned with the study of the composition, structure, and properties of substances and the transformations they undergo.

Solids, Liquids, and Gases

The Greek philosophers of some 2,500 years ago thought of all matter as being made up of four elements that were identified as earth, air, water, and fire. It is interesting that, as we know now, the ancient Greeks were not describing elements. They were describing the general forms in which all matter exists and one form of energy. Earth, air, and water are the most common examples of the solid, gaseous, and liquid phases of matter. Fire is the most common example of heat, the energy involved in changes of matter such as nonmetallic ores to metals. Thus, the ancient Greeks were actually describing the basics that make up the universe, matter and energy.

Solid, liquid, and gas are called the **phases of matter,** and these phases represent another way to classify matter. On the earth, all matter generally belongs to one of the three groups, or phases, which are defined by two general properties. These properties are how well a sample of matter maintains (1) its shape and (2) its volume (figure 10.3). For example, the *gaseous* phase of matter is not able to maintain a definite shape or a definite volume. A sample of gas released into a completely evacuated, rigid container will disperse throughout the entire space inside the container, taking the shape and volume of the container. Since density is mass per unit volume, the density of the gas will depend on how much gas is placed in the container as well as the volume of the container.

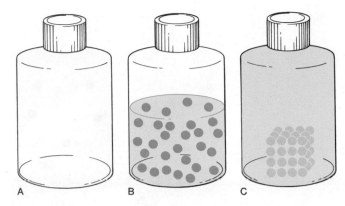

Figure 10.3

(*a*) A gas disperses throughout a container, taking the shape and volume of the container. (*b*) A liquid takes the shape of the container but retains its own volume. (*c*) A solid retains its own shape and volume.

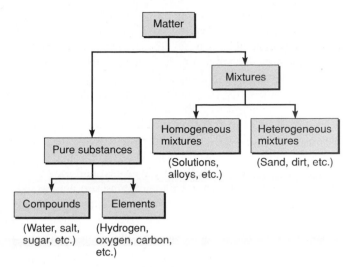

Figure 10.4

A classification scheme for matter.

Liquids, like gases, will flow and can be poured from one container to another. Both liquids and gases are fluids. Both have an indefinite shape and take the shape of their container. A liquid, however, has a definite volume and does not fill any container as a gas does. A 500 cm³ sample of gas will disperse to fill a 1,000 cm³ container completely. A 500 cm³ sample of liquid will still have that volume when placed into a 1,000 cm³ container. Liquids have a much greater density than gases and are not very compressible in comparison. Most common liquids have a density not much different from water at 1 g/cm³ (1 kg/L) at 4° C.

Solids, like liquids, have a definite volume. Unlike gases and liquids, a solid has a definite shape. A solid with a cubic shape placed in a container of some other shape maintains both its volume and its original shape. Most solids are more dense when in the solid state than in the liquid state. Ice is the common exception to this generalization as ice floats on water, which is more dense. This exception will be explained when water is discussed in a later chapter.

Mixtures and Pure Substances

Still another way to classify matter is to consider its composition, that is, how it is put together (figure 10.4). In this classification scheme all common matter occurs in one of two groups, mixtures or pure substances. **Mixtures** are made of unlike parts that have a variable composition. Common sand, for example, is a mixture of bits of rocks, minerals, and perhaps hard parts from plants and animals. Since sand is a mixture, any given sample might have more of some things and less of others. The black sand of Hawaii is mostly tiny pieces of volcanic rocks. The white sand of Florida is mostly tiny pieces of ground-up seashells. The composition and proportions of a sand mixture depend on the source of the components and how far the sand has been transported. Solutions, gases, and alloys are also mixtures since they are made of unlike parts with a variable composition.

Mixtures can be separated into their component parts by physical means. For example, you can physically separate the parts making up a sand mixture by using a magnifying glass and tweezers to move and isolate each part. A solution of salt in water is a mixture since the amount of salt dissolved in water can have a variable composition. But how do you separate the parts of a solution? One way is to evaporate the water, leaving the salt behind. There are many methods for separating mixtures but all involve a **physical change.** A physical change does not alter the *identity* of matter. When water is evaporated from the salt solution, for example, it changes to a different state of matter (water vapor) but is still recognized as water. Physical changes involve physical properties only and no new substances are formed. Examples of physical changes include evaporation, condensation, melting, freezing, and dissolving.

Mixtures can be physically separated into **pure substances,** materials that are the same throughout and have a fixed, definite composition. If you closely examine a sample of table salt you will see that it is made up of hundreds of tiny cubes. Any one of these cubes will have the identical properties as any other cube, including a salty taste. Sugar, like table salt, has all of its parts alike. Unlike salt, sugar grains have no special shape or form. But each grain has the same sweet taste and other properties as any other sugar grain.

If you heat salt and sugar in separate containers, you will find very different results. Some pure substances, such as salt, undergo a physical change and melt, changing back to a solid upon cooling with the same properties that it had originally. Sugar, however, changes to a black material upon heating, while it gives off water vapor. The black material does not change back to sugar upon cooling. The sugar has *decomposed* to a new substance while the salt did not. The new substance is carbon and it has properties completely different from sugar. The original substance (sugar) and its properties have changed to a new substance (carbon) with new properties. The sugar has gone through a **chemical change.** A chemical change alters the identity of matter, producing new substances with different properties. In this case the chemical change was one of decomposition. Heat produced a chemical change by decomposing the sugar into carbon and water vapor.

Elements and the Periodic Table

The decomposition of sugar always produces the same mass ratio of carbon to water, so sugar has a fixed, definite composition. A **compound** is a pure substance that can be decomposed by a chemical change into simpler substances with a fixed mass ratio. This means that sugar is a compound (figure 10.5).

A pure substance that cannot be broken down into anything simpler by chemical or physical means is an **element.** Sugar is decomposed by heating into carbon and water vapor. Carbon cannot be broken down further, so carbon is an element. It has been known since about 1800 that water is a compound that can be broken down by electrolysis into hydrogen and oxygen, two gases that cannot be broken down to anything simpler. So sugar is a compound made from the elements of carbon, hydrogen, and oxygen.

But what about the table salt? Is table salt a compound? Table salt is a stable compound that is not decomposed by heating. It melts at a temperature of about 1,000° C, then returns to the solid form with the same salty properties upon cooling. Electrolysis was used in the early 1800s to decompose table salt into the elements sodium and chlorine, positively proving that it is a compound. Heat brings about a chemical change and decomposes some compounds, such as sugar. Heat will not bring about a chemical change in table salt, so other means are needed.

Pure substances are either compounds or elements. Decomposition through heating and decomposition through electrolysis are two means of distinguishing between compounds and elements. If a substance can be decomposed into something simpler, you know for sure that it is a compound. If the substance cannot be decomposed it might be an element or it might be a stable compound that resists decomposition. More testing would be necessary before you can be confident that you have identified an element. Most pure substances are compounds. There are millions of different compounds but only 109 known elements at the present time. These elements are the fundamental materials of which all matter is made.

Elements

Modern science is usually identified as beginning just after the time of Galileo and Newton, about three hundred years ago. Understandings about matter at this time were still based on the writings of Aristotle of some two thousand years earlier. There were alternative ideas about the number of elements, but consideration of earth, air, fire, and water as the basic elements was generally accepted.

Reconsidering the Fire Element

During the early 1700s the use of fuels such as coal, the process of burning, and the production of steam became topics of general interest. The first working steam engine was invented by Thomas Newcomen in 1711, and the time was ripe for new ideas, new theories, and answers to questions about burning and what was going on during the burning process.

About 1700 a new theory about burning was introduced by a German physician. This theory considered all burnable materials to contain a substance called *phlogiston,* a word from the Greek meaning "fire." Burning was considered to be the

Figure 10.5

Sugar is a compound that can be easily decomposed to simpler substances by heating. One of the simpler substances is the black element carbon, which cannot be further decomposed by chemical or physical means.

escape of phlogiston from fuels into the air. Materials that did not burn were considered not to contain phlogiston, either because they never had it or because they had already lost it (such as ashes). So far, phlogiston might remind you of Aristotle's "fire element" with a different name. But this theory continued on in detail to explain other observations. For example, a candle in a closed container would burn for a few minutes, then go out. The explanation for this observation was that the air could hold only so much phlogiston, which was released during burning. When the air was saturated it could accept no more phlogiston. The escape of more phlogiston was thus prevented and the fire would go out.

Other observations explained by the theory involved (1) the conversion of nonmetallic, earthlike materials into metals by fire and (2) the rusting of metals, returning them to nonmetallic, earthlike matter over a period of time. The earthlike materials were considered to be phlogiston poor. Placing the materials in a fire permitted them to absorb phlogiston, becoming metals. The metals could not hold on to the phlogiston and, over time, it leaked away, returning the metals to their nonmetallic form (figure 10.6). The phlogiston theory, with its convincing explanations, became the accepted understanding of burning, metal smelting, rusting, and the role of air in these processes during the 1700s.

Discovery of Modern Elements

A series of events would lead to the downfall of the phlogiston theory and the discovery of modern elements in the 1770s. The first of these events was an experiment in gas chemistry conducted by an English minister named Joseph Priestley. Priestley was an amateur chemist with a natural flair for experimental research. One of his first discoveries was a method for producing carbon dioxide by reacting chalk and sulfuric acid, then forcing the gas into a container of water. Priestley had invented soda water, which soon would become modern-day cola. During

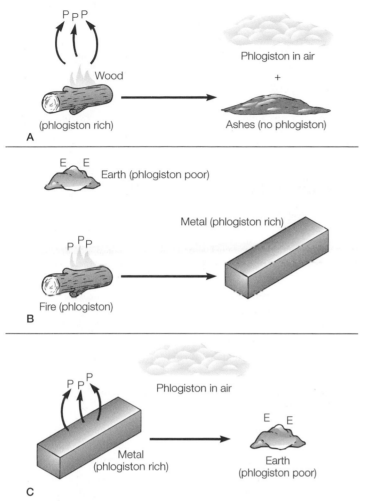

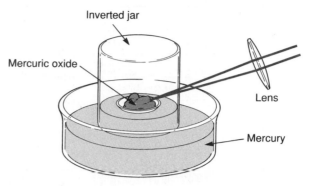

Figure 10.7

Priestley produced a gas (oxygen) by using sunlight to heat mercuric oxide kept in a closed container. The oxygen forced some of the mercury out of the jar as it was produced, increasing the volume about five times.

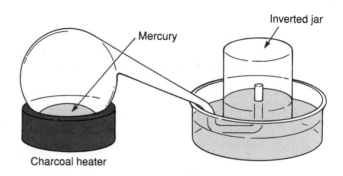

Figure 10.8

Lavoisier heated a measured amount of mercury to form the red oxide of mercury. He measured the amount of oxygen removed from the jar and the amount of red oxide formed. When the reaction was reversed, he found the original amounts of mercury and oxygen.

Figure 10.6

The phlogiston theory. (*a*) In this theory, burning was considered to be the escape of phlogiston into the air. (*b*) Smelting combined phlogiston-poor earth with phlogiston from a fire to make a metal. (*c*) Metal rusting was considered to be the slow escape of phlogiston from a metal into the air.

Priestley's time, a navy officer won approval to supply ships with soda water, believing it might help with scurvy. The officer, coincidentally, was named Lord Sandwich.

Priestley is recognized today for the discovery of oxygen. In one of his experiments, Priestley heated a nonmetallic red powder that was then called mercurius calcinatus (mercuric oxide) in a jar inverted in a bowl of mercury (figure 10.7). The powder gave off a gas, forcing some of the mercury out of the inverted jar. From this, Priestley could tell that the red powder gave off about five times its own volume of gas. He found that this gas caused a candle to burn vigorously, a glowing piece of wood to sparkle, and a mouse to jump about with vigor.

Priestley had discovered oxygen, but he interpreted his findings in terms of the then-current phlogiston theory. He believed that the red powder combined with phlogiston from the air. That was why things burned vigorously in the air. The air was depleted of phlogiston, Priestley thought, which rapidly "pulled" it from flaming candles and glowing pieces of wood. The phlogiston theory provided an explanation for what was observed, and Priestley was not much concerned with measurements.

Antoine Lavoisier was already a recognized French chemist in 1772 when, at the age of twenty-nine, he began to experiment with the role of phlogiston in burning and the rusting of metals. He had observed that sulfur and phosphorus *gained* weight when they burned, not lost weight as the phlogiston theory predicted. He supposed that the sulfur and phosphorus combined with air to make the additional weight. When he read about Priestley's "dephlogisticated air," Lavoisier repeated Priestley's work with a carefully measured analysis. Following through with his earlier observations about burning and weight gain, Lavoisier proved that Priestley's "dephlogisticated air" was actually a substance he called *oxygen* (figure 10.8). He replaced the phlogiston theory with a theory that burning is a chemical combination of some material with oxygen.

In the meantime, Henry Cavendish had isolated a very light, highly flammable gas by reacting metals with acids. Cavendish found that when this gas burned, pure water was produced. Lavoisier repeated the Cavendish experiment and concluded that the light gas combined with oxygen to produce water. Lavoisier named the gas *hydrogen* from Greek words meaning "water former." This proved to be the final blow to the ancient Greek concepts of water and air as elements as well as the end of the phlogiston theory. Lavoisier recognized the need for a whole new concept of elements, compounds, and chemical

Elements and the Periodic Table

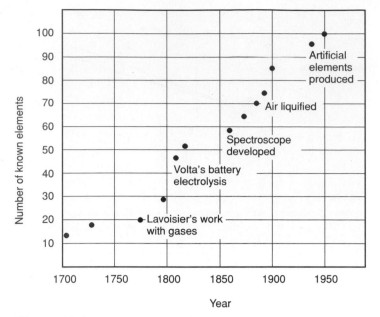

Figure 10.9

The number of known elements increased as new chemical and analytical techniques were developed.

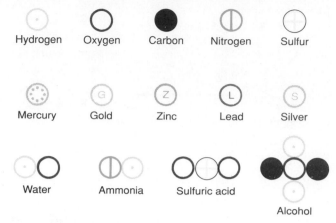

Figure 10.10

Here are some of the symbols Dalton used for atoms of elements and molecules of compounds. He probably used a circle for each because he thought of atoms as the ancient Greeks did, as tiny, round, hard spheres.

change. The concept of elements was reconsidered, and by working with other leading chemists of the time, a whole new list of elements was developed.

Lavoisier published the results of his experiments and a list of the known elements in 1789. Lavoisier's list of thirty-three elements included twenty-three that are recognized as elements today and eight that are recognized as compounds along with light and heat. This was the beginning of modern chemistry and of understanding elements as they are known today.

Figure 10.9 shows the number of known elements since the time of Lavoisier. New elements were discovered as new chemical and analytical techniques were developed. The increase just before 1800 resulted from an interest in rocks, minerals, and the materials of the earth. Just after the 1800s the invention of the electric battery was followed by experiments in which compounds were decomposed into elements by electrolysis. These experiments resulted in the discovery of more elements. By 1809, about half of the natural elements had been discovered. The development of the spectroscope in the late 1850s led to the identification of more elements by their spectral lines. Mendeleev found patterns in elements in 1869, which resulted in even more discoveries. By the 1890s, technology was developed that made it possible to produce very cold, liquified gases, and this resulted in even more discoveries. All the natural elements were discovered by 1940. The naturally occurring elements generally range from hydrogen (atomic number 1) to uranium (atomic number 92). The exceptions are:

• Technetium (43) and promethium (61) are not found anywhere on earth unless they are artificially produced.

• Francium (87) and astatine (85) exist only a short time before undergoing radioactive decay, so they do not exist in any significant quantities. Half a sample of francium decays in 21 minutes. The most stable form of astatine decays in 7 1/2 hours.

• Plutonium (94) has been found occurring naturally in small amounts.

The other elements that do not occur naturally were artificially made during nuclear physics experiments and are not found on the earth. Thus, there are 89 naturally occurring elements that occur in significant quantities and 20 artificially prepared ones that are known at the present time for a total of 109.

Names of the Elements

Elements such as sulfur, zinc, tin, and iron have been known since ancient times but were not recognized as elements. The original meanings of their names are ancient and have become obscured with time. More recently, the right to name an element belonged to the discoverer, who could call it anything as long as the name of a metal ended with "-um." The first 103 elements have internationally accepted names. Sources of these names have included the following:

1. the compound or substance in which the element was discovered;
2. an unusual or identifying property of the element;
3. places, cities, and countries;
4. famous scientists;
5. Greek mythology or some other mythology;
6. astronomical objects.

Chemical Symbols

When John Dalton introduced his atomic theory in the early 1800s he introduced a system of symbols to represent atoms and how they formed compounds. Each element had its own special symbol, which stood for one atom of that element. Some of the symbols Dalton used to explain his atomic theory are shown in figure 10.10. Each atom was a circle, probably because Dalton

Table 10.1

Elements with symbols from Latin or German names

Atomic Number	Name	Source of Symbol	Symbol
11	Sodium	Latin: Natrium	Na
19	Potassium	Latin: Kalium	K
26	Iron	Latin: Ferrum	Fe
29	Copper	Latin: Cuprum	Cu
47	Silver	Latin: Argentum	Ag
50	Tin	Latin: Stannum	Sn
51	Antimony	Latin: Stibium	Sb
74	Tungsten	German: Wolfram	W
79	Gold	Latin: Aurum	Au
80	Mercury	Latin: Hydrargyrum	Hg
82	Lead	Latin: Plumbum	Pb

Table 10.2

Elements making up 99 percent of earth's crust

Element (Symbol)	Percent by Weight
Oxygen (O)	46.6
Silicon (Si)	27.7
Aluminum (Al)	8.1
Iron (Fe)	5.0
Calcium (Ca)	3.6
Sodium (Na)	2.8
Potassium (K)	2.6
Magnesium (Mg)	2.1

thought of atoms as tiny, indivisible spheres. He indicated the different elements by symbols or letters within each circle, for example, a dot for hydrogen and a "G" for gold. Dalton represented compounds with combinations of element symbols as shown in the figure.

Letter symbols came into use about 1913 after an earlier recommendation by Jons Berzelius, a Swedish chemist. Berzelius recommended that the letter symbol should be the capitalized first letter of the name of an element, for example, the symbol H for hydrogen. Today, there are about a dozen common elements that have single capitalized first letters as their symbols. All the rest have two letters with the first letter *always* capitalized and the second letter *never* capitalized. The second, lowercase letter is either (1) the second letter in the name of the element or (2) a letter representing a strong consonant heard when the name of the element is spoken. Examples of using the first two letters of the name are Ca for calcium and Si for silicon. Examples of using the first letter and the letter that is heard when the name is spoken are Cl for chlorine and Cr for chromium.

Some elements have symbols from the earlier use of Latin names for the elements. For example, the symbol Au is used for gold because the metal was earlier known by its Latin name of *aurum,* meaning "shining dawn." There are ten elements with symbols from Latin names and one with a symbol from a German name. These eleven elements are listed in table 10.1, together with the sources of their names and their symbols.

Chemical symbols are like the vocabulary of a new language and are used to describe chemical changes with clarity and precision. The first key to understanding this language is to understand that a chemical symbol identifies a specific element and represents one atom of that element. Thus, the symbol H means one atom of the element hydrogen. The symbol Hg means one atom of the element mercury and so on.

You usually learn the symbols by using them and looking up a symbol as needed from a table such as the one located on the inside back cover of this text. This is really not as big a task as it might seem at first, because the elements are not equally abundant and only a few are common. In table 10.2, for example, you can see that only eight elements make up about 99 percent of the solid surface of the earth. Oxygen is most abundant, making up about 50 percent of the weight of the earth's crust. Silicon makes up more than 25 percent, so these two nonmetals alone make up about 75 percent of the earth's solid surface. Almost all the rest is made up of just six metals, as shown in the table.

The number of common elements is limited elsewhere, too. Only two elements make up about 99 percent of the atmospheric air around the earth. Air is mostly nitrogen (about 78 percent) and oxygen (about 21 percent) with traces of five other elements and compounds. Water on the earth is hydrogen and oxygen, of course, but seawater also contains elements in solution. These elements are chlorine (55 percent), sodium (31 percent), sulfur (8 percent), and magnesium (4 percent). Only three elements make up about 97 percent of your body. These elements are hydrogen (60 percent), oxygen (26 percent), and carbon (11 percent). Generally, all of this means that the elements are not equally distributed or equally abundant in nature (figure 10.11).

Symbols and Atomic Structures

When Dalton developed his atomic theory he pictured atoms as tiny, hard spheres. He considered these spheres to differ only in their masses, with all the atoms of one element having the same mass, which was different from the atomic masses of the other elements. An important contribution of Dalton's theory was the attempt to determine a comparison of the atomic masses. He used the fixed mass ratios of combining elements to determine the relative atomic masses. For example, hydrogen always combines with oxygen in a mass ratio of 1:8 to make water. This means that one gram of hydrogen combines with eight grams of oxygen to form nine grams of water. Dalton assumed that one atom of hydrogen combined with one atom of oxygen to

Elements and the Periodic Table

Figure 10.11

Just imagine, the elements of aluminum, iron, oxygen, and silicon make up about 88 percent of the earth's solid surface. Water on the surface and in the air as clouds and fog is made up of hydrogen and oxygen. The air is 99 percent nitrogen and oxygen. Hydrogen, oxygen, and carbon make up 97 percent of the person. Thus, almost everything you see in this picture is made up of just 6 elements.

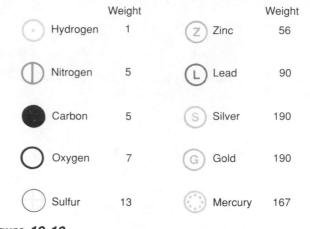

	Weight		Weight
Hydrogen	1	Zinc	56
Nitrogen	5	Lead	90
Carbon	5	Silver	190
Oxygen	7	Gold	190
Sulfur	13	Mercury	167

Figure 10.12

Using information from the fixed mass ratios of combining elements, Dalton was able to calculate the relative atomic masses of some of the elements. Many of his findings were wrong, as you can see from this sample of his table.

form a particle of water. Today, the particle of water is called a **molecule.** For now, consider a molecule to be a particle composed of two or more atoms held together by an attractive force called a chemical bond.

From the assumption that one atom of hydrogen joined with one atom of oxygen to form a molecule of water, Dalton reasoned that the mass ratio of 1:8 could be explained if one atom of oxygen is eight times more massive than one atom of hydrogen. Measuring mass ratios of hydrogen, oxygen, and other elements, Dalton was able to construct the table of relative atomic masses (figure 10.12).

Dalton's relative atomic mass table was a step in the right direction, but his findings were all wrong. They were wrong because of measurement errors and the assumption that one atom of an element combines with one atom of another element. Today, for example, it is known that two atoms of hydrogen combine with one atom of oxygen to form water. Chemists eventually worked out techniques for determining the number of atoms of elements that combine to form compounds, and these techniques will be discussed in later chapters. It took about one hundred years of carefully measured chemical experimentation before accurate values of relative atomic masses were established.

Radioactivity was discovered in the late 1890s, and before long new but puzzling data were observed. Recall that Dalton considered all atoms of an element to have the same mass and all to act the same when combining with other elements. That is, all the atoms of an element were considered to have the same physical properties and the same chemical properties. Analysis of radioactive elements found that the masses of atoms were *not* always the same. One sample of the metal lead, for example, was radioactive but another sample was not radioactive. Further analysis found that the two samples had different masses

even though the two samples were the same element with identical chemical behaviors. In 1910 Frederic Soddy called such varieties of an element an **isotope** because they were in the same place chemically (from the Greek *iso* meaning "same" and *tope* meaning "place"). So isotopes are atoms of an element with identical chemical properties but with different masses.

The masses and abundance of isotopes were measured by Francis William Aston in the 1910s. Aston had been an assistant to J. J. Thomson and was familiar with his method of measuring the charge to mass ratio of the electron. But Aston now had a new device called a *mass spectrometer* (figure 10.13). This instrument uses a beam of electrons to ionize, for example, vaporized atoms of an element. Positive ions are formed, then accelerated by high voltage through a tube. In the tube they are deflected by a magnetic field, and the amount of deflection depends on their masses. A wide range of mass values is scanned by varying the magnetic field and accelerating voltage. A detector is thus not only able to measure the presence of various masses but also the abundance of each mass from the intensity. A plot of the intensity measured by the detector of the mass is called a *mass spectrum.*

Aston was able to provide the first clear, undisputed evidence for the existence of isotopes with the mass spectrograph. He confirmed that suspected isotopes existed for some elements and discovered new ones as well. Figure 10.14, for example, shows that a sample of chlorine gas has two chlorine isotopes in different proportions, and this was unknown before Aston analyzed chlorine in the mass spectrograph. In 1919, Aston reported that the atomic mass of each substance was very close to, but not exactly, a whole number. Chlorine, for example, has one isotope with an atomic mass very close to 35 (34.969 units). Thus, the atomic mass of an element with more than one isotope must be the *average* of the atomic masses of the isotopes. The contribution of a particular isotope to this average value depends on its abundance. For example, if the chlorine isotope with an atomic mass of 34.969 units makes up 75.53 percent of a

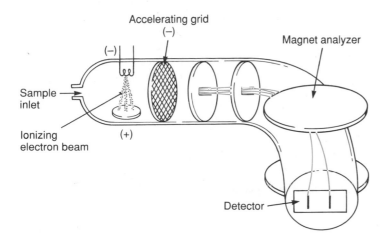

Figure 10.13

A schematic of a mass spectrometer. The atoms of a sample of gas become positive ions after being bombarded by a beam of electrons. The ions are deflected into a curved path by a magnetic field, which separates them according to their charge to mass ratio. Less massive ions are deflected the most, so the device identifies different groups of particles with different masses.

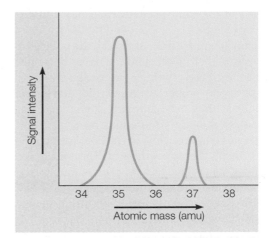

Figure 10.14

A mass spectrum of chlorine from a mass spectrometer. Note that two separate masses of chlorine atoms are present, and their abundance can be measured from the signal intensity. The greater the signal intensity, the more abundant the isotope.

Table 10.3

Selected atomic weights calculated from mass and abundance of isotopes

Stable Isotopes	Mass of Isotope Compared to C-12	Abundance	Atomic Weight
$^{1}_{1}H$	1.007	99.985%	
$^{2}_{1}H$	2.0141	0.015%	1.0079
$^{9}_{4}Be$	9.01218	100.%	9.01218
$^{14}_{7}N$	14.00307	99.63%	
$^{15}_{7}N$	15.00011	0.37%	14.0067
$^{16}_{8}O$	15.99491	99.759%	
$^{17}_{8}O$	16.99914	0.037%	
$^{18}_{8}O$	17.00016	0.204%	15.9994
$^{19}_{9}F$	18.9984	100.%	18.9984
$^{20}_{10}Ne$	19.99244	90.92%	
$^{21}_{10}Ne$	20.99395	0.257%	
$^{22}_{10}Ne$	21.99138	8.82%	20.179
$^{27}_{13}Al$	26.9815	100.%	26.9815

sample of chlorine gas, and the chlorine isotope with an atomic mass of 36.966 units makes up the other 24.47 percent, then

$$\text{average atomic mass} = (75.53\% \times 34.969 \text{ units})$$
$$+ (24.47\% \times 36.966 \text{ units})$$
$$= 26.41 \text{ units} + 9.05 \text{ units}$$
$$= 35.46 \text{ units}$$

The average atomic mass of chlorine in this example is 35.46 units, which is a *weighted average* of the masses of the chlorine isotopes as they occur. **Atomic weight** is the name given to the weighted average of the masses of stable isotopes of an element as they occur in nature.

Dalton had originally assigned hydrogen a mass of one unit, then compared the mass of other atoms to this standard. Today, the mass of any isotope is compared to the mass of a particular isotope of carbon. This isotope is *assigned* a mass of exactly 12.00 units called **atomic mass units** (u). This isotope, called carbon-12, provides the standard to which the masses of all other isotopes are compared. The mass of any isotope is based on the mass of a carbon-12 isotope. A table of atomic weights, such as the one on the inside back cover of this text, lists the atomic weight of carbon as slightly more than 12 u. Carbon occurs naturally as two isotopes, one with an atomic mass of exactly 12.00 u (98.9 percent) and one with an atomic mass of about 13 u (1.11 percent). Thus, the weighted average, or *atomic weight,* of carbon is slightly greater than about 12.01 u. Other examples are illustrated in table 10.3.

The abundance of each isotope making up a naturally occurring element is always the same, no matter where the sample is measured. While isotopes have different masses, they have the *same chemical behavior.* Isotopes occur because of varying numbers of neutrons in the nuclei of atoms with the same atomic number. The atomic number is the number of protons in

Elements and the Periodic Table

the nucleus and the number of electrons in the atom. What element an atom belongs to is identified by the number of protons, and the chemical nature is identified by the number of electrons. Since the isotopes of an element all have the same number of protons and electrons with the same configuration, they all have the same chemical properties. Since they have different numbers of neutrons they have different atomic masses.

A neutron is slightly more massive than a proton, but an electron has a comparatively trivial mass (about 1/1,840 the mass of a proton). Thus, neutrons and protons make up almost all the mass of an atom. The sum of the number of protons and neutrons in a nucleus is called the **mass number** of an atom. Mass numbers are used to identify isotopes. A chlorine atom with 17 protons and 20 neutrons has a mass number of 17 + 20, or 37, and is referred to as "chlorine-37." A chlorine atom with 17 protons and 18 neutrons has a mass number of 17 + 18, or 35, and is referred to as "chlorine-35." Using symbols, chlorine-37 is written as

$$^{37}_{17}Cl$$

where Cl is the chemical symbol for chlorine, the subscript to the bottom left is the atomic number, and the superscript to the top left is the mass number. Consider the following rules when you use a chemical symbol to identify isotopes with a mass number:

1. The mass number is the closest whole number to the atomic mass of an isotope. Only carbon-12 has an atomic mass with a whole number (by definition).
2. The number of protons in the nucleus of an atom equals the atomic number.
3. The number of neutrons in the nucleus of an atom equals the mass number minus the atomic number.

Atomic numbers are always whole numbers because they represent the number of protons in the nucleus of an atom. *Mass numbers* are always whole numbers because they represent the number of protons and neutrons in the nucleus of an isotope. The *atomic mass of an isotope* is *not* a whole number, with the exception of carbon-12, which was assigned a mass of exactly 12 units to set up a relative scale from which the masses of all other atoms are compared. Why are the atomic masses of other isotopes not whole numbers? The answer is found in a mass contribution from the internal energy of the nucleus. There is a relationship between energy and mass, and this relationship will be explained in a future chapter on nuclear energy. *Atomic weights* are also *not* whole numbers, but you might expect this because atomic weight is a weighted average of the masses of the isotopes. Compare these values in table 10.3.

Example 10.1

Identify the number of protons, neutrons, and electrons in an atom of $^{16}_{8}O$.

Solution

The subscript to the bottom left is the atomic number. Atomic number is defined as the number of protons in the nucleus, so this number identifies the number of protons as 8. Any atom with 8 protons is an

atom of oxygen, which is identified with the symbol O. The superscript to the top left identifies the mass number of this isotope of oxygen, which is 16. The mass number is defined as the sum of the number of protons and the number of neutrons in the nucleus. Since you already know the number of protons is 8 (from the atomic number), then the number of neutrons is 16 minus 8, or 8 neutrons. Since a neutral atom has the same number of electrons as protons, an atom of this oxygen isotope has 8 protons, 8 neutrons, and 8 electrons.

Example 10.2

How many protons, neutrons, and electrons are found in an atom of $^{17}_{8}O$? (Answer: 8 protons, 9 neutrons, and 8 electrons)

The Periodic Law

Many new elements were discovered in a short period of time in the early 1800s, nearly doubling the number of known elements to 60 by the 1840s. As the list grew, information about the behavior of elements and their compounds created a large body of apparently unrelated facts. Seeking to organize and make sense out of this mass of information, chemists made attempts to classify the elements according to their properties. They were looking for some underlying order, a pattern that would not only organize the large body of facts but also perhaps form the basis of a theory that would account for the facts. The process began with attempts to relate somehow the behaviors of elements to their atomic weights.

As information about the behavior of elements increased, chemists began to recognize certain patterns of similarities. Lithium, sodium, and potassium were found to be shiny, soft metals that reacted vigorously with water to form an alkaline solution. Calcium, strontium, and barium were found to be another group of soft metals with similar properties, but properties that were different from the sodium group. Iron, cobalt, and nickel were similar hard metals. Chlorine, bromine, and iodine were similar nonmetals. By 1829, a German chemist named Johann Dobereiner described the existence of *triads*, groups of three elements with similar chemical properties. Dobereiner found that when the elements of a triad were listed in order of atomic weights, the atomic weight of the middle element was almost equal to the average atomic weight of the other two elements. In most cases the values for density, melting point, and other properties of the middle element were also midway between values for the other two elements. Three of these triads are identified in table 10.4. There was no explanation that would account for the occurrence of triads at this time, and they were considered an interesting curiosity.

A workable classification scheme of the elements was developed independently by two scientists. The Russian chemist Dmitri Mendeleev and the German physicist Lothar Meyer published similar classification schemes in 1869. Mendeleev's scheme was based mostly on the chemical properties of elements and their atomic weight, and Meyer's was based on physical properties and atomic weight. But both arranged the elements in order of increasing atomic weight and observed that

Table 10.4

Examples of element triads

Element	Atomic Weight (u)	Average Atomic Weight (First and Third)	Density (g/cm³)
Lithium (Li)	6.9		0.53
Sodium (Na)	23.0	23.0	0.97
Potassium (K)	39.1		0.86
Calcium (Ca)	40.1		1.55
Strontium (Sr)	87.6	88.7	2.54
Barium (Ba)	137.3		3.50
Chlorine (Cl)	35.5		1.56
Bromine (Br)	79.9	81.2	3.12
Iodine (I)	126.9		4.93

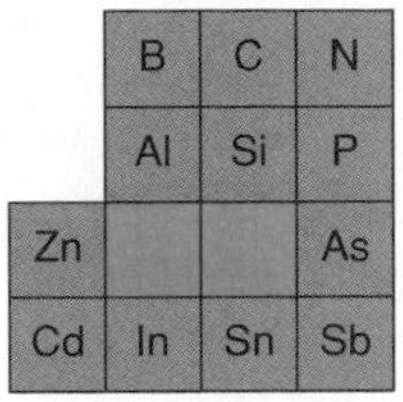

the properties recur periodically. Both had devised schemes that would systematize the study of chemistry and lead to the modern periodic table.

Mendeleev and Meyer both arranged the elements in rows of increasing atomic weights, arranged so that elements with similar properties made vertical columns. These vertical columns contained Dobereiner's triads of elements with similar properties. Mendeleev and Meyer were not restricted by trying to fit their individual schemes to some perceived "law" as others before them. Both left blank spaces if the element with the next highest atomic weight did not fit with a vertical family. For example, the next known element following zinc (Zn) in the sequence of increasing atomic weight was arsenic (As). But placing As after Zn in the table would place it in the same vertical column with aluminum (Al). The properties of As suggested that it belonged in the column with phosphorus (P), not Al. So two blank spaces were left after Zn, suggesting undiscovered elements.

Mendeleev demonstrated his understanding and daring by predicting that elements would be discovered to fill the gaps in his table and predicting the physical and chemical properties of the yet to be discovered elements. The gaps were directly below boron, aluminum, and silicon in the 1871 version of his table. The gaps below aluminum and silicon are illustrated in figure 10.15. Mendeleev named the elements eka-boron, eka-aluminum, and eka-silicon ("eka" is Sanskrit for "one" and in this context means "next one"). Eka-aluminum was discovered in 1875 and is now called gallium (Ga), eka-boron was discovered in 1879 and is now called scandium (Sc), and eka-silicon was discovered in 1886 and is now called germanium (Ge). There was an impressive correspondence between Mendeleev's predicted properties of these elements and the properties that were observed after their discovery. Mendeleev is usually given credit for developing the periodic table, probably because of his dramatic and highly publicized predictions about the unknown elements.

Figure 10.15

Mendeleev left blank spaces in his table when the properties of the elements above and below did not seem to match. The existence of unknown elements was predicted by Mendeleev on the basis of the blank spaces. When the unknown elements were discovered, it was found that Mendeleev had closely predicted the properties of the elements as well as their discovery.

There were some problems with Mendeleev's original periodic table because not all atomic weights were quite right for the proper placement. And there was no theoretical accounting for similar chemical and physical properties in terms of atomic weight. After the work of Rutherford and Moseley on the atomic nucleus, it became clear that the *atomic number,* not the atomic weight, was the fundamental factor. The atomic number, that is, the number of protons in the nucleus and the number of electrons around the nucleus, is the significant, essential basis for the modern periodic table. Thus the modern **periodic law** is:

Similar physical and chemical properties recur periodically when the elements are listed in order of increasing atomic number.

The Modern Periodic Table

The periodic table is made up of squares, with each element having its own square in a specific location. The squares are arranged in rows and columns but not symmetrically. The arrangement has a meaning, both about atomic structure and about chemical behaviors. The key to meaningful, satisfying use of the table is to understand the code of this structure. The following explains some of what the code means. It will facilitate your understanding of the code if you refer frequently to a periodic table during the following discussion (see the inside back cover of this text and figure 10.22).

An element is identified in each square with its chemical symbol. The number above the symbol is the atomic number of the element, and the number below the symbol is the rounded atomic weight of the element. Horizontal *rows* of elements run from left to right with increasing atomic numbers. Each row is called a **period.** The periods are numbered from 1 to 7 on the left side. Period 1, for example, has only two elements, H (hydrogen) and He (helium). Period 2 starts with Li (lithium) and ends with Ne (neon). The two rows at the bottom of the table are actually part of periods 6 and 7 (between atomic numbers 57 and 72, 89 and 104). They are moved so the table is not so wide.

A vertical *column* of elements is called a **family** (or *group*) of elements. Elements in families have similar properties, but this is true of some families more than others. Note that the families are identified with Roman numerals and letters at the top of each column. Group IIA, for example, begins with Be

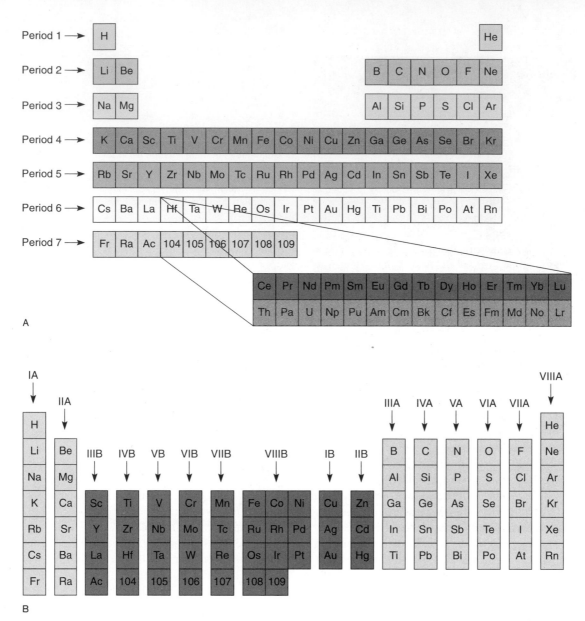

Figure 10.16

(a) Periods of the periodic table, and (b) families of the periodic table.

(beryllium) at the top and has Ra (radium) at the bottom. The A families are in sequence from left to right. The B families are not in sequence, and one group contains more elements than the others (figure 10.16).

Example 10.3

Identify the periodic table (a) period and (b) family of the element silicon.

Solution

According to the list of elements on the inside back cover of this text, silicon has the symbol Si and an atomic number of 14. The square with the symbol Si and the atomic number 14 is located in the third period (third row) and in the column identified as IVA.

Example 10.4

Identify the (a) period and (b) family of the element iron. (Answer: Iron has the symbol Fe and is located in period 4, family VIIIB)

Periodic Patterns

Hydrogen is a colorless, odorless, light gas that burns explosively, combining with oxygen to form water. Helium is a colorless, odorless, light gas that does not burn and, in fact, will not react with other elements at all. Why are hydrogen and helium different? Why does one have atoms that are very reactive while the other has atoms that do not react at all? The answer to these questions is found in the atomic structures of the two gases. Recall that the number of protons determines the identity of an atom. Any atom that has one proton is an atom of hydrogen. Any atom that has two protons is an atom of helium. The chemical behavior of the two gases is determined by their electron configuration, which, in turn, is determined by the number of electrons they have.

Since the energy levels of electrons are quantized, they roughly correspond to a series of distances that resemble the spherical layers of an onion. These spherical layers are sometimes referred to as **shells.** The term *shell* is used to represent all the electrons with the same value of n. The shell concept first came from studies of X-ray spectra, and X-ray vocabulary is used to identify the shells. The shortest X rays are called K X rays, and they were experimentally found to be produced by electrons closest to the nucleus. This energy level closest to the nucleus, $n = 1$, came to be called the K shell, and the shells further out were called the L shell, M shell, N shell, and so on. As noted in chapter 11, the number of electrons that can occupy a given orbital is limited, so *there is a limit to how many electrons can occupy a given shell.* We will return to this important key concept shortly.

As already noted, the first period contains just hydrogen (H) and helium (He). Hydrogen has an atomic number of one, so it has one proton and one electron. This electron is the lowest possible energy level, in the K shell. Helium has two protons in the nucleus so it has two electrons. Both electrons can occupy the K shell, which means that the helium atom has a filled outside shell since a maximum of two electrons is allowed in the K shell. Note that period 1 ends with the filling of the orbital in the K shell.

After helium, lithium (Li) is the next element, and it begins the second period. Li has three electrons since it has an atomic number of three. Two of these electrons are accommodated in the K shell, but the third must go to the next higher level, in the L shell. Lithium is followed by beryllium, boron, carbon, nitrogen, oxygen, fluorine, and neon in order. Each element adds one additional electron to the outermost shell, in this case, the L shell. Note that period 2 ends with the filling of the orbitals in the L shell (table 10.5).

The third period also contains eight elements, from sodium (Na) to argon (Ar). The outer shell is filled just as the elements in the second period, but this time at the third energy level. The third period ends with the filling of the orbitals in the M shell.

By now a couple of patterns are becoming apparent. First, note that the first three periods contain just A families. Each period *begins* with a single electron in a new outer shell. Second, each period *ends* with the filling of an orbital in an outer shell, completing the maximum number of electrons that can occupy that shell. Since the first A family is identified as IA, this means that all the atoms of elements in this family have one electron in their outer shells. All the atoms of elements in family IIA have two electrons in their outer shells. This pattern continues on to family VIIIA, in which all the atoms of elements have

eight electrons in their outer shells except helium. Thus, the number identifying the A families *also identifies the number of electrons in the outer shells* with the exception of helium. Helium is nonetheless similar to the others in this family since all have filled orbitals in their outer shells. The electron theory of chemical bonding, which is discussed in the next chapter, states that only the electrons in the outermost shell of an atom are involved in chemical reactions. Thus, *the outer shell electrons are mostly responsible for the chemical properties of an element.* Therefore, since the members of a family all have similar outer configurations, you would expect them to have similar chemical behaviors, and they do.

The members of the "A-group" families are called the *main-group* or **representative elements.** The members of the "B-group" families are called the **transition elements** (or metals). How the members of the families (figure 10.17) in the representative elements resemble one another will be discussed next.

Table 10.5

Electron configurations for periods 2 and 3

Period 2 From the End of Period 1 Where He: $1s^2$

Element (Atomic Number and Symbol)	Electron Configuration	Number of Electrons in K (First) Shell	Number of Electrons in L (Second) Shell
Lithium ($_3$Li)	[He] $2s^1$	2	1
Beryllium ($_4$Be)	[He] $2s^2$	2	2
Boron ($_5$B)	[He] $2s^2 2p^1$	2	3
Carbon ($_6$C)	[He] $2s^2 2p^2$	2	4
Nitrogen ($_7$N)	[He] $2s^2 2p^3$	2	5
Oxygen ($_8$O)	[He] $2s^2 2p^4$	2	6
Fluorine ($_9$F)	[He] $2s^2 2p^5$	2	7
Neon ($_{10}$Ne)	[He] $2s^2 2p^6$	2	8

Period 3 From the End of Period 2 Where Ne: $1s^2 2s^2 2p^6$

Element (Atomic Number and Symbol)	Electron Configuration	Electrons in K Shell	Electrons in L Shell	Electrons in M Shell
Sodium ($_{11}$Na)	[Ne] $3s^1$	2	8	1
Magnesium ($_{12}$Mg)	[Ne] $3s^2$	2	8	2
Aluminum ($_{13}$Al)	[Ne] $3s^2 3p^1$	2	8	3
Silicon ($_{14}$Si)	[Ne] $3s^2 3p^2$	2	8	4
Phosphorus ($_{15}$P)	[Ne] $3s^2 3p^3$	2	8	5
Sulfur ($_{16}$S)	[Ne] $3s^2 3p^4$	2	8	6
Chlorine ($_{17}$Cl)	[Ne] $3s^2 3p^5$	2	8	7
Argon ($_{18}$Ar)	[Ne] $3s^2 3p^6$	2	8	8

Elements and the Periodic Table

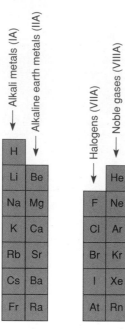

Figure 10.17

Four chemical families of the periodic table: the alkali metals (IA), alkaline earth metals (IIA), halogens (VIIA), and the noble gases (VIIIA).

Example 10.5

How many outer shell electrons are found in an atom of (a) oxygen, (b) calcium, and (c) aluminum?

Solution

(a) According to the list of elements on the inside back cover of this text, oxygen has the symbol O and an atomic number of 8. The square with the symbol O and the atomic number 8 is located in the column identified as VIA. Since the A family number is the same as the number of electrons in the outer shell, oxygen therefore has six outer shell electrons. (b) Calcium has the symbol Ca (atomic number 20) and is located in column IIA, so a calcium atom has two outer shell electrons. (c) Aluminum has the symbol Al (atomic number 13) and is located in column IIIA, so an aluminum atom has three outer shell electrons.

Chemical Families

As shown in table 10.6, all of the elements in group IA have an outside electron configuration of one electron. With the exception of hydrogen, the IA elements are shiny, low-density metals that are so soft you can cut them easily with a knife. These IA metals are called the **alkali metals** because they react violently with water to form an alkaline solution. The alkali metals do not occur in nature as free elements because they are so reactive. Hydrogen is a unique element in the periodic table. It is not an alkali metal and is placed in the IA group because it seems to fit there with one outer shell electron.

The elements in group IIA all have an outside configuration of two electrons and are called the **alkaline earth metals.**

Figure 10.18

Halogen lamps are high-intensity discharge (HID) devices in which a metal is vaporized and ionized between two electrodes. There are three basic types of HID lamps—mercury, sodium, and metal halide (halogen). As you can see, this headlight contains a halogen lamp.

Table 10.6

Electron structures of the alkali metal family

Element	Electron Configuration	Number of Electrons in Shell						
		1st	2nd	3rd	4th	5th	6th	7th
Lithium (Li)	[He] 2s^1	2	1	—	—	—	—	—
Sodium (Na)	[Ne] 3s^1	2	8	1	—	—	—	—
Potassium (K)	[Ar] 4s^1	2	8	8	1	—	—	—
Rubidium (Rb)	[Kr] 5s^1	2	8	18	8	1	—	—
Cesium (Cs)	[Xe] 6s^1	2	8	18	18	8	1	—
Francium (Fr)	[Rn] 7s^1	2	8	18	32	18	8	1

The alkaline earth metals are soft, reactive metals but not as reactive or soft as the alkali metals. Calcium and magnesium are familiar examples of this group.

The elements in group VIIA all have an outside configuration of seven electrons, needing only one more electron to completely fill the outer shell. These elements are called the **halogens.** The halogens are very reactive nonmetals. The halogens fluorine and chlorine are greenish colored gases. Bromine is a reddish brown liquid and iodine is a dark purple solid. Halogens are used as disinfectants, bleaches, and combined with a metal, as a source of light in halogen headlights (figure 10.18). Halogens react with metals to form a group of chemicals called salts, such as sodium chloride. In fact, the word halogen is Greek, meaning "salt former."

As shown in table 10.7, the elements in group VIIIA have orbitals that are filled to capacity in the outside shells. These elements are colorless, odorless gases that almost never react

Table 10.7

Electron structures of the noble gas family

Element	Electron Configuration	Number of Electrons in Shell						
		1st	2nd	3rd	4th	5th	6th	7th
Helium (He)	$1s^2$	2	—	—	—	—	—	—
Neon (Ne)	[He] $2s^2 2p^8$	2	8	—	—	—	—	—
Argon (Ar)	[Ne] $3s^2 3p^6$	2	8	8	—	—	—	—
Krypton (Kr)	[Ar] $4s^2 3d^{10} 4p^6$	2	8	18	8	—	—	—
Xenon (Xe)	[Kr] $5s^2 4d^{10} 5p^6$	2	8	18	18	8	—	—
Radon (Rn)	[Xe] $6s^2 4f^{14} 5d^{10} 6p^6$	2	8	18	32	18	8	—

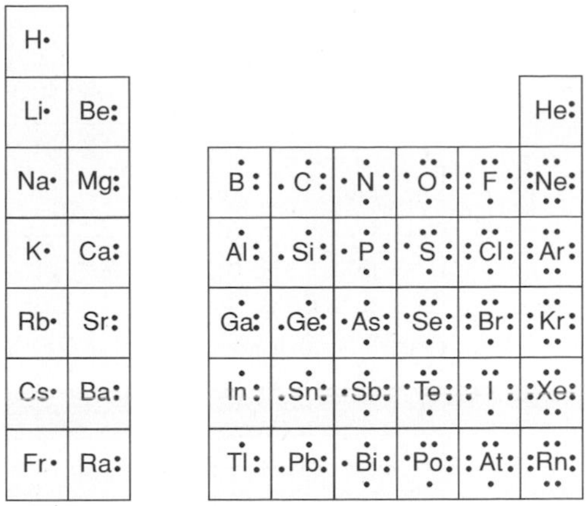

Figure 10.19

Electron dot notation for the representative elements

with other elements to form compounds. Sometimes they are called the **noble gases** because they are chemically inert, perhaps indicating they are above the other elements. They have also been called the *rare gases* because of this scarcity and *inert gases* because they are mostly chemically inert, not forming compounds. The noble gases are inert because they have filled outer electron configurations, a particularly stable condition.

Example 10.6

(a) To what chemical family does chlorine belong? (b) How many electrons does an atom of chlorine have in its outer shell?

Solution

According to the list of elements on the inside back cover of this text, chlorine has the symbol Cl and an atomic number of 17. The square with the symbol Cl and the atomic number 17 is located in the third period and in the column identified as VIIA. (a) Column VIIA is the chemical family known as the halogens. (b) Each A family number is the same as the number of electrons in the outer shell, so an atom of chlorine has seven electrons in its outer shell.

Metals, Nonmetals, and Semiconductors

As indicated earlier, chemical behavior is mostly concerned with the outer shell electrons. The outer shell electrons, that is, the highest energy level electrons, are conveniently represented with an **electron dot notation.** This notation is made by writing the chemical symbol with dots around it indicating the number of outer shell electrons. Electron dot notations are shown for the representative elements in figure 10.19. Again, note the pattern in figure 10.19—all the noble gases are in group VIIIA and all (except helium) have eight outer electrons. All the group IA elements (alkali metals) have one dot, all the IIA elements have two dots, and so on. This pattern will explain the difference in metals, nonmetals, and a third group of in-between elements called semiconductors.

This chapter began with a discussion of several ways to classify matter according to properties. One example given was to group substances according to the physical properties of metals and nonmetals—luster, conductivity, malleability, and

ductility. Metals and nonmetals also have certain chemical properties that are related to their positions in the periodic table. Figure 10.20 shows where the *metals, nonmetals,* and *semiconductors* are located. Note that about 80 percent of all the elements are metals.

The noble gases have completely filled outer orbitals in their highest energy levels, and this is a particularly stable arrangement. Other elements react chemically, either *gaining or losing electrons to attain a filled outermost energy level like the noble gases.* When an atom loses or gains electrons it acquires an unbalanced electron charge called an **ion.** An atom of lithium, for example, has three protons (plus charges) and three electrons (negative charges). If it loses the outermost electron it now has an outer filled orbital structure like helium, a noble gas. It is also now an ion since it has three protons $(3+)$ and two electrons $(2-)$, for a net charge of $1+$. A lithium ion thus has a $1+$ charge.

Example 10.7

(a) Is strontium a metal, nonmetal, or semiconductor? (b) What is the charge on a strontium ion?

Solution

(a) The list of elements inside the back cover identifies the symbol for strontium as Sr (atomic number 38). In the periodic table, Sr is located in family IIA, which means that an atom of strontium has two electrons in its outer shell. For several reasons, you know that strontium is a metal: (1) An atom of strontium has two electrons in its outer shell and atoms with one, two, or three outer electrons are identified as metals; (2) Strontium is located in the IIA family, which is named the alkaline earth metals, and; (3) Strontium is located on the left side of the periodic table and, in general, elements located in the left two-thirds of the table are metals. (b) Elements with one, two, or three outer electrons tend to lose electrons to form positive ions. Since strontium has an atomic number of 38, you know that it has thirty-eight protons $(38+)$ and thirty-eight electrons $(38-)$. When it loses its two outer shell electrons, it has $38+$ and $36-$ for a charge of $2+$.

Figure 10.20

The location of metals, nonmetals, and semiconductors in the periodic table.

1 H				Metals													2 He
3 Li	4 Be			Nonmetals							5 B	6 C	7 N	8 O	9 F	10 Ne
11 Na	12 Mg			Semiconductors							13 Al	14 Si	15 P	16 S	17 Cl	18 Ar
19 K	20 Ca	21 Sc	22 Ti	23 V	24 Cr	25 Mn	26 Fe	27 Co	28 Ni	29 Cu	30 Zn	31 Ga	32 Ge	33 As	34 Se	35 Br	36 Kr
37 Rb	38 Sr	39 Y	40 Zr	41 Nb	42 Mo	43 Tc	44 Ru	45 Rh	46 Pd	47 Ag	48 Cd	49 In	50 Sn	51 Sb	52 Te	53 I	54 Xe
55 Cs	56 Ba	57 La	72 Hf	73 Ta	74 W	75 Re	76 Os	77 Ir	78 Pt	79 Au	80 Hg	81 Tl	82 Pb	83 Bi	84 Po	85 At	86 Rn
87 Fr	88 Ra	89 Ac	104	105	106	107	108	109									

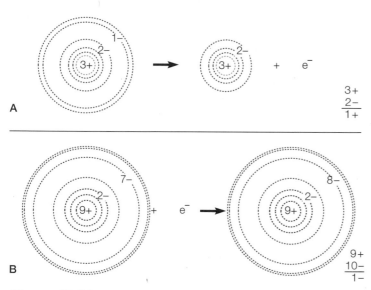

Figure 10.21

(a) Metals lose their outer electrons to acquire a noble gas structure and become positive ions. Lithium becomes a 1+ ion as it loses its one outer electron. (b) Nonmetals gain electrons to acquire an outer noble gas structure and become negative ions. Fluorine gains a single electron to become a 1− ion.

Elements with one, two, or three outer electrons tend to lose electrons to form positive ions. The metals lose electrons like this and the *metals are elements that lose electrons to form positive ions* (figure 10.21). Nonmetals, on the other hand, are elements with five to seven outer electrons that tend to acquire electrons to fill their outer orbitals. *Nonmetals are elements that gain electrons to form negative ions.* In general, elements lo-

Example 10.8

(a) Is iodine a metal, nonmetal, or semiconductor? (b) What is the charge on an iodine ion? (Answer: Nonmetal with a charge of 1−)

cated in the left two-thirds or so of the periodic table are metals. The nonmetals are on the right side of the table (figure 10.22).

The dividing line between the metals and nonmetals is a steplike line from the left top of group IIIA down to the bottom left of group VIIA. This is not a line of sharp separation between the metals and nonmetals, and elements *along* this line sometimes act like metals, sometimes act like nonmetals, and sometimes act like both. These ill-defined elements are called **semiconductors** (or *metalloids*). Silicon, germanium, and arsenic have physical properties of nonmetals; for example, they are brittle materials that cannot be hammered into a new shape. Yet these elements conduct electric currents under certain conditions. The ability to conduct an electric current is a property of a metal, and nonmalleability is a property of nonmetals, so as you can see these semiconductors have the properties of both metals and nonmetals.

Some of the elements near the semiconductors exhibit **allotropic forms,** which means the same element can have several different structures with very different physical properties. Carbon atoms, for example, are joined differently in diamond, graphite, and charcoal. All three are made of the element carbon, but all three have different physical properties. They are allotropic forms of carbon. Phosphorus, sulfur, and tin are other elements near the dividing line that also have several allotropic forms.

The Periodic Table

IA																	VIIIA
1 H 1.01	IIA											IIIA	IVA	VA	VIA	VIIA	2 He 4.0
3 Li 6.94	4 Be 9.01											5 B 10.8	6 C 12.0	7 N 14.0	8 O 16.0	9 F 19.0	10 Ne 20.2
11 Na 23.0	12 Mg 24.3	IIIB IVB VB VIB VIIB — VIIIB — IB IIB										13 Al 27.0	14 Si 28.1	15 P 31.0	16 S 32.1	17 Cl 35.5	18 Ar 39.9
19 K 39.1	20 Ca 40.1	21 Sc 45.0	22 Ti 47.9	23 V 50.9	24 Cr 52.0	25 Mn 54.9	26 Fe 55.8	27 Co 58.9	28 Ni 58.7	29 Cu 63.5	30 Zn 65.4	31 Ga 69.7	32 Ge 72.6	33 As 74.9	34 Se 79.0	35 Br 79.9	36 Kr 83.8
37 Rb 85.5	38 Sr 87.6	39 Y 88.9	40 Zr 91.2	41 Mb 92.9	42 Mo 95.9	43 Tc 98.9	44 Ru 101.1	45 Rh 102.9	46 Pd 106.4	47 Ag 107.9	48 Cd 112.4	49 In 114.8	50 Sn 121.8	51 Sb 121.8	52 Te 127.6	53 I 126.9	54 Xe 131.3
55 Cs 132.9	56 Ba 137.3	57 La 138.9	72 Hf 168.5	73 Ta 180.9	74 W 183.9	75 Re 186.2	76 Os 190.2	77 Ir 192.2	78 Pt 195.1	79 Au 197.0	80 Hg 200.6	81 Ti 204.4	82 Pb 207.2	83 Bi 209.0	84 Po (210)	85 At (210)	86 Rn (222)
87 Fr (223)	88 Ra 226.0	89 Ac (227)	104 (261)	105 (262)	106 (263)	107 (262)	108 (265)	109 (267)									

Box key: 1 — Atomic number / H — Symbol / 1.01 — Atomic weight (rounded value)

58 Ce 140.1	59 Pr 140.9	60 Nd 144.2	61 Pm (145)	62 Sm 150.4	63 Eu 152.0	64 Gd 157.3	65 Tb 158.9	66 Dy 162.5	67 Ho 164.9	68 Er 167.3	69 Tm 168.9	70 Yb 173.0	71 Lu 175.0
90 Th 232.0	91 Pa (231)	92 U 238.0	93 Np (237)	94 Pu (244)	95 Am (243)	96 Cm (247)	97 Bk (247)	98 Cf (251)	99 Es (252)	100 Fm (257)	101 Md (258)	102 No (259)	103 Lr (260)

( ) represents an isotope

Figure 10.22
The periodic table of the elements.

Activities

Make a periodic table. List the elements in boxes on a roll of adding tape. Spiral the tape so that the noble gases appear one below the other. Use cellophane tape to hold the noble gas family together, then cut the tape just to the right of this group. When this cut spiral is spread flat, a long form of the periodic table will be the result. Moving the inner transition elements will produce the familiar short form of the periodic table.

The transition elements, which are all metals, are located in the B group families. Unlike the representative elements, which form vertical families of similar properties, the transition elements tend to form horizontal groups of elements with similar properties. Iron (Fe), cobalt (Co), and nickel (Ni) in group VIIIB; for example, are three horizontally arranged metallic elements that show magnetic properties.

A family of representative elements all form ions with the same charge. Alkali metals, for example, all lose an electron to form a 1+ ion. The transition elements have *variable charges*. Some transition elements, for example, lose their one outer electron to form 1+ ions (copper, silver, gold). Copper, because of its special configuration, can also lose an additional electron to form a 2+ ion. Thus, copper can form either a 1+ ion or a 2+ ion. Some transition elements have two outer electrons and lose them both to form 2+ ions (iron, cobalt, nickel). But some of these elements also have special configurations that permit them to lose more of their electrons. Thus, iron and cobalt, for example, can form either a 2+ ion or a 3+ ion. Much more can be interpreted from the periodic table, and more generalizations will be made as the table is used in the following chapters.

The Rare Earths

Compounds of the rare earths were first identified when they were isolated from uncommon minerals in the late 1700s. The elements are very reactive and have similar chemical properties, so they were not recognized as elements until some fifty years later. Thus, they were first recognized as earths, that is, nonmetal substances, when in fact they are metallic elements. They were also considered to be rare since, at that time, they were known to occur only in uncommon minerals. Today, these metallic elements are known to be more abundant in the earth than gold, silver, mercury, or tungsten. The rarest of the rare earths, thulium, is twice as abundant as silver. The rare earth elements are neither rare nor earths, and they are important materials in glass, electronic, and metallurgical industries.

You can identify the rare earths in the two lowermost rows of the periodic table. These rows contain two series of elements that actually belong in the periods 6 and 7, but they are moved below so the entire table is not so wide. Together, the two series are called the inner transition elements. The top series is fourteen elements wide from elements 58 through 71. Since this series belongs next to element 57, lanthanum, it is sometimes called the *lanthanide series*. This series is also known as the rare earths. The second series of fourteen elements is likewise called the *actinide series*. These are mostly the artificially prepared elements that do not occur naturally.

You may have never heard of the rare earth elements, but they are key materials in many advanced or high-technology products. Lanthanum, for example, gives glass special refractive properties and is used in optic fibers and expensive camera lenses. Samarium, neodymium, and dysprosium are used to manufacture crystals used in lasers. Samarium, ytterbium, and terbium have special magnetic properties that have made possible new electric motor designs, magnetic-optical devices in computers, and the creation of a ceramic superconductor. Other rare earth metals are also being researched for use in possible high-temperature superconductivity materials. Many rare earths are also used in metal alloys, for example, an alloy of cerium is used to make heat resistant jet engine parts. Erbium is also used in high-performance metal alloys. Dysprosium and holmium have neutron absorbing properties and are used in control rods to control nuclear fission. Europium should be mentioned because of its role of making the red color of color television screens. The rare earths are relatively abundant metallic elements that play a key role in many common and high-technology applications. They may also play a key role in superconductivity research.

Summary

Matter can be *classified,* that is, mentally grouped into sets with common properties to make thinking about the wide variety of matter a little less complex. One classification scheme identifies matter as *metals* or *nonmetals* according to the *metallic properties* of *luster, conductivity, malleability, and ductility.* A second classification scheme identifies matter as *solids, liquids, and gases,* according to how well it maintains its shape and volume. A third scheme identifies groups of matter according to how it is put together. *Mixtures* are made up of *unlike parts* with a *variable composition. Pure substances* are the *same throughout* and have a *definite composition.* Mixtures can be separated into their components by *physical changes,* changes that do not alter the identity of matter. Some pure substances can be broken down into simpler substances by a *chemical change,* a change that *alters the identity of matter as it produces new substances with different properties.* A pure substance that can be decomposed by chemical change into simpler substances with a definite composition is a *compound.* A pure substance that cannot be broken down into anything simpler is an *element.* There are ninety-one naturally occurring elements, eighteen that have been made artificially, and millions of known compounds.

Elements are identified by *letter symbols,* and each symbol for an element identifies the element and *represents one atom* of that element. Studies of radioactive elements found that elements have identical chemical behaviors with different masses. The name *isotope* was given to atoms of the same element with different masses. The masses of isotopes were accurately measured with a *mass spectroscope,* and new isotopes were discovered for many elements. These isotopes were found to occur in different proportions with a mass very near a whole number.

The mass of each isotope was compared to the mass of the carbon-12 isotope, which was assigned a mass of exactly 12.00 *atomic mass units* (u). The mass contribution of isotopes according to their abundance is used to determine a *weighted average* of all the isotopes of an element. The weighted average is called the *atomic weight* of that element. Isotopes are identified by their *mass number,* defined as the sum of the number of protons and neutrons in the nucleus. The mass number is the closest whole number to the actual atomic mass of an isotope. Isotopes are identified by a chemical symbol with the atomic number (the number of protons) shown as a subscript and the mass number (the number of protons and neutrons) shown as a superscript.

The *periodic law* states that the properties of elements recur periodically when the elements are listed in order of increasing atomic number. The periodic table has horizontal rows of elements called *periods* and vertical columns of elements called *families.* Families have the same outer shell electron configurations, and it is the electron configuration that is mostly responsible for the chemical properties of an element. The *chemical families* of *alkali metals, alkaline earth metals, halogens,* and *noble gases* all have elements with similar properties and identical outer electron arrangements. The outer electrons are represented by *electron dot notations,* which use dots around a chemical symbol to represent the outer electrons.

Elements react chemically to *gain or lose electrons to attain a filled outer orbital structure like the noble gases.* When an atom gains or loses electrons it acquires an imbalanced charge and is called an *ion.* Metals *lose electrons to form positive ions.* Nonmetals *gain electrons to form negative ions.* Elements on the dividing line between metals and nonmetals are *semiconductors.* Elements near the dividing line exhibit *allotropic forms* with different physical properties.

Key Terms

Applying the Concepts

1. Matter is usually defined as anything that
   a. has a consistent weight unless subdivided.
   b. has a specific volume, but in any shape or size.
   c. occupies space and emits energy in the form of visible light.
   d. occupies space and has mass.

2. One of the physical properties of a metal, compared to a nonmetal, is that the metal is malleable. This means you can
   a. see your reflection in it.
   b. use it to conduct electricity.
   c. pound it into a thin sheet.
   d. pull it into a wire that will conduct electricity.

3. Chemistry is the science concerned with
   a. how the world around you moves and changes.
   b. the study of matter and the changes it undergoes.
   c. atomic theory and the structure of an atom.
   d. the making of synthetic materials.

4. A sample of matter has an indefinite shape and takes the shape of its container. This sample must be in which phase?
   a. gas
   b. liquid
   c. either gaseous or liquid
   d. None of the above.

5. A sample of a salt water solution is homogeneous throughout. Is this sample a mixture or a pure substance?
   a. a pure substance since it is the same throughout
   b. a mixture because it can have a variable composition
   c. a pure substance because it has a definite composition
   d. a mixture because it has a fixed, definite composition

6. Which of the following represents a chemical change?
   a. heating a sample of ice until it melts
   b. tearing a sheet of paper into tiny pieces
   c. burning a sheet of paper
   d. All of the above are correct.

7. Which of the following represents a physical change?
   a. the fusion of hydrogen in the sun
   b. making an iron bar into a magnet
   c. burning a sample of carbon
   d. All of the above are correct.

8. A pure substance that can be decomposed by a chemical change into simpler substances with a fixed mass ratio is called a (an)
   a. element.
   b. compound.
   c. mixture.
   d. isotope.

9. A pure substance that cannot be decomposed into anything simpler by chemical or physical means is called a (an)
   a. element.
   b. compound.
   c. mixture.
   d. isotope.

10. If you cannot decompose a pure substance into anything simpler, you know for sure that the substance is a (an)
    a. element.
    b. compound.
    c. element or stable compound.
    d. element, compound, or mixture.

11. Priestley discovered oxygen when he heated a red powder, but interpreted his findings in terms of the phlogiston theory. He would not have made this interpretation if, in his experiment, he would have
    a. made measurements.
    b. smelled the air.
    c. forced the gas into a container of water.
    d. poured the gas onto a flaming candle.

12. How many elements are found naturally on the earth, not including the ones that were artificially produced in nuclear physics experiments?
    a. 81
    b. 89
    c. 103
    d. 109

13. You see the symbols "CO" in a newspaper article. According to the list of elements on the back cover of this book, these symbols mean
    a. one atom of cobalt.
    b. one atom of copper.
    c. one atom of carbon and one atom of oxygen.
    d. one atom of copper and one atom of oxygen.

14. The chemical symbol for sodium is
    a. S.
    b. So.
    c. Na.
    d. Nd..

15. The chemical symbol "Fe" means
    a. the element fluorine.
    b. one atom of the element fermium.
    c. one atom of the element iron.
    d. the fifth isotope of the element fluorine.

16. About 75 percent of the earth's solid crust is made up of how many different kinds of atoms, not including isotopes?
    a. 2
    b. 8
    c. 91
    d. 109

17. Dalton was able to construct a table of relative atomic masses by
    a. using an instrument called a mass spectrometer.
    b. comparing the fixed mass ratios of combining elements.
    c. weighing the molecules of a compound before and after decomposition.
    d. finding the masses of combining elements, then dividing by the number of atoms.

18. Two isotopes of the same element have
    a. the same number of protons, neutrons, and electrons.
    b. the same number of protons and neutrons, but different numbers of electrons.
    c. the same number of protons and electrons, but different numbers of neutrons.
    d. the same number of neutrons and electrons, but different numbers of protons.

19. Atomic weight is
    a. the weight of an atom in grams.
    b. the average atomic mass of the isotopes as they occur in nature.
    c. the number of protons and neutrons in the nucleus.
    d. All of the above are correct.

20. The mass of any isotope is based on the mass of
    a. hydrogen, which is assigned the number 1 since it is the lightest element.
    b. oxygen, which is assigned a mass of 16.
    c. an isotope of carbon, which is assigned a mass of 12.
    d. its most abundant isotope as found in nature.

21. The isotopes of a given element always have
    a. the same mass and the same chemical behavior.
    b. the same mass and a different chemical behavior.
    c. different masses and different chemical behaviors.
    d. different masses and the same chemical behavior.

22. If you want to know the number of protons in an atom of a given element, you would look up the
    a. mass number.
    b. atomic number.
    c. atomic weight.
    d. abundance of isotopes compared to the mass number.

23. If you want to know the number of neutrons in an atom of a given element, you would
    a. round the atomic weight to the nearest whole number.
    b. add the mass number and the atomic number.
    c. subtract the atomic number from the mass number.
    d. add the mass number and the atomic number, then divide by two.

24. Which of the following is always a whole number?
    a. atomic mass of an isotope
    b. mass number of an isotope
    c. atomic weight of an element
    d. None of the above are correct.

25. The chemical family of elements called the noble gases are found in what column of the periodic table?
    a. IA
    b. IIA
    c. VIIA
    d. VIIIA

26. A particular element is located in column IVA of the periodic table. How many dots would be placed around the symbol of this element in its electron dot notation?
    a. 1
    b. 3
    c. 4
    d. 8

Answers
1. d   2. c   3. b   4. c   5. b   6. c   7. b   8. b   9. a
10. c   11. a   12. b   13. c   14. c   15. c   16. a   17. b
18. c   19. b   20. c   21. d   22. b   23. c   24. b   25. d
26. c

Questions for Thought

1. How was Mendeleev able to predict the chemical and physical properties of elements that were not yet discovered? __

2. What is an isotope? Are *all* atoms isotopes? Explain.

3. Which of the following are whole numbers and which are not whole numbers? Explain why for each.
   (a) atomic number
   (b) isotope mass
   (c) mass number
   (d) atomic weight

4. Why does the carbon-12 isotope have a whole number mass but not the other isotopes?

5. What two things does a chemical symbol represent?

6. What do the members of the noble gas family have in common? What are the differences?

7. How are the isotopes of an element similar? How are they different?

8. What is the difference between a chemical change and a physical change? Give three examples of each.

9. What is an ion? How are ions formed?

10. What patterns are noted in the electron structures of elements found in a period and a family in the periodic table?

11. What is a semiconductor?

12. Why do chemical families exist?

Exercises

1. Write the chemical symbols for the following chemical elements:
   (a) Silicon
   (b) Silver
   (c) Helium
   (d) Potassium
   (e) Magnesium
   (f) Iron

2. Lithium has two naturally occurring isotopes, lithium-6 and lithium-7. Lithium-6 has a mass of 6.01512 relative to carbon-12 and makes up 7.42 percent of all naturally occurring lithium. Lithium-7 has a mass of 7.016 compared to carbon-12 and makes up the remaining 92.58 percent. According to this information, what is the atomic weight of lithium?

3. Identify the number of protons, neutrons, and electrons in the following isotopes:
   (a) $^{12}_{6}C$
   (b) $^{1}_{1}H$
   (c) $^{40}_{18}Ar$
   (d) $^{2}_{1}H$
   (e) $^{197}_{79}Au$
   (f) $^{235}_{92}U$

4. Identify the period and family in the periodic table for the following elements:
   (a) Radon
   (b) Sodium
   (c) Copper
   (d) Neon
   (e) Iodine
   (f) Lead

5. How many outer shell electrons are found in an atom of:
   (a) Li
   (b) N
   (c) F
   (d) Cl
   (e) Ra
   (f) Be

6. Write electron dot notations for the following elements:
   (a) Boron
   (b) Bromine
   (c) Calcium
   (d) Potassium
   (e) Oxygen
   (f) Sulfur

7. Identify the charge on the following ions:
   (a) Boron
   (b) Bromine
   (c) Calcium
   (d) Potassium
   (e) Oxygen
   (f) Nitrogen

1. Write the chemical symbols for the following chemical elements:
   (a) Argon
   (b) Gold
   (c) Neon
   (d) Sodium
   (e) Calcium
   (f) Tin

2. Boron has two naturally occurring isotopes, boron-10 and boron-11. Boron-10 has a mass of 10.0129 relative to carbon-12 and makes up 19.78 percent of all naturally occurring boron. Boron-11 has a mass of 11.00931 compared to carbon-12 and makes up the remaining 80.22 percent. What is the atomic weight of boron?

3. Identify the number of protons, neutrons, and electrons in the following isotopes:
   (a) $^{14}_{7}N$
   (b) $^{7}_{3}Li$
   (c) $^{35}_{17}Cl$
   (d) $^{48}_{20}Ca$
   (e) $^{63}_{29}Cu$
   (f) $^{230}_{92}U$

4. Identify the period and the family in the periodic table for the following elements:
   (a) Xenon
   (b) Potassium
   (c) Chromium
   (d) Argon
   (e) Bromine
   (f) Barium

5. How many outer shell electrons are found in an atom of
   (a) Na
   (b) P
   (c) Br
   (d) I
   (e) Te
   (f) Sr

6. Write electron dot notations for the following elements:
   (a) Aluminum
   (b) Fluorine
   (c) Magnesium
   (d) Sodium
   (e) Carbon
   (f) Chlorine

7. Identify the charge on the following ions:
   (a) Aluminum
   (b) Chlorine
   (c) Magnesium
   (d) Sodium
   (e) Sulfur
   (f) Hydrogen

8. Use the periodic table to identify if the following are metals, nonmetals, or semiconductors:
   (a) Krypton
   (b) Cesium
   (c) Silicon
   (d) Sulfur
   (e) Molybdenum
   (f) Plutonium

9. From their charges, predict the periodic table family number for the following ions:
   (a) Br^{-1}
   (b) K^{+1}
   (c) Al^{+3}
   (d) S^{-2}
   (e) Ba^{+2}
   (f) O^{-2}

10. Use chemical symbols and numbers to identify the following isotopes:
   (a) Oxygen-16
   (b) Sodium-23
   (c) Hydrogen-3
   (d) Chlorine-35

8. Use the periodic table to identify if the following are metals, nonmetals, or semiconductors:
   (a) Radon
   (b) Francium
   (c) Arsenic
   (d) Phosphorus
   (e) Hafnium
   (f) Uranium

9. From their charges, predict the periodic table family number for the following ions:
   (a) F^{-1}
   (b) Li^{+1}
   (c) B^{+3}
   (d) O^{-2}
   (e) Be^{+2}
   (f) Si^{+4}

10. Use chemical symbols and numbers to identify the following isotopes:
   (a) Potassium-39
   (b) Neon-22
   (c) Tungsten-184
   (d) Iodine-127

Chapter

11

Outline

Compounds and Chemical Change
Valence Electrons and Ions
Chemical Bonds
    Ionic Bonds
        Energy and Electrons in Ionic
           Bonding
        Ionic Compounds and Formulas
    Covalent Bonds
        Covalent Compounds and
           Formulas
        Multiple Bonds
        Coordinate Covalent Bonds
Bond Polarity
Composition of Compounds
    Ionic Compound Names
    Ionic Compound Formulas
    Covalent Compound Names
    Covalent Compound Formulas

**Feature: Microwave Ovens and
Molecular Bonds**

Compounds and
Chemical Change

A chemical change occurs when iron rusts.

223

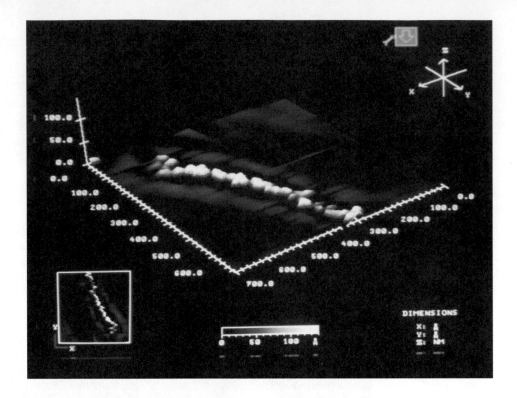

Figure 11.1

This is a scanning tunneling electron microscope image of DNA, showing the actual grooves of the double helix. The scale is in Angstroms (1 Angstrom = 10^{-10} m), and each twist of the helix is about 35 Angstroms. A molecule like this one determined the characteristics that other people recognize as you.

*I*N the previous two chapters you learned how the modern atomic theory is used to describe the structures of atoms of different elements. The electron structures of different atoms successfully account for the position of elements in the periodic table as well as for groups of elements with similar properties. On a large scale all metals were found to have a similarity in electron structure, as were nonmetals. On a smaller scale, chemical families such as the alkali metals were found to have the same outer electron configurations. Thus, the modern atomic theory accounts for observed similarities between elements in terms of atomic structure.

So far, only individual, isolated atoms have been discussed without considering how atoms of elements join together to produce compounds. There is a relationship between the electron structure of atoms and the reactions they undergo to produce specific compounds. Understanding this relationship will explain the changes that matter itself undergoes. For example, hydrogen is a highly flammable, gaseous element that burns with an explosive reaction. Oxygen, on the other hand, is a gaseous element that supports burning. As you know, hydrogen and oxygen combine to form water. Water is a liquid that neither burns nor supports burning. What happens when atoms of elements such as hydrogen and oxygen join to form molecules such as water? Why do such atoms join and why do they stay together? Why does water have different properties from the elements that combine to produce it? And finally, why is water H_2O and not H_3O or H_4O?

Answers to questions about why and how atoms join together in certain numbers are provided by considering the electronic structures of the atoms. Chemical substances are formed from the interactions of electrons as their structures merge, forming new patterns that result in molecules with new properties (figure 11.1). It is the new electron pattern of the water molecule that gives water different properties than the oxygen or hydrogen from which it formed. Understanding how electron structures of atoms merge to form new patterns is understanding the changes that matter itself undergoes, the topic of this chapter.

Compounds and Chemical Change

The air you breathe, the liquids you drink, and all the things around you are elements, compounds, or mixtures. Most are compounds, however, and very few are pure elements. Water, sugar, gasoline, and chalk are examples of compounds. Each can be broken down into the elements that make it up. Recall that elements are basic substances that cannot be broken down into simpler substances. Examples of elements are hydrogen, carbon, and calcium. Why and how these elements join together in different ways to form different compounds is the subject of this chapter.

You have already learned that elements are made up of atoms that can be described by the modern atomic theory. You can also consider an **atom** to be *the smallest unit of an element that can exist alone or in combination with other elements.* Compounds are formed when atoms are held together by an attractive force called a *chemical bond.* The chemical bond binds

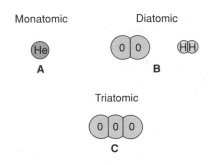

Monatomic   Diatomic

A   B

Triatomic

C

Figure 11.2

(a) The noble gases are monatomic, occurring as single atoms. (b) Many gases, such as hydrogen and oxygen, are diatomic, with two atoms per molecule. (c) Ozone is a form of oxygen that is triatomic, occurring with three atoms per molecule.

individual atoms together in a compound. A molecule is generally thought of as a tightly bound group of atoms that maintains its identity. More specifically, a **molecule** is defined as *the smallest particle of a compound, or a gaseous element, that can exist and still retain the characteristic properties of a substance.* Compounds with one type of chemical bond, as you will see, have molecules that are electrically neutral groups of atoms held together strongly enough to be considered independent units. For example, water is a compound. The smallest unit of water that can exist alone is an electrically neutral unit made up of two hydrogen atoms and one oxygen atom held together by chemical bonds. The concept of a molecule will be expanded as chemical bonds are discussed.

Compounds occur naturally as gases, liquids, and solids. Many common gases occur naturally as molecules made up of two or more atoms. For example, at ordinary temperatures hydrogen gas occurs as molecules of two hydrogen atoms bound together. Oxygen gas also usually occurs as molecules of two oxygen atoms bound together. Both hydrogen and oxygen occur naturally as *diatomic molecules* ("di-" means "two"). Oxygen sometimes occurs as molecules of three oxygen atoms bound together. These *triatomic molecules* ("tri-" means "three") are called *ozone*. The noble gases are unique, occurring as single atoms called *monatomic* ("mon-" or "mono-" means "one") (figure 11.2). These monatomic particles are sometimes called monatomic molecules since they are the smallest unit of the noble gases that can exist alone. Helium and neon are examples of the monatomic noble gases.

When multiatomic molecules of any size are formed or broken down into simpler substances, new materials with new properties are produced. This kind of a change in matter is called a chemical change, and the process is called a chemical reaction. A **chemical reaction** is defined as

a change in matter where different chemical substances are created by forming or breaking chemical bonds.

In general, chemical bonds are formed when atoms of elements are bound together to form compounds. Chemical bonds are broken when a compound is decomposed into simpler substances.

Chemical reactions happen all the time, all around you. A growing plant, burning fuels, and your body's utilization of food all involve chemical reactions. These reactions produce different chemical substances with greater or smaller amounts of internal potential energy (see chapter 4 for a discussion of internal potential energy). Energy is *absorbed* to produce new chemical substances with more internal potential energy. Energy is *released* when new chemical substances are produced with less internal potential energy. In general, changes in internal potential energy are called **chemical energy.** For example, new chemical substances are produced in green plants through the process called photosynthesis. A green plant uses radiant energy (sunlight), carbon dioxide, and water to produce new chemical materials and oxygen. These new chemical materials, the stuff that leaves, roots, and wood are made of, contain more chemical energy than the carbon dioxide and water they were made from.

A **chemical equation** is a way of describing what happens in a chemical reaction. Later, you will learn how to use formulas in a chemical reaction. For now, the chemical reaction of photosynthesis will be described by using words in an equation:

$$\text{energy (sunlight)} + \text{carbon dioxide molecules} + \text{water molecules} \rightarrow \text{plant material molecules} + \text{oxygen molecules}$$

The substances that are changed are on the left side of the word equation and are called *reactants*. The reactants are carbon dioxide molecules and water molecules. The equation also indicates that energy is absorbed, since the term *energy* appears on the left side. The arrow means *yields*. The new chemical substances are on the right side of the word equation and are called *products*. Reading the photosynthesis reaction as a sentence, "Carbon dioxide and water use energy to react, yielding plant materials and oxygen."

The plant materials produced by the reaction have more internal potential energy, also known as chemical energy, than the reactants. You know this from the equation because the term *energy* appears on the left side but not the right. This means that the energy on the left went into internal potential energy on the right. You also know this because the reaction can be reversed to release the stored energy (figure 11.3). When plant materials (such as wood) are burned, the materials react with oxygen and chemical energy is released in the form of radiant energy (light) and high kinetic energy of the newly formed gases and vapors. In words,

$$\text{plant material molecules} + \text{oxygen molecules} \rightarrow \text{carbon dioxide molecules} + \text{water molecules} + \text{energy}$$

If you compare the two equations, you will see that burning is the opposite process of photosynthesis! The energy released in burning is the exact same amount of solar energy that was stored as internal potential energy by the plant. Such chemical changes where chemical energy is stored in one reaction and released by another reaction are the result of the making, then the breaking, of chemical bonds. Chemical bonds were formed

Compounds and Chemical Change

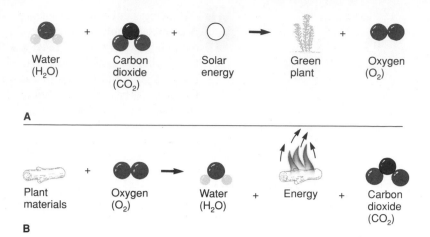

A

B

Figure 11.3

(a) New chemical bonds are formed as a green plant makes new materials and stores solar energy through the photosynthesis process. (b) The chemical bonds are later broken and the same amount of energy and the same original materials are released. The same energy and the same materials are released rapidly when the plant materials burn, and they are released slowly when the plant decomposes.

by utilizing energy to produce new chemical substances. Energy was released when these bonds were broken to produce the original substances (figure 11.4). Chemical reactions and energy flow can be explained by the making and breaking of chemical bonds. Chemical bonds can be explained in terms of changes in the electron structures of atoms. Thus, the place to start in seeking understanding about chemical reactions is the electron structure of the atoms themselves.

Valence Electrons and Ions

As discussed in chapter 10, it is the number of electrons in the outermost shell that usually determines the chemical properties of an atom. These outer electrons are called **valence electrons,** and it is the valence electrons that participate in chemical bonding. The inner electrons are in stable, fully occupied orbitals and do not participate in chemical bonds. The representative elements have valence electrons in the outermost orbitals, which contain from one to eight valence electrons. Recall that you can easily find the number of valence electrons by referring to a periodic table. The number at the top of each representative family is the same as the number of outer shell electrons (with the exception of helium).

The noble gases have filled outer orbitals and do not normally form compounds. Apparently half-filled and filled orbitals are particularly stable arrangements. Atoms have a tendency to seek such a stable, filled outer orbital arrangement such as the one found in the noble gases. For the representative elements, this tendency is called the **octet rule.** The octet rule states that *atoms attempt to acquire an outer orbital with eight electrons* through chemical reactions. This rule is a generalization, and a few elements do not meet the requirement of "eight" but do seek the same general trend of stability. There are a few

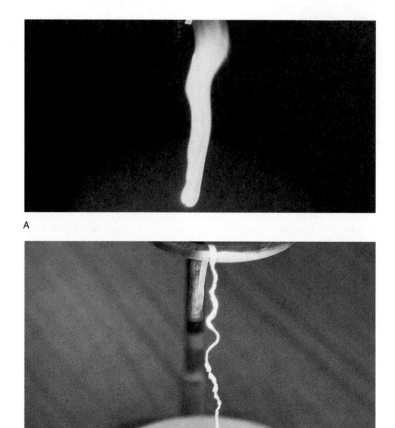

A

B

Figure 11.4

Magnesium is an alkaline earth *metal* that (a) burns brightly in air, releasing heat and light. As chemical energy is released, a new chemical substance (b) is formed. The new chemical material is magnesium oxide, a soft powdery material that forms an alkaline solution in water (called "milk of magnesia").

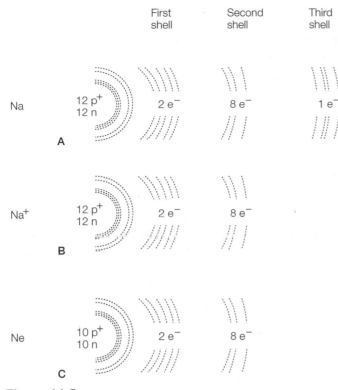

	First shell	Second shell	Third shell
Na	12 p$^+$ 12 n	2 e$^-$	8 e$^-$	1 e$^-$

A

Na$^+$	12 p$^+$ 12 n	2 e$^-$	8 e$^-$

B

Ne	10 p$^+$ 10 n	2 e$^-$	8 e$^-$

C

Figure 11.5

(a) A sodium atom has two electrons in the first energy level, eight in the second energy level, and one in the third level. (b) When it loses its one outer, or valence, electron, it becomes a sodium ion with the same electron structure as an atom of neon (c).

other exceptions, and the octet rule should be considered a generalization that helps keep track of the valence electrons in most representative elements.

The periodic table representative element family number tells you the number of valence electrons and what the atom must do to reach the stability suggested by the octet rule. For example, consider sodium (Na). Sodium is in family IA, so sodium has one valence electron. If the sodium atom can get rid of this outer valence electron through a chemical reaction, it will have the same outer electron configuration as an atom of the noble gas neon (Ne) (compare figures 11.5b and 11.5c).

When a sodium atom (Na) loses an electron to form a sodium ion (Na$^+$) it has the same, stable outer electron configuration as a neon atom (Ne). The sodium ion (Na$^+$) is still a form of sodium since it still has eleven protons. But it is now a sodium *ion*, not a sodium *atom*, since it has eleven protons (eleven positive charges) and now has ten electrons (ten negative charges) for a total of

$$11+ \text{ (protons)}$$
$$\underline{10- \text{ (electrons)}}$$
$$1+ \text{ (net charge on sodium ion)}$$

This charge is shown on the chemical symbol of Na$^+$ and is called a *sodium ion.* Note that the sodium nucleus and the inner orbitals do not change when the sodium atom is ionized. The

sodium ion is formed when a sodium atom loses its valence electron, and the process can be described by

$$\text{energy} + \text{Na}\cdot \rightarrow \text{Na}^+ + e^- \qquad \textbf{equation 11.1}$$

where Na$\cdot$ is the electron dot symbol for sodium and the e$^-$ is the electron that has been pulled off the sodium atom.

Example 11.1

What is the symbol and charge for a calcium ion?

Solution

From the list of elements on the inside back cover, the symbol for calcium is Ca and the atomic number is 20. The periodic table tells you that Ca is in family IIA, which means that calcium has 2 valence electrons. According to the octet rule, the calcium ion must lose 2 electrons to acquire the stable outer arrangement of the noble gases. Since the atomic number is 20 a calcium atom has 20 protons (20+) and 20 electrons (20−). When it is ionized, the calcium ion will lose 2 electrons for a total charge of (20+) + (18−), or 2+. The calcium ion is represented by the chemical symbol for calcium and the charge shown as a superscript: Ca^{2+}.

Example 11.2

What is the symbol and charge for an aluminium ion? (Answer: Al^{3+})

Chemical Bonds

Atoms gain or lose electrons through a chemical reaction to achieve a state of lower energy, the stable electron arrangement of the noble gas atoms. Such a reaction results in a **chemical bond,** an *attractive force that holds atoms together in a compound.* There are three general classes of chemical bonds: (1) ionic bonds, (2) covalent bonds, and (3) metallic bonds.

Ionic bonds are formed when atoms *transfer* electrons to achieve the noble gas electron arrangement. Electrons are given up or acquired in the transfer, forming positive and negative ions. The electrostatic attraction between oppositely charged ions forms ionic bonds, and ionic compounds are the result. In general, ionic compounds are formed when a metal from the left side of the periodic table reacts with a nonmetal from the right side.

Covalent bonds result when atoms achieve the noble gas electron structure by *sharing* electrons. Covalent bonds are generally formed between the nonmetallic elements on the right side of the periodic table.

Metallic bonds are formed in solid metals such as iron, copper, and the other metallic elements that make up about 80 percent of all the elements. The atoms of metals are closely packed and share many electrons in a "sea" that is free to move throughout the metal, from one metal atom to the next. Metallic bonding accounts for metallic properties such as high electrical conductivity.

Ionic, covalent, and metallic bonds are attractive forces that hold atoms or ions together in molecules and crystals. There are two ways to describe what happens to the electrons when

Compounds and Chemical Change

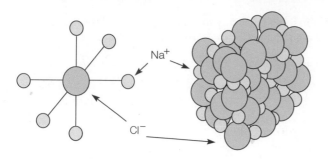

Figure 11.6
Sodium chloride crystals are composed of sodium and chlorine ions held together by electrostatic attraction. Each sodium ion is surrounded by six chlorine ions, and each chlorine ion is surrounded by six sodium ions. A crystal builds up like this, giving the sodium chloride crystal a cubic structure.

one of these bonds is formed, by considering (1) the new patterns formed when atomic orbitals overlap to form a combined orbital called a *molecular orbital* or (2) the atoms in a molecule as *isolated atoms* with changes in their outer shell arrangements. The molecular orbital description considers that the electrons belong to the whole molecule and form a molecular orbital with its own shape, orientation, and energy levels. The isolated atom description considers the electron energy levels as if the atoms in the molecule were alone, isolated from the molecule. The isolated atom description is less accurate than the molecular orbital description, but it is less complex and more easily understood. Thus, the following details about chemical bonding will mostly consider individual atoms and ions in compounds.

Ionic Bonds

An **ionic bond** is defined as the *chemical bond of electrostatic attraction* between negative and positive ions. Ionic bonding forms when an atom of a metal reacts with an atom of a nonmetal. The reaction results in a transfer of one or more valence electrons from the metal atom to the valence shell of the nonmetal atom. The atom that loses electrons becomes a positive ion, and the atom that gains electrons becomes a negative ion. Oppositely charged ions attract one another, and when pulled together, they form an ionic solid with the ions arranged in an orderly geometric structure. This results in a crystalline solid that is typical of salts such as sodium chloride (figure 11.6).

As an example of ionic bonding, consider the reaction of sodium (a soft reactive metal) with chlorine (a pale yellow-green gas). When an atom of sodium and an atom of chlorine collide, they react violently as the valence electron is transferred from the sodium to the chlorine atom. This produces a sodium ion and a chlorine ion. The reaction can be illustrated with electron dot symbols as follows:

$$ Na \cdot \quad + \quad \cdot \overset{\displaystyle ..}{\underset{\displaystyle ..}{Cl}} \colon \quad \longrightarrow \quad Na^+ \left(\colon \overset{\displaystyle ..}{\underset{\displaystyle ..}{Cl}} \colon \right)^- \qquad \textbf{equation 11.2} $$

As you can see, the sodium ion transferred its valence electron and the resulting ion now has a stable electron configuration. The chlorine atom accepted the electron in its outer

orbital to acquire a stable electron configuration. Thus, a stable positive ion and a stable negative ion are formed. Because of opposite electrical charges, the ions attract each other to produce an ionic bond. When many ions are involved, each Na^+ ion is surrounded by six Cl^- ions and each Cl^- ion is surrounded by six Na^+ ions. This gives the resulting solid NaCl its crystalline cubic structure as shown in figure 11.6. In the solid state all the sodium ions and all the chlorine ions are bound together in one giant unit. Thus, the term "molecule" is not really appropriate for ionic solids such as sodium chloride. But the term is sometimes used anyway since any given sample will have the same number of Na^+ ions as Cl^- ions.

Energy and Electrons in Ionic Bonding
The sodium-chloride reaction can be represented with electron dot notation as occurring in three steps:

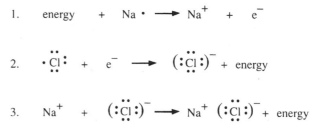

The energy released in steps 2 and 3 is greater than the energy absorbed in step 1, and an ionic bond is formed. The energy released is called the **heat of formation.** It is also the amount of energy required to decompose the compound (sodium chloride) into its elements. The reaction does not take place in steps as described, however, but occurs all at once. Note again, as in the photosynthesis-burning reactions described earlier, that the total amount of chemical energy is conserved. The energy released by the formation of the sodium chloride compound is the *same* amount of energy needed to decompose the compound.

Ionic bonds are formed by electron transfer, and electrons are conserved in the process. This means that electrons are not created or destroyed in a chemical reaction. The same total number of electrons exists after a reaction that existed before the reaction. There are two rules you can use for keeping track of electrons in ionic bonding reactions:

1. Ions are formed as atoms gain or lose valence electrons to achieve the stable noble gas structure.
2. There must be a balance between the number of electrons lost and the number of electrons gained by atoms in the reaction.

The sodium-chloride reaction follows these two rules. The loss of one valence electron from a sodium atom formed a stable sodium ion. The gain of one valence electron by the chlorine atom formed a stable chlorine ion. Thus, both ions have noble gas configurations (rule 1), and one electron was lost and one was gained, so there is a balance in the number of electrons lost and the number gained (rule 2).

Ionic Compounds and Formulas
The **formula** of a compound *describes what elements are in the compound and in what proportions.* Sodium chloride contains one positive sodium ion for each negative chlorine ion. The formula of the compound sodium chloride is NaCl. If there are no

subscripts at the lower right part of each symbol, it is understood that the symbol has a number "1." Thus, NaCl indicates a compound made up of the elements sodium and chlorine, and there is one sodium atom for each chlorine atom.

Calcium (Ca) is an alkaline metal in family IIA, and fluorine (F) is a halogen in family VIIA. Since calcium is a metal and fluorine is a nonmetal, you would expect calcium and fluorine atoms to react, forming a compound with ionic bonds. Calcium has two valence electrons to lose to acquire a noble gas configuration. Fluorine needs one valence electron to acquire a noble gas configuration. So calcium needs to lose two electrons and fluorine needs to gain one electron to achieve a stable configuration (rule 1). Two fluorine atoms, each acquiring one electron, are needed to balance the number of electrons lost and the number of electrons gained. The compound formed from the reaction, calcium fluoride, will therefore have a calcium ion with a charge of plus two for every fluorine ion with a charge of minus one. Recalling that electron dot symbols show only the outer valence electrons, you can see that the reaction is

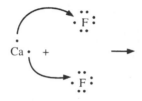

which shows that a calcium atom transfers two electrons, one each to two fluorine atoms. Now showing the results of the reaction, a calcium ion is formed from the loss of two electrons (charge 2+) and two fluorine ions are formed by gaining one electron each (charge 1−):

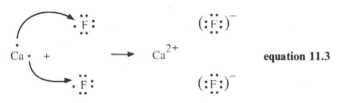

equation 11.3

The formula of the compound is therefore CaF_2, with the subscript 2 for fluorine and the understood subscript 1 for calcium. This means that there are two fluorine atoms for each calcium atom in the compound.

Sodium chloride (NaCl) and magnesium fluoride (MgF_2) are examples of compounds held together by ionic bonds. Compounds held together by ionic bonds are called **ionic compounds.** Ionic compounds of the representative elements are generally white, crystalline solids that form colorless solutions. Sodium chloride, the most common example, is common table salt. Many of the transition elements form colored compounds that make colored solutions. Ionic compounds dissolve in water, producing a solution of ions that can conduct an electric current.

In general, the elements in families IA and IIA of the periodic table tend to form positive ions by losing electrons. The ion charge for these elements equals the family number of these elements. The elements in families VIA and VIIA tend to form negative ions by gaining electrons. The ion charge for these elements equals their family number minus eight. The elements in

families IIIA and VA have less of a tendency to form ionic compounds except those in higher periods. Common ions of representative elements are given in table 11.1. The transition elements form positive ions of several different charges. Some common ions of the transition elements are listed in table 11.2.

Table 11.1
Common ions of representative elements

Element	Symbol	Ion
Lithium	Li	1+
Sodium	Na	1+
Potassium	K	1+
Magnesium	Mg	2+
Calcium	Ca	2+
Barium	Ba	2+
Aluminum	Al	3+
Oxygen	O	2−
Sulfur	S	2−
Hydrogen	H	1+, 1−
Fluorine	F	1−
Chlorine	Cl	1−
Bromine	Br	1−
Iodine	I	1−

Table 11.2
Common ions of transition elements

Single-Charge Ions		
Element	Symbol	Charge
Zinc	Zn	2+
Tungsten	W	6+
Silver	Ag	1+
Cadmium	Cd	2+
Variable-Charge Ions		
Element	Symbol	Charge
Chromium	Cr	2+, 3+, 6+
Manganese	Mn	2+, 4+, 7+
Iron	Fe	2+, 3+
Cobalt	Co	2+, 3+
Nickel	Ni	2+, 3+
Copper	Cu	1+, 2+
Tin	Sn	2+, 4+
Gold	Au	1+, 3+
Mercury	Hg	1+, 2+
Lead	Pb	2+, 4+

Compounds and Chemical Change

The single-charge representative elements and the variable-charge transition elements form single, monatomic negative ions. There are also many polyatomic ("poly" means "many") negative ions, charged groups of atoms that act like a single unit in ionic compounds. Polyatomic ions are held together by covalent bonds, which will be discussed in the next section.

Example 11.3

Use electron dot notation to predict the formula of a compound formed when aluminum (Al) combines with fluorine (F).

Solution
Aluminum, atomic number 13, is in family IIIA so it has three valence electrons and an electron dot notation of

$$\overset{\cdot}{Al}\,\cdot$$

According to the octet rule, the aluminum atom would need to lose three electrons to acquire the stable noble gas configuration. Fluorine, atomic number 9, is in family VIIA so it has seven valence electrons and an electron dot notation of

$$\cdot\overset{\cdot\cdot}{\underset{\cdot\cdot}{F}}:$$

Fluorine would acquire a noble gas configuration by accepting one electron. Three fluorine atoms, each acquiring one electron, are needed to balance the three electrons lost by aluminum. The reaction can be represented as

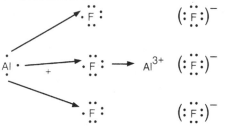

The ratio of aluminum atoms to fluorine atoms in the compound is 1:3. The formula for aluminum fluoride is therefore AlF_3.

Example 11.4

Predict the formula of the compound formed between aluminum and oxygen using electron dot notation. (Answer: Al_2O_3)

Covalent Bonds

Most substances do not have the properties of ionic compounds since they are not composed of ions. Most substances are molecular, composed of electrically neutral groups of atoms that are tightly bound together. As noted earlier, many gases are diatomic, occurring naturally as two atoms bound together as an electrically neutral molecule. Hydrogen, for example, occurs as molecules of H_2 and no ions are involved. The hydrogen atoms are held together by a covalent bond. A **covalent bond** is a *chemical bond formed by the sharing of a pair of electrons.* In the diatomic hydrogen molecule each hydrogen atom contributes a single electron to the shared pair. Both hydrogen atoms count

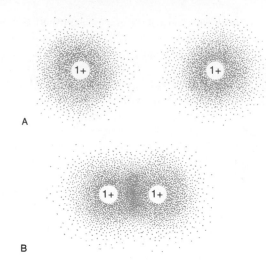

Figure 11.7
(a) Two hydrogen atoms, each with its own probability distribution of electrons about the nucleus. (b) When the hydrogen atoms bond, a new electron distribution pattern forms around the entire molecule and both electrons occupy the molecular orbital.

the shared pair of electrons in achieving their noble gas configuration. Hydrogen atoms both share one pair of electrons, but other elements might share more than one pair to achieve a noble gas structure.

Consider how the covalent bond forms between two hydrogen atoms by imagining two hydrogen atoms moving toward one another. Each atom has a single electron. As the atoms move closer and closer together, their orbitals begin to overlap. Each electron is attracted to the oppositely charged nucleus of the other atom and the overlap tightens. Then the repulsive forces from the like-charged nuclei will halt the merger. A state of stability is reached between the two nuclei and two electrons, and a H_2 molecule has been formed. The two electrons are now shared by both atoms, and the attraction of one nucleus for the other electron and vice versa holds the atoms together (figure 11.7).

Covalent Compounds and Formulas
Electron dot notation can be used to represent the formation of covalent bonds. For example, the joining of two hydrogen atoms to form a H_2 molecule can be represented as:

$$H\cdot \;+\; H\cdot \longrightarrow H:H$$

Since an electron pair is *shared* in a covalent bond the two electrons move throughout the entire molecular orbital. Since each hydrogen atom now has both electrons on an equal basis, each can be considered to now have the noble gas configuration of helium. A dashed circle around each symbol shows that both atoms have two electrons:

$$H\cdot \;+\; H\cdot \longrightarrow (H:H) \qquad \textbf{equation 11.4}$$

Hydrogen and fluorine react to form a covalent molecule (how this is known will be discussed shortly), and this bond can be represented with electron dots. Fluorine is in the VIIA family, so you know an atom of fluorine has seven valence electrons in the outermost energy level. The reaction is

$$H \cdot \ + \ \cdot \overset{\cdot \cdot}{\underset{\cdot \cdot}{F}} : \ \longrightarrow \ (H \overset{\cdot \cdot}{\underset{\cdot \cdot}{F}} :) \qquad \textbf{equation 11.5}$$

Each atom shares a pair of electrons to achieve a noble gas configuration. Hydrogen achieves the helium configuration and fluorine achieves the neon configuration. All the halogens have seven valence electrons like this, and all need to gain one electron (ionic bond) or share an electron pair (covalent bond) to achieve a noble gas configuration. This also explains why the halogen gases occur as diatomic molecules. Two fluorine atoms can achieve a noble gas configuration by sharing a pair of electrons:

$$\cdot \overset{\cdot \cdot}{\underset{\cdot \cdot}{F}} : \ + \ \cdot \overset{\cdot \cdot}{\underset{\cdot \cdot}{F}} : \ \longrightarrow \ (: \overset{\cdot \cdot}{\underset{\cdot \cdot}{F}} \overset{\times}{\underset{\times}{:}} \overset{\cdot \cdot}{\underset{\cdot \cdot}{F}} :) \qquad \textbf{equation 11.6}$$

Each fluorine atom thus achieves the neon configuration by bonding together. Note that there are two types of electron pairs: (1) orbital pairs and (2) bonding pairs. Orbital pairs are not shared since they are the two electrons in an orbital, each with a separate spin. Orbital pairs are also called *lone pairs*, since they are not shared. *Bonding pairs,* as the name implies, are the electron pairs shared between two atoms. Considering again the F_2 molecule,

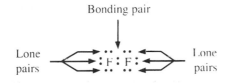

Often, the number of bonding pairs that are formed by an atom is the same as the number of single, *unpaired* electrons in the atomic electron dot notation. For example, hydrogen has one unpaired electron and oxygen has two unpaired electrons. Hydrogen and oxygen combine to form an H_2O molecule as

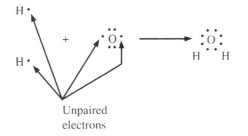

The diatomic hydrogen (H_2) and fluorine (F_2), hydrogen fluoride (HF), and water (H_2O) are examples of compounds held together by covalent bonds. A compound held together by covalent bonds is called a **covalent compound.** In general, covalent compounds form from nonmetal elements on the right side of the periodic table. For elements in families IVA through VIIA, the number of unpaired electrons (and thus the number of covalent bonds formed) is eight minus the family number. You can get a lot of information from the periodic table from generalizations like this one. For another generalization, compare table 11.3 with the periodic table. The table gives the structures of nonmetals combined with hydrogen and the resulting compounds.

Table 11.3

Structures and compounds of nonmetal elements combined with hydrogen

Nonmetallic Elements	Element (E Represents Any Element of Family)	Compound
Family IVA: C, Si, Ge	$\cdot \overset{\cdot}{\underset{\cdot}{E}} \cdot$	$H : \overset{H}{\underset{H}{E}} : H$
Family VA: N, P, As, Sb	$\cdot \overset{\cdot \cdot}{\underset{\cdot}{E}} \cdot$	$H : \overset{\cdot \cdot}{\underset{H}{E}} : H$
Family VIA: O, S, Se, Te	$\cdot \overset{\cdot \cdot}{\underset{\cdot \cdot}{E}} \cdot$	$H : \overset{\cdot \cdot}{\underset{\cdot \cdot}{E}} : H$
Family VIIA: F, Cl, Br, I	$\cdot \overset{\cdot \cdot}{\underset{\cdot \cdot}{E}} :$	$H : \overset{\cdot \cdot}{\underset{\cdot \cdot}{E}} :$

Multiple Bonds

Two dots can represent a lone pair of valence electrons or it can represent a bonding pair, a single pair of electrons being shared by two atoms. Bonding pairs of electrons are often represented by a simple line between two atoms. For example,

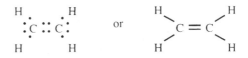

Note that the line between the two hydrogen atoms represents an electron pair, so each hydrogen atom has two electrons in the outer shell like helium. In the water molecule each hydrogen atom has two electrons as before. The oxygen atom has two lone pairs (a total of four electrons) and two bonding pairs (a total of four electrons) for a total of eight electrons. Thus, oxygen has acquired a stable octet of electrons.

A covalent bond in which a single pair of electrons is shared by two atoms is called a *single covalent bond* or simply a **single bond.** Some atoms have two unpaired electrons and can share more than one electron pair. A **double bond** is a covalent bond formed when *two pairs* of electrons are shared by two atoms. This happens mostly in compounds involving atoms of the elements C, N, O, and S. Ethylene, for example, is a gas given off from ripening fruit. The electron dot formula for ethylene is

$$H \quad \quad H \\ \quad \overset{\cdot \cdot}{\underset{}{C}} :: \overset{\cdot \cdot}{\underset{}{C}} \quad \text{or} \quad \overset{H}{\underset{H}{}} C = C \overset{H}{\underset{H}{}} \\ H \quad \quad H$$

Compounds and Chemical Change

Figure 11.8

Acetylene is a hydrocarbon consisting of two carbon atoms and two hydrogen atoms held together by a triple covalent bond between the two carbon atoms. When mixed with oxygen gas (the tank to the right), the resulting flame is hot enough to cut through most metals.

The ethylene molecule has a double bond between two carbon atoms. Since each line represents two electrons, you can simply count the lines around each symbol to see if the octet rule has been satisfied. Each H has one line, so each H atom is sharing two electrons. Each C has four lines so each C atom has eight electrons, satisfying the octet rule.

A **triple bond** is a covalent bond formed when *three pairs* of electrons are shared by two atoms. Triple bonds occur mostly in compounds with atoms of the elements C and N. Acetylene, for example, is a gas often used in welding torches (figure 11.8). The electron dot formula for acetylene is

$$H \!:\! C \!:\!:\!:\! C \!:\! H \quad \text{or} \quad H - C \equiv C - H$$

The acetylene molecule has a triple bond between two carbon atoms. Again, note that each line represents two electrons. Each C atom has four lines, so the octet rule is satisfied.

Coordinate Covalent Bonds

The single, double, and triple covalent bonds are formed when an atom shares one, two, or three pairs of electrons. For each pair each atom contributes one electron and another atom shares one of its electrons in return. There is another type of covalent bond that is a "hole and plug" kind of sharing. A **coordinate covalent bond** is formed when *the shared electron pair is contributed by one atom.* The coordinate covalent bond is like the other covalent bonds since a pair of electrons is shared between two atoms. The difference is that both of these electrons come from one atom, not one each from two atoms. Ammonia, for example, has an electron dot structure of

Table 11.4

Some common polyatomic ions

Ion Name	Formula
Acetate	$(C_2H_3O_2)^{1-}$
Ammonium	$(NH_4)^{1+}$
Borate	$(BO_3)^{3-}$
Carbonate	$(CO_3)^{2-}$
Chlorate	$(ClO_3)^{1-}$
Chromate	$(CrO_4)^{2-}$
Cyanide	$(CN)^{1-}$
Dichromate	$(Cr_2O_7)^{2-}$
Hydrogen carbonate (or bicarbonate)	$(HCO_3)^{1-}$
Hydrogen sulfate (or bisulfate)	$(HSO_4)^{1-}$
Hydroxide	$(OH)^{1-}$
Hypochlorite	$(ClO)^{1-}$
Nitrate	$(NO_3)^{1-}$
Nitrite	$(NO_2)^{1-}$
Perchlorate	$(ClO_4)^{1-}$
Permanganate	$(MnO_4)^{1-}$
Phosphate	$(PO_4)^{3-}$
Phosphite	$(PO_3)^{3-}$
Sulfate	$(SO_4)^{2-}$
Sulfite	$(SO_3)^{2-}$

which appears to be a stable structure since the octet rule is satisfied. However, notice the lone pair of electrons on the ammonia molecule. Ammonia can contribute its pair of nonbonding electrons to, say, a hydrogen ion, which shares the pair with ammonia

equation 11.7

forming an ammonium ion. All of the covalent bonds with hydrogen are now identical, with each hydrogen atom sharing a pair of electrons. The molecule is now an ion, however, since it has a net charge. An ion made up of many atoms is called a **polyatomic ion.** Coordinate covalent bonding is common in many polyatomic ions. These ions are important in many common chemicals, which will be discussed later. Some of the common polyatomic ions are listed in table 11.4. Note that (1) all have negative charges except the ammonium ion, (2) all are formed exclusively of nonmetals except three that contain metals, and (3) some are similar with different "-ite" and "-ate" endings. The "-ate" ion always has one more oxygen than the "-ite" ion.

Bond Polarity

How do you know if a bond between two atoms will be ionic or covalent? In general, ionic bonds form between metal atoms and nonmetal atoms, especially those from the opposite sides of the

Figure 11.9

Electronegativities of the elements. These values are comparative only, assigned an arbitrary scale to indicate the relative tendency of atoms to attract shared electrons.

periodic table. Also in general, covalent bonds form between the atoms of nonmetals. If an atom has a much greater electron-pulling ability than another atom, the electron is pulled completely away from the atom with lesser pulling ability and an ionic bond is the result. If the electron-pulling ability is more even between the two atoms, the electron is shared and a co-valent bond results. As you can imagine, all kinds of reactions are possible between atoms with different combinations of electron-pulling abilities. The result is that it is possible to form many gradations of bonding between completely ionic and completely covalent bonding. Which type of bonding will result can be found by comparing the electronegativity of the elements involved. **Electronegativity** is the *comparative ability of atoms of an element to attract bonding electrons*. The assigned numerical values for electronegativities are given in figure 11.9. Elements with higher values have the greatest attraction for bonding electrons, and elements with the lowest values have the least attraction for bonding electrons.

The absolute ("absolute" means without plus or minus signs) difference in the electronegativity of two bonded atoms can be used to predict if a bond is ionic or covalent. A large difference means that one element has a much greater attraction for bonding electrons than the other element. *If the absolute difference in electronegativity is 1.7 or more,* one atom pulls the bonding electron completely away and *an ionic bond results*. For example, sodium (Na) has an electronegativity of 0.9. Chlorine (Cl) has an electronegativity of 3.0. The difference is 2.1, so you can expect sodium and chloride to form ionic bonds.

Table 11.5

The meaning of absolute differences in electronegativity

Absolute Difference	→	Type of Bond Expected
1.7 or greater	means	ionic bond
between 0.5 and 1.7	means	polar covalent bond
0.5 or less	means	covalent bond

If the absolute difference in electronegativity is 0.5 or less, both atoms have about the same ability to attract bonding electrons. The result is that the electron is shared and *a covalent bond results*. A given hydrogen atom (H) has an electronegativity of another hydrogen atom, so the difference is 0. Zero is less than 0.5 so you can expect a molecule of hydrogen gas to have a covalent bond.

An ionic bond can be expected when the difference in electronegativity is 1.7 or more, and a covalent bond can be expected when the difference is less than 0.5. What happens when the difference is between 0.5 and 1.7? A covalent bond is formed, but there is an inequality since one atom has a greater bonding electron attraction than the other atom. Thus, the bonding electrons are shared unequally. A **polar covalent bond** is *a covalent bond in which there is an unequal sharing of bonding electrons*. Thus, the bonding electrons spend more time around one atom than the other. The term "polar" means "poles," and that is

Compounds and Chemical Change

Electron distribution and kinds of bonding

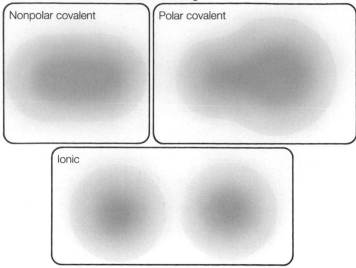

Figure 11.10

The absolute difference in electronegativities determines the kind of bond formed.

what forms in a polar molecule. Since the bonding electrons spend more time around one atom than the other, one end of the molecule will have a negative pole and the other end will have a positive pole. Since there are two poles, the molecule is sometimes called a *dipole*. Note that the molecule as a whole still contains an equal number of electrons and protons, so it is overall electrically neutral. The poles are created by an uneven charge distribution, not an imbalance of electrons and protons. Figure 11.10 shows this uneven charge distribution for a polar covalent compound. The bonding electrons spend more time near the atom on the right, giving this side of the molecule a negative pole.

Figure 11.10 also shows a molecule that has an even charge distribution. The electron distribution around one atom is just like the charge distribution around the other. This molecule is thus a *nonpolar molecule* with a *nonpolar bond*. Thus, a polar bond can be viewed as an intermediate type of bond between a nonpolar covalent bond and an ionic bond. Many gradations are possible between the transition from a purely nonpolar covalent bond and a purely ionic bond.

Example 11.5

Predict if the following bonds are nonpolar covalent, polar covalent, or ionic: (a) H-O; (b) C-Br; and, (c) K-Cl

Solution
From the electronegativity values in figure 11.9, the absolute differences are

(a) H-O, 1.4
(b) C-Br, 0.3
(c) K-Cl, 2.2

Since an absolute difference of less than 0.5 means nonpolar covalent, between 0.5 and 1.7 means polar covalent, and greater than 1.7 means ionic, then

(a) H-O, polar covalent
(b) C-Br, nonpolar covalent
(c) K-Cl, ionic

Example 11.6

Predict if the following bonds are nonpolar covalent, polar covalent, or ionic: (a) Ca-O; (b) H-Cl, and; (c) C-O.
(Answer: (a) ionic; (b) polar covalent; (c) polar covalent)

Composition of Compounds

As you can imagine, there are literally millions of different chemical compounds from all the possible combinations of over ninety natural elements held together by ionic or covalent bonds. Each of these compounds has its own name, so there are millions of names and formulas for all the compounds. In the early days, compounds were given *common names* according to how they were used, where they came from, or some other means of identifying them. Thus, sodium carbonate was called "soda," and closely associated compounds were called baking soda (sodium bicarbonate), washing soda (sodium carbonate), caustic soda (sodium hydroxide), and the bubbly drink made by reacting "soda" with acid was called "soda water," later called "soda pop" (figure 11.11). Potassium carbonate was extracted from charcoal by soaking in water and came to be called "potash." Such common names are colorful, and some are descriptive, but it was impossible to keep up with the names as the number of known compounds grew. So a systematic set of rules was developed to determine the name and formula of each compound. Once you know the rules you can write the formula when you hear the name. Conversely, seeing the formula will tell you the systematic name of the compound. This can be an interesting intellectual activity and can also be important when reading the list of ingredients to understand the composition of a product.

There is a different set of systematic rules to be used with ionic compounds and covalent compounds, but there are a few rules in common. For example, a compound made of only two different elements always ends with the suffix "-ide." So when you hear the name of a compound ending with "-ide" you automatically know that the compound is made up of only two elements. Sodium chlor*ide* is an ionic compound made up of sodium and chlorine ions. Carbon diox*ide* is a covalent compound with carbon and oxygen atoms. Thus, the systematic name tells you what elements are present in a compound with an "-ide" ending.

Ionic Compound Names

Ionic compounds formed by representative metal ions are named by stating the name of the metal (positive ion) first, then the name of the nonmetal (negative ion). Ionic compounds formed

Figure 11.11
These substances are made up of sodium and some form of a carbonate ion. All have common names with the term "soda" for this reason. Soda water (or "soda pop") was first made by reacting soda (sodium carbonate) with an acid, so it was called "soda water."

by variable-charge ions of the transition elements have an additional rule to identify which variable-charged ion is involved. There was an old way of identifying the charge on the ion by adding either "-ic" or "-ous" to the name of the metal. The suffix "-ic" meant the higher of two possible charges, and the suffix "-ous" meant the lower of two possible charges. For example, iron has two possible charges, $2+$ or $3+$. The old system used the Latin name for the root. The Latin name for iron is ferrum, so a higher charged iron ion ($3+$) was named a "ferric ion." The lower charged iron ion ($2+$) was called a "ferrous ion."

You still hear the old names sometimes, but chemists now have a better way to identify the variable-charge ion. The newer system uses the English name of the metal with Roman numerals in parentheses to indicate the charge number. Thus, an iron ion with a charge of $2+$ is called an "iron(II) ion" and an iron ion with a charge of $3+$ is an "iron(III) ion." Table 11.6 gives some of the modern names for variable-charge ions. These names are used with the name of a nonmetal ending in "-ide" just like the single-charge ions in ionic compounds made up of two different elements.

Some ionic compounds are more complex than a combination of a metal and a nonmetal ion, containing three or more elements. This is possible because they have *polyatomic ions,* groups of two or more atoms that are bound together tightly and behave very much like a single monatomic ion. For example, the OH^{1-} ion is an oxygen atom bound to a hydrogen

Table 11.6
Modern names of some variable-charged ions

Ion	Name of Ion
Fe^{2+}	Iron(II) ion
Fe^{3+}	Iron(III) ion
Cu^{+}	Copper(I) ion
Cu^{2+}	Copper(II) ion
Pb^{2+}	Lead(II) ion
Pb^{4+}	Lead(IV) ion
Sn^{2+}	Tin(II) ion
Sn^{4+}	Tin(IV) ion
Cr^{2+}	Chromium(II) ion
Cr^{3+}	Chromium(III) ion
Cr^{6+}	Chromium(VI) ion

atom with a net charge of $1-$. This polyatomic ion is called a *hydroxide ion.* The hydroxide compounds make up one of the main groups of ionic compounds, the *metal hydroxides.* A metal hydroxide is an ionic compound consisting of a metal with the hydroxide ion. Another main group consists of the salts with polyatomic ions.

Compounds and Chemical Change

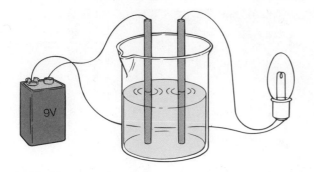

Figure 11.12

A battery and bulb will tell you if a solution contains ion. See activities below.

The metal hydroxides are named by identifying the metal first and the term *hydroxide* second. Thus, NaOH is named sodium hydroxide and KOH is potassium hydroxide. The salts are similarly named with the metal (or ammonium ion) identified first, then the name of the polyatomic ion. So NaNO$_3$ is named sodium nitrate and NaNO$_2$ is sodium nitrite. Note that the suffix "-ate" means the polyatomic ion with one more oxygen atom than the "-ite" ion. For example, the chlor*ate* ion is $(ClO_3)^{1-}$ and the chlor*ite* ion is $(ClO_2)^{1-}$. Sometimes more than two possibilities exist, and more oxygen atoms are identified with the prefix "per-" and less with the prefix "hypo-". Thus the *per*chlor*ate* ion is $(ClO_4)^{1-}$ and the *hypo*chlor*ite* ion is $(ClO)^{1-}$.

Activities

Dissolving an ionic compound in water results in ions being pulled from the crystal lattice to form free ions. A solution that contains ions will conduct an electric current, so electrical conductivity is one way to test a dissolved compound to see if it is an ionic compound or not.

Make a conductivity tester like the one illustrated in figure 11.12. This one is a 9 V radio battery and a miniature Christmas tree bulb with two terminal wires. Sand these wires to make a good electrical contact. Try testing dry (1) baking soda, (2) table salt, and (3) sugar. Test solutions of these substances dissolved in distilled water. Test vinegar and rubbing alcohol. Explain which contains ions and why.

Ionic Compound Formulas

The formulas for ionic compounds are easy to write. There are two rules: (1) the symbol for the positive element is written first, followed by the symbol for the negative element just as in the order in the name, and (2) subscripts are used to indicate the numbers of ions needed to produce an electrically neutral compound. For example, the name "calcium chloride" tells you that this compound consists of positive calcium ions and negative chlorine ions. Again, "-ide" means only two elements are present. The calcium ion is Ca^{2+} and the chlorine ion is Cl^- (you know this by applying the atomic theory, knowing their positions in the periodic table, or by using a table of ions and their charges). To be electrically neutral, the compound must have an equal number of pluses and minuses. Thus, two negative chlorine ions

are needed for every calcium ion with its 2+ charge. The formula is CaCl$_2$. The total charge of two chlorines is thus 2−, which balances the 2+ charge on the calcium ion.

One easy way to write a formula showing that a compound is electrically neutral is to cross over the absolute charge numbers (without plus or minus signs) and use them as subscripts. For example, the symbols for the calcium ion and the chlorine ion are

$$Ca^{2+} \quad Cl^{1-}$$

Crossing the absolute numbers as subscripts,

$$Ca_1^{2+} \qquad Cl_2^{1-}$$

then dropping the charge numbers gives

$$Ca_1 Cl_2$$

No subscript is written for 1; it is understood. The formula for calcium chloride is thus

$$CaCl_2$$

When using the crossover technique it is sometimes necessary to reduce the ratio to the lowest common multiple. Thus, Mg$_2$O$_2$ means an equal ratio of magnesium and oxygen ions, so the correct formula is MgO. The crossover technique works because ionic bonding results from a transfer of electrons and the net charge is conserved. A calcium ion has a 2+ charge because the atom lost two electrons and two chlorine atoms gain one electron each for a total of two electrons gained. Two electrons lost equals two electrons gained, and the net charge on calcium chloride is zero as it has to be.

The formulas for variable-charge ions are easy to write since the Roman numeral tells you the charge number. The formula for tin(II) fluoride is written by crossing over the charge numbers (Sn^{2+}, F^{1-}) and the formula is SnF$_2$.

Example 11.7

Name the following compounds: (a) LiF and (b) PbF$_2$. Write the formulas for the following compounds: (c) potassium bromide and (d) copper (I) sulfide.

Solution

(a) The formula LiF means that the positive metal ions are lithium, the negative nonmetal ions are fluorine, and there are only two elements in the compound. Lithium ions are Li^{1+} (family IA) and fluorine ions are F^{1-} (family VIIA). The name is lithium fluoride.

(b) Lead is a variable-charge transition element (table 11.6) and fluorine ions are F^{1-}. The lead ion must be Pb^{2+} because the compound PbF$_2$ is electrically neutral. Therefore, the name is lead(II) fluoride.

(c) The ions are K^{1+} and Br^{1-}. Crossing over the charge numbers and dropping the signs gives the formula KBr.

(d) The Roman numeral tells you the charge on the copper ion so the ions are Cu^{1+} and S^{2-}. The formula is Cu$_2$S.

The formulas for ionic compounds with polyatomic ions are written from combinations of positive metal ions or the ammonium ion with the polyatomic ions as listed in table 11.4. Since the polyatomic ion is a group of atoms that has a charge and stays together in a unit, it is sometimes necessary to indicate this with parentheses. For example, magnesium hydroxide is composed of Mg^{2+} ions and $(OH)^{1-}$ ions. Using the crossover technique to write the formula, you get

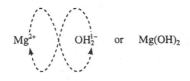

$$Mg^{2+} \quad OH_2^{1-} \quad \text{or} \quad Mg(OH)_2$$

The parentheses are used and the subscript is written *outside* the parenthesis to show that the entire hydroxide unit is taken twice. The formula $Mg(OH)_2$ means

$$Mg^{2+} \Big\langle \begin{array}{l} (OH)^{1-} \\ \\ (OH)^{1-} \end{array}$$

which shows that the pluses equal the minuses. Parentheses are not used, however, when only one polyatomic ion is present. Sodium hydroxide is NaOH, not $Na(OH)_1$.

Example 11.8

Name the following compounds: (a) Na_2SO_4 and (b) $Cu(OH)_2$. Write formulas for the following compounds: (c) calcium carbonate and (d) calcium phosphate.

Solution

(a) The ions are Na+ (sodium ion) and $(SO_4)^{2-}$ (sulfate ion). The name of the compound is sodium sulfate.

(b) Copper is a variable-charge transition element (table 11.6) and the hydroxide ion $(OH)^{1-}$ has a charge of 1−. Since the compound $Cu(OH)_2$ must be electrically neutral, the copper ion must be Cu^{2+}. The name is copper (II) hydroxide.

(c) The ions are Ca^{2+} and $(CO_3)^{2-}$. Crossing over the charge numbers and dropping the signs gives the formula $Ca_2(CO_3)_2$. Reducing the ratio to the lowest common multiple gives the correct formula of $CaCO_3$.

(d) The ions are Ca^{2+} and $(PO_4)^{3-}$ (from table 11.4). Using the crossover technique gives the formula $Ca_3(PO_4)_2$. The parentheses indicate that the entire phosphate unit is taken twice.

Covalent Compound Names

Covalent compounds are molecular and the molecules are composed of two *nonmetals,* as opposed to the metal and nonmetal elements that make up ionic compounds. The combinations of nonmetals alone do not present simple names as the ionic compounds did, so a different set of rules for naming and formula writing is needed.

Table 11.7

Prefixes and element stem names

Prefixes		Stem Names	
Prefix	Meaning	Element	Stem
Mono-	1	Hydrogen	Hydr-
Di-	2	Carbon	Carb-
Tri-	3	Nitrogen	Nitr-
Tetra-	4	Oxygen	Ox-
Penta-	5	Fluorine	Fluor-
Hexa-	6	Phosphorus	Phosph-
Hepta-	7	Sulfur	Sulf-
Octa-	8	Chlorine	Chlor-
Nona-	0	Bromine	Brom-
Deca-	10	Iodine	Iod-

Note: the *a* or *o* ending on the prefix is often dropped if the stem name begins with a vowel, e.g., "tetroxide," not "tetraoxide."

Ionic compounds were named by stating the name of the positive metal ion, then the name of the negative nonmetal ion with an "-ide" ending. This system is not adequate for naming the covalent compounds. To begin, covalent compounds are composed of two or more nonmetal atoms that form a molecule. It is possible for some atoms to form single, double, or even triple bonds with other atoms, including atoms of the same element, and coordinate covalent bonding is also possible in some compounds. The net result is that the same two elements can form more than one kind of covalent compound. Carbon and oxygen, for example, can combine to form the gas released from burning and respiration, carbon dioxide (CO_2). Under certain conditions the very same elements combine to produce a different gas, the poisonous carbon monoxide (CO). Similarly, sulfur and oxygen can combine differently to produce two different covalent compounds. A successful system for naming covalent compounds must therefore provide a means of identifying different compounds made of the same elements. This is accomplished by using a system of Greek prefixes (see table 11.7). The rules are as follows:

1. The first element in the formula is named first with a prefix indicating the number of atoms if the number is greater than one.
2. The stem name of the second element in the formula is next. A prefix is used with the stem if two elements form more than one compound. The suffix "-ide" is again used to indicate a compound of only two elements.

For example, CO is carbon monoxide and CO_2 is carbon dioxide. The compound BF_3 is boron trifluoride and N_2O_4 is dinitrogen tetroxide. Knowing the formula and the prefix and stem information in table 11.7, you can write the name of any covalent compound made up of two elements by ending it with "-ide." Conversely, the name will tell you the formula. However, there are a few polyatomic ions that have "-ide" endings that are compounds made up of more than just two elements (hydroxide and cyanide). Compounds formed with the ammonium

Compounds and Chemical Change

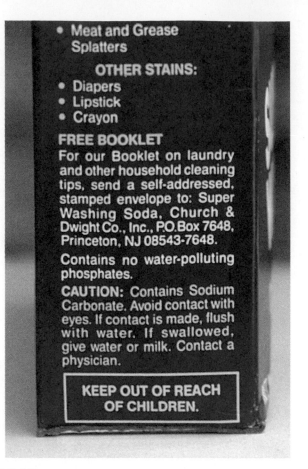

Figure 11.13

Once you understand chemical names and formulas, you can figure out what chemical compounds are contained in different household products. For example, (a) washing soda is sodium carbonate (Na_2CO_3) and (b) oven cleaner is sodium hydroxide (NaOH), which is also known as lye.

will also have an "-ide" ending, and these are also made up of more than two elements.

Covalent Compound Formulas

The systematic name tells you the formula for a covalent compound. The gas that dentists use as an anesthetic, for example, is dinitrogen monoxide. This tells you there are two nitrogen atoms and one oxygen atom in the molecule, so the formula is N_2O. A different molecule composed of the very same elements is nitrogen dioxide. Nitrogen dioxide is the pollutant responsible for the brownish haze of smog. The formula for nitrogen dioxide is NO_2. Other examples of formulas from systematic names are carbon dioxide (CO_2) and carbon tetrachloride (CCl_4).

Formulas of covalent compounds indicate a pattern of how many atoms of one element combine with atoms of another. Carbon, for example, combines with no more than two oxygen atoms to form carbon dioxide. Carbon combines with no more than four chlorine atoms to form carbon tetrachloride. Electron dot formulas show these two molecules as

Using a dash to represent bonding pairs,

$$O - C - O \qquad\qquad Cl - \underset{\displaystyle |}{\overset{\displaystyle Cl}{\underset{\displaystyle Cl}{C}}} - Cl$$

In both of these compounds the carbon atom forms four covalent bonds with another atom. The number of covalent bonds that an atom can form is called its **valence.** Carbon has a valence of four and can form single, double, or triple bonds. Here are the possibilities for a single carbon atom (combining elements not shown):

Hydrogen has only one unshared electron so the hydrogen atom has a valence of one. Oxygen has a valence of two and

Microwave Ovens and Molecular Bonds

A microwave oven rapidly cooks foods that contain water, but paper, glass, and plastic products remain cool in the oven. If they are warmed at all it is from the heat conducted from the food. The explanation of how the microwave oven heats water, but not most other substances, begins with the nature of the chemical bond.

A chemical bond acts much like a stiff spring, resisting both compression and stretching as it maintains an equilibrium distance between the atoms. As a result, a molecule tends to vibrate when energized or buffeted by other molecules. The rate of vibration depends on the "stiffness" of the spring, which is determined by the bond strength, and the mass of the atoms making up the molecule. Each kind of molecule therefore has its own set of characteristic vibrations, a characteristic natural frequency.

Disturbances with a wide range of frequencies can impact a vibrating system. When the frequency of a disturbance matches the natural frequency, energy is transferred very efficiently and the system undergoes a large increase in amplitude. Such a frequency match is called resonance. When the disturbance is visible light or some other form of radiant energy, a resonant match results in absorption of the radiant energy and an increase in the molecular kinetic energy of vibration. Thus a resonant match results in a temperature increase.

The natural frequency of a water molecule matches the frequency of infrared radiation, so resonant heating occurs when infrared radiation strikes water molecules. It is the water molecules in your skin that absorb infrared radiation from the sun, a fire, or some hot object, resulting in the warmth that you feel. Because of this match between the frequency of infrared radiation and the natural frequency of a water molecule, infrared is often called "heat radiation." Since infrared radiation is absorbed by water molecules, it is mostly absorbed on the surface of an object, penetrating only a short distance.

The frequency ranges of visible light, infrared radiation, and microwave radiation are given in box table 11.1. Most microwave ovens operate at the lower end of the microwave frequency range, between 1×10^9 Hz to 3×10^9 Hz, or 1 to 3 gigahertz. This range is too low for a resonant match with water molecules, so something else must transfer energy from the microwaves to heat the water. This something else is a result of another characteristic of the water molecule, the type

Box Table 11.1

Approximate ranges of visible light, infrared radiation, and microwave radiation

Radiation	Frequency Range (Hz)
Visible light	4×10^{14} to 8×10^{14}
Infrared radiation	3×10^{11} to 4×10^{14}
Microwave radiation	1×10^9 to 3×10^{11}

of covalent bond holding the molecule together.

The difference in electronegativity between a hydrogen and oxygen atom is 1.4, meaning the water molecule is held together by a polar covalent bond. The electrons are strongly shifted toward the oxygen end of the molecule, creating a negative pole at the oxygen end and a positive pole at the hydrogen ends. The water molecule is thus a dipole as shown in box figure 11.1A.

The dipole of water molecules has two effects: (1) the molecule can be rotated by the electric field of a microwave (see box figure 11.1B) and (2) groups of individual molecules are held together by an electrostatic attraction between the positive hydrogen ends of a water molecule with the negative oxygen end of another molecule (see box figure 11.1C). This attraction is called a hydrogen bond (see chapter 13).

One model to explain how microwaves heat water involves a particular group of three molecules, arranged so that the end molecules of the group are aligned with the microwave electric field, with the center molecule not aligned. The microwave torques the center molecule, breaking its hydrogen bond. The energy of the microwave goes into doing the work of breaking the hydrogen bond, and the molecule now has increased potential energy as a consequence. The detached water molecule reestablishes its hydrogen bond, giving up its potential energy, which goes into the vibration of the group of molecules. Thus, the energy of the microwaves is converted into a temperature increase of the water. The temperature increase is high enough to heat and cook most foods.

Microwave cooking is different from conventional cooking because the heating results from energy transfer in polar water molecules, not conduction and convection. The surface of the food never reaches a temper-

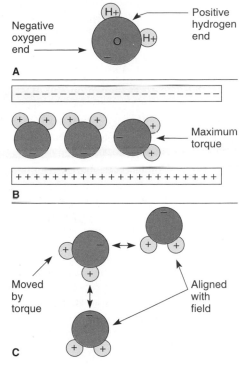

A

B

C

Box Figure 11.1

(a) A water molecule is polar, with a negative pole on the oxygen end and positive poles on the hydrogen ends. (b) An electric field aligns the water dipoles, applying a maximum torque at right angles to the dipole vector.
(c) Electrostatic attraction between the dipoles holds groups of water molecules together.

ature over the boiling point of water, so a microwave oven does not brown food (a conventional oven may reach temperatures almost twice as high). Large food items continue to cook for a period of time after being in a microwave oven as the energy is conducted from the water molecules to the food. Most recipes allow for this continued cooking by specifying a waiting period after removing the food from the oven.

Microwave ovens are able to defrost frozen foods because ice always has a thin layer of liquid water (which is what makes it slippery). To avoid "spot cooking" of small pockets of liquid water, many microwave ovens cycle on and off in the defrost cycle. The electrons in metals, like the dipole water molecules, are affected by the electric field of a microwave. A piece of metal near the wall of a microwave oven can result in sparking, which can ignite paper. Metals also reflect microwaves, which can damage the radio tube that produces the microwaves.

nitrogen has a valence of three. Here are the possibilities for hydrogen, oxygen, and nitrogen:

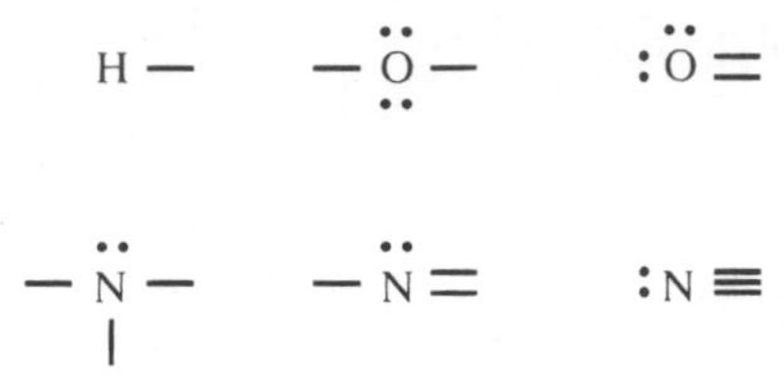

$$H- \qquad -\ddot{\underset{..}{O}}- \qquad :\ddot{O}=$$

$$-\ddot{\underset{|}{N}}- \qquad -\ddot{N}= \qquad :N\equiv$$

Note the lone pairs shown on the oxygen and nitrogen atoms. Such lone pairs create the possibility of forming a coordinate covalent bond with another atom. The number of bonds and the number of lone pairs also determine the *shape* of a molecule. Molecules are not flat like the formulas on paper. Molecules have three dimensional shapes as illustrated in figure 11.14. More will be said about molecular shapes later.

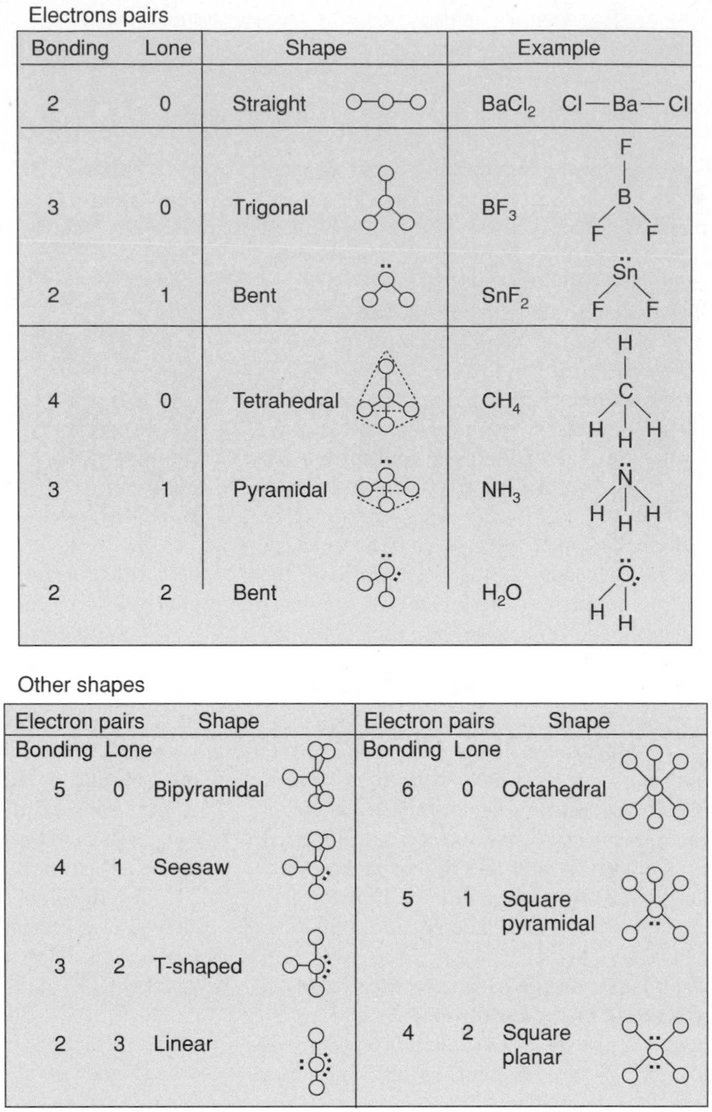

Summary

Elements are basic substances that cannot be broken down into anything simpler, and an *atom* is the smallest unit of an element. *Compounds* are combinations of two or more elements and can be broken down into simpler substances. Compounds are formed when atoms are held together by an attractive force called a *chemical bond*. A *molecule* is the smallest unit of a compound, or a gaseous element, that can exist and still retain the characteristic properties of a substance.

A *chemical change* produces new substances with new properties, and the new materials are created by making or breaking chemical bonds. The process of chemical change in which different chemical substances are created by forming or breaking chemical bonds is called a *chemical reaction*. During a chemical reaction different chemical substances with greater or lesser amounts of internal potential energy are produced. *Chemical energy* is the change of internal potential energy during a chemical reaction, and other reactions absorb energy. A *chemical equation* is a shorthand way of describing a chemical reaction. An equation shows the substances that are changed, the *reactants* on the left side, and the new substances produced, the *products* on the right side.

Chemical reactions involve *valence electrons,* the electrons in the outermost shell of an atom. Atoms tend to lose or acquire electrons to achieve the configuration of the noble gases with stable, filled outer orbitals. This tendency is generalized as the *octet rule,* that atoms lose or gain electrons to acquire the noble gas structure of eight electrons in the outer orbital. Atoms form negative or positive *ions* in the process.

A chemical bond is an attractive force that holds atoms together in a compound. Chemical bonds that are formed when atoms transfer electrons to become ions are *ionic bonds*. An ionic bond is an electrostatic attraction between oppositely charged ions. Chemical bonds formed when ions share electrons are *covalent bonds*.

Ionic bonds result in *ionic compounds* with a crystalline structure. The energy released when an ionic compound is formed is called the *heat of formation*. It is the same amount of energy that is required to decompose the compound into its elements. A *formula* of a compound uses symbols to tell what elements are in a compound and in what proportions. Ions of representative elements have a single, fixed charge but many transition elements have variable charges. Electrons are conserved when ionic compounds are formed and the ionic compound is electrically neutral. The formula shows this overall balance of charges.

Electrons pairs

Bonding	Lone	Shape		Example	
2	0	Straight	O—O—O	$BaCl_2$	Cl—Ba—Cl
3	0	Trigonal		BF_3	
2	1	Bent		SnF_2	
4	0	Tetrahedral		CH_4	
3	1	Pyramidal		NH_3	
2	2	Bent		H_2O	

Other shapes

Electron pairs		Shape		Electron pairs		Shape	
Bonding	Lone			Bonding	Lone		
5	0	Bipyramidal		6	0	Octahedral	
4	1	Seesaw					
3	2	T-shaped		5	1	Square pyramidal	
2	3	Linear		4	2	Square planar	

Figure 11.14

As you can see by studying these two charts, there is a relationship between the number of bonding electron pairs and the number of lone electron pairs and the shape of a molecule.

Covalent compounds are molecular, composed of electrically neutral groups of atoms bound together by *covalent bonds*. A *single covalent bond* is formed by the sharing of a pair of electrons, with each atom contributing a single electron to the shared pair. Covalent bonds formed when two pairs of electrons are shared are called *double bonds*. A *triple bond* is the sharing of three pairs of electrons. If a shared electron pair comes from a single atom, the bond is called a *coordinate covalent bond*. Coordinate covalent bonding sometimes results in a group of atoms with a charge that acts together as a unit. The charged unit is called a *polyatomic ion*.

The electron-pulling ability of an atom in a bond is compared with arbitrary values of *electronegativity*. A high electronegative value means a greater attraction for bonding electrons. If the absolute difference in electronegativity of two bonded atoms is 1.7 or more, one atom pulls the bonding electron away and an ionic bond results. If the difference is less than 0.5, the electrons are equally shared in a covalent bond. Between 0.5 and 1.7, the electrons are shared unequally in a *polar covalent bond*. A polar covalent bond results in electrons spending more time around the atom or atoms with the greater pulling ability, creating a negative pole at one end and a positive pole at the other. Such a molecule is called a *dipole* since it has two poles, or centers, of charge.

Compounds are named with systematic rules for ionic and covalent compounds. Both ionic and covalent compounds that are made up of only two different elements always end with an "-ide" suffix, but there are a few "-ide" names for compounds that have more than just two elements.

The modern systematic system for naming variable-charge ions states the English name and gives the charge with Roman numerals in parentheses. Ionic compounds are electrically neutral and formulas must show a balance of charge. The *crossover technique* is an easy way to write formulas that show a balance of charge.

Covalent compounds are molecules of two or more nonmetal atoms held together by a covalent bond. The system for naming covalent compounds uses Greek prefixes to identify the numbers of atoms since more than one compound can form from the same two elements (CO and CO_2, for example).

Summary of Equations

11.1

Ionization of a metal atom:

$$\text{energy} \quad + \quad Na \cdot \quad \longrightarrow \quad Na^+ \quad + \quad e^-$$

11.2

Ionic bonding reaction (single charges):

$$Na \cdot \quad + \quad \cdot \ddot{\underset{\cdot\cdot}{Cl}} : \quad \longrightarrow \quad Na^+ (: \ddot{\underset{\cdot\cdot}{Cl}} :)^-$$

11.3

Ionic bonding reaction (single and double charges):

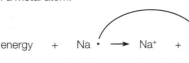

11.4

Covalent bonding (hydrogen-hydrogen):

$$H \cdot \quad + \quad H \cdot \quad \longrightarrow \quad (H \overset{\cdot\cdot}{\cdot} H)$$

11.5

Covalent bonding (hydrogen-fluorine):

$$H \cdot \quad + \quad \cdot \ddot{\underset{\cdot\cdot}{F}} : \quad \longrightarrow \quad (H \overset{\cdot\cdot}{\cdot} \ddot{\underset{\cdot\cdot}{F}} :)$$

11.6

Covalent bonding (fluorine-fluorine):

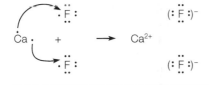

11.7

Coordinate covalent bonding (ammonium ion):

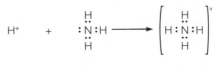

Key Terms

atom (p. 224)
chemical bond (p. 227)
chemical energy (p. 225)
chemical equation (p. 225)
chemical reaction (p. 225)
coordinate covalent bond (p. 232)
covalent bond (p. 230)
covalent compound (p. 231)
double bond (p. 231)
electronegativity (p. 233)
formula (p. 228)

heat of formation (p. 228)
ionic bond (p. 228)
ionic compounds (p. 229)
molecule (p. 225)
octet rule (p. 226)
polar covalent bond (p. 233)
polyatomic ion (p. 232)
single bond (p. 231)
triple bond (p. 232)
valence (p. 238)
valence electrons (p. 226)

Applying the Concepts

1. The smallest unit of an element that can exist alone or in combination with other elements is the
   a. electron.
   b. atom.
   c. molecule.
   d. chemical bond.

2. The smallest unit of a covalent compound that can exist while retaining the properties of the compound is the
   a. electron.
   b. atom.
   c. molecule.
   d. ionic bond.

3. You know that a chemical reaction is taking place if
   a. the temperature of a substance increases.
   b. electrons move in a steady current.
   c. chemical bonds are formed or broken.
   d. All of the above are correct.

4. Chemical reactions that involve changes in the internal potential energy of molecules always involve changes of
   a. the mass of the reactants as compared to the products.
   b. chemical energy.
   c. radiant energy.
   d. the weight of the reactants.

5. The energy released in burning materials produced by photosynthesis has what relationship to the solar energy that was absorbed in making the materials? It is
   a. less than the solar energy absorbed.
   b. the same as the solar energy absorbed.
   c. more than the solar energy absorbed.
   d. variable, having no relationship to the energy absorbed.

6. The electrons that participate in chemical bonding are (the)
   a. valance electrons.
   b. electrons in fully occupied orbitals.
   c. stable inner electrons.
   d. all of the above.

7. Atoms of the representative elements have a tendency to seek stability through
   a. acquiring the noble gas structure.
   b. filling or emptying their outer orbitals.
   c. any situation that will satisfy the octet rule.
   d. all of the above.

8. An ion is formed when an atom of a representative element
   a. gains or loses protons.
   b. shares electrons to achieve stability.
   c. loses or gains electrons to satisfy the octet rule.
   d. All of the above are correct.

9. An atom of an element that is in family VIA will have what charge when it is ionized?
   a. 2+
   b. 6+
   c. 6−
   d. 2−

10. Which type of chemical bond is formed by a transfer of electrons?
    a. ionic
    b. covalent
    c. metallic
    d. coordinate covalent

11. Which type of chemical bond is formed by the sharing of two electrons?
    a. ionic
    b. covalent
    c. metallic
    d. coordinate covalent

12. Salts, such as sodium chloride, are what type of compounds?
    a. ionic compounds
    b. covalent compounds
    c. polar compounds
    d. Any of the above are correct.

13. If there are two bromine atoms for each barium ion in a compound the chemical formula is
    a. $_2Br_1Ba$.
    b. Ba_2Br.
    c. $BaBr_2$.
    d. none of the above.

14. Which combination of elements forms crystalline solids that will dissolve in water, producing a solution of ions that can conduct an electric current?
    a. metal and metal
    b. metal and nonmetal
    c. nonmetal and nonmetal
    d. All of the above are correct.

15. In a single covalent bond between two atoms,
    a. a single electron from one of the atoms is shared.
    b. a pair of electrons from one of the atoms is shared.
    c. a pair of electrons, one from each atom is shared.
    d. a single electron is transferred from one atom.

16. The number of pairs of shared electrons in a covalent compound is often the same as the number of
    a. unpaired electrons in the electron dot notation.
    b. valence electrons.
    c. orbital pairs.
    d. protons in the nucleus.

17. Sulfur and oxygen are both in the VIA family of the periodic table. If element X combines with oxygen to form the compound X_2O, element X will combine with sulfur to form the compound
    a. XS_2.
    b. X_2S.
    c. X_2S_2.
    d. It is impossible to say without more information.

18. One element is in the IA family of the periodic table and a second is in the VIIA family. What type of compound will the two elements form?
    a. ionic
    b. covalent
    c. They will not form a compound.
    d. More information is needed to answer this question.

19. One element is in the VA family of the periodic table and a second is in the VIA family. What type of compound will these two elements form?
    a. ionic
    b. covalent
    c. They will not form a compound.
    d. More information is needed to answer this question.

20. A covalent bond in which there is an unequal sharing of bonding electrons is a
    a. single covalent bond.
    b. double covalent bond.
    c. triple covalent bond.
    d. polar covalent bond.

21. A compound made of only two different elements has a name that always ends with the suffix
    a. -ite.
    b. -ate.
    c. -ide.
    d. -ous.

22. Dihydrogen monoxide is the systematic name for a compound that has the common name of
    a. laughing gas.
    b. water.
    c. smog.
    d. rocket fuel.

Answers

1. b 2. c 3. c 4. b 5. b 6. a 7. d 8. c 9. d
10. a 11. b 12. a 13. c 14. b 15. c 16. a 17. b
18. a 19. b 20. d 21. c 22. b

Questions for Thought

1. Describe how the following are alike and how they are different (a) a sodium atom and a sodium ion and (b) a sodium ion and a neon atom.

2. What is the difference between a polar covalent bond and a nonpolar covalent bond?

3. What is the difference between an ionic and covalent bond? What do atoms forming the two bond types have in common?

4. What is the octet rule?

5. Is there a relationship between the number of valence electrons and how many covalent bonds an atom can form? Explain.

6. Write electron dot formulas for molecules formed when hydrogen combines with (a) chlorine, (b) oxygen, and (c) carbon.

7. Sodium fluoride is often added to water supplies to strengthen teeth. Is sodium fluoride ionic, nonpolar covalent, or polar covalent? Explain the basis of your answer.

8. What is the modern systematic name of a compound with the formula (a) SnF_2? (b) PbS?

9. What kinds of elements are found in (a) Ionic compounds with a name ending with an "-ide" suffix? (b) Covalent compounds with a name ending with an "-ide" suffix?

10. Why is it necessary to use a system of Greek prefixes to name binary covalent compounds?

11. What are variable charge ions? Explain how variable charge ions are identified in the modern systematic system of naming compounds.

12. What is a polyatomic ion? Give the names and formulas for several common polyatomic ions.

13. Write the formula for magnesium hydroxide. Explain what the parentheses mean.

14. What is a double bond? A triple bond?

Exercises

Group A—Solutions Provided in Appendix

1. Use electron dot symbols in equations to predict the formula of the ionic compound formed between the following:
   (a) K and I
   (b) Sr and S
   (c) Na and O
   (d) Al and O

2. Name the following ionic compounds formed from variable charge transition elements:
   (a) CuS
   (b) Fe_2O_3
   (c) CrO
   (d) PbS

3. Name the following polyatomic ions:
   (a) $(OH)^{1-}$
   (b) $(SO_3)^{2-}$
   (c) $(ClO)^{1-}$
   (d) $(NO_3)^{1-}$
   (e) $(CO_3)^{2-}$
   (f) $(ClO_4)^{1-}$

4. Use the crossover technique to write formulas for the following compounds:
   (a) Iron(III) hydroxide
   (b) Lead(II) phosphate
   (c) Zinc carbonate
   (d) Ammonium nitrate NH_4NO_3
   (e) Potassium hydrogen carbonate
   (f) Potassium sulfite

5. Write formulas for the following covalent compounds:
   (a) Carbon tetrachloride
   (b) Dihydrogen monoxide
   (c) Manganese dioxide
   (d) Sulfur trioxide
   (e) Dinitrogen pentoxide
   (f) Diarsenic pentasulfide

6. Name the following covalent compounds:
   (a) CO
   (b) CO_2
   (c) CS_2
   (d) N_2O
   (e) P_4S_3
   (f) N_2O_3

7. Predict if the bonds formed between the following pairs of elements are ionic, polar covalent, or nonpolar covalent:
   (a) Si and O
   (b) O and O
   (c) H and Te
   (d) C and H
   (e) Li and F
   (f) Ba and S

Group B—Solutions Not Given

1. Use electron dot symbols in equations to predict the formulas of the ionic compound formed between the following:
   (a) Li and F
   (b) Be and S
   (c) Li and O
   (d) Al and S

2. Name the following ionic compounds formed from variable charge transition elements:
   (a) $PbCl_2$
   (b) FeO
   (c) Cr_2O_3
   (d) PbO

3. Name the following polyatomic ions:
   (a) $(C_2H_3O_2)^{1-}$
   (b) $(HCO_3)^{1-}$
   (c) $(SO_4)^{2-}$
   (d) $(NO_2)^{1-}$
   (e) $(MnO_4)^{1-}$
   (f) $(CO_3)^{2-}$

4. Use the crossover technique to write formulas for the following compounds:
   (a) Aluminum hydroxide
   (b) Sodium phosphate
   (c) Copper (II) chloride
   (d) Ammonium sulfate
   (e) Sodium hydrogen carbonate
   (f) Cobalt(II) chloride

5. Write formulas for the following covalent compounds:
   (a) Silicon dioxide
   (b) Dihydrogen sulfide
   (c) Boron trifluoride
   (d) Dihydrogen dioxide
   (e) Carbon tetrafluoride
   (f) Nitrogen trihydride

6. Name the following covalent compounds:
   (a) N_2O
   (b) SO_2
   (c) SiC
   (d) PF_5
   (e) $SeCl_6$
   (f) N_2O_4

7. Predict if the bonds formed between the following pairs of elements are ionic, polar covalent, or nonpolar covalent:
   (a) Si and C
   (b) Cl^+ and Cl
   (c) S and O
   (d) Sr and F
   (e) O and H
   (f) K and F

Chemical Formulas and Equations

These microscopic bubbles can be any size with no definite mass, but elements always react in definite mass ratios.

W_E live in a chemical world that has been partly manufactured through controlled chemical change. Consider all of the synthetic fibers and plastics that are used in clothing, housing, and cars. Consider all the synthetic flavors and additives in foods, how these foods are packaged, and how they are preserved. Consider also the synthetic drugs and vitamins that keep you healthy. There are millions of such familiar products that are the direct result of chemical research. Most of these products simply did not exist sixty years ago.

Many of the products of chemical research have remarkably improved the human condition. For example, synthetic fertilizers have made it possible to supply food in quantities that would not otherwise be possible. Chemists learned how to take nitrogen from the air and convert it into fertilizers on an enormous scale. Other chemical research resulted in products such as weed killers, insecticides, and mold and fungus inhibitors. The fertilizers and these products have made it possible to supply food for millions of people who would have otherwise starved (figure 12.1).

Yet, we also live in a world with concerns about chemical pollutants, the greenhouse effect, acid rain, and a disappearing ozone shield. The very nitrogen fertilizers that have increased food supplies also wash into rivers, polluting the waterways and bays. Such dilemmas require an understanding of chemical products and the benefits and hazards of possible alternatives. Understanding requires a knowledge of chemistry, since the benefits, and risks, are chemical in nature.

The previous chapters were about the modern atomic theory and how it explains elements and how compounds are formed in chemical change. This chapter is concerned with describing chemical changes and the different kinds of chemical reactions that occur. These reactions are explained with balanced chemical equations, which are concise descriptions of reactions that produce the products used in our chemical world.

Figure 12.1

The products of chemical research have substantially increased food supplies but have also increased the possibilities of pollution. Balancing the benefits and hazards of the use of chemicals requires a knowledge of chemistry and a knowledge of the alternatives.

Chemical Formulas

In chapter 11 you learned how to name and write formulas for ionic and covalent compounds, including the ionic compound of table salt and the covalent compound of ordinary water. Recall that a formula is a shorthand way of describing the elements or ions that make up a compound. There are basically three kinds of formulas that describe compounds: (1) *empirical* formulas, (2) *molecular* formulas, and (3) *structural* formulas. Empirical and molecular formulas, and their use, will be considered in this chapter. Structural formulas will be considered in chapter 14.

An **empirical formula** identifies the elements present in a compound and describes the *simplest whole number ratio* of atoms of these elements with subscripts. For example, the empirical formula for ordinary table salt is $NaCl$. This tells you that the elements sodium and chlorine make up this compound and there is one atom of sodium for each chlorine atom. The empirical formula for water is H_2O, meaning there are two atoms of hydrogen for each atom of oxygen.

Covalent compounds exist as molecules. A chemical formula that identifies the *actual numbers* of atoms in a molecule is known as a **molecular formula.** Figure 12.2 shows the

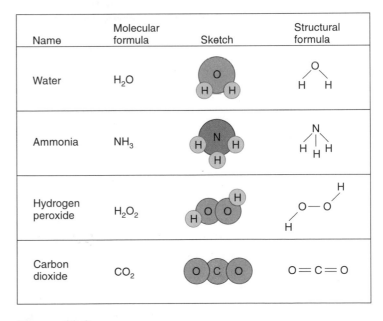

Name	Molecular formula	Sketch	Structural formula
Water	H_2O		
Ammonia	NH_3		
Hydrogen peroxide	H_2O_2		
Carbon dioxide	CO_2		$O=C=O$

Figure 12.2

The name, molecular formula, sketch, and structural formula of some common molecules. Compare the kinds and numbers of atoms making up each molecule in the sketch to the molecular formula.

Chemical Formulas and Equations

structure of some common molecules and their molecular formulas. Note that each formula identifies the elements and numbers of atoms in each molecule. The figure also indicates how molecular formulas can be written to show how the atoms are arranged in the molecule. Formulas that show the relative arrangements are called structural formulas. Compare the structural formulas in the illustration with the three-dimensional representations and the molecular formulas.

How do you know if a formula is empirical or molecular? First, you need to know if the compound is ionic or covalent. You know that ionic compounds are usually composed of metal and nonmetal atoms with an electronegativity difference greater than 1.7. Formulas for ionic compounds are *always* empirical formulas. Ionic compounds are composed of many positive and negative ions arranged in an electrically neutral array. There is no discrete unit, or molecule, in an ionic compound so it is only possible to identify ratios of atoms with an empirical formula.

Covalent compounds are generally nonmetal atoms bonded to nonmetal atoms in a molecule. You could therefore assume that a formula for a covalent compound is a molecular formula unless it is specified otherwise. You can be certain it is a molecular formula if it is not the simplest whole number ratio. Glucose, for example, is a simple sugar (also known as dextrose) with the formula $C_6H_{12}O_6$. This formula is divisible by six, yielding a formula with the simplest whole number ratio of CH_2O. Therefore, CH_2O is the empirical formula for glucose and $C_6H_{12}O_6$ is the molecular formula.

Molecular and Formula Weights

The **formula weight** of a compound is the sum of the atomic weights of all the atoms in a chemical formula. For example, the formula for water is H_2O. Hydrogen and oxygen are both nonmetals, so the formula means that one atom of oxygen is bound to two hydrogen atoms in a molecule. From the periodic table, you know that the approximate (rounded) atomic weight of hydrogen is 1.0 u and oxygen is 16.0 u. Adding the atomic weights for *all* the atoms,

Atoms	Atomic Weight		Totals
2 of H	2 × 1.0 u	=	2.0 u
1 of O	1 × 16.0 u	=	16.0 u
	Formula weight	=	18.0 u

Thus, the formula weight of a water molecule is 18.0 u.

The formula weight of an ionic compound is found the same way, by adding the rounded atomic weights of atoms (or ions) making up the compound. Sodium chloride is NaCl, so the formula weight is 23.0 u plus 35.5 u, or 58.5 u. The *formula weight* can be calculated for an ionic or molecular substance. The **molecular weight** is the formula weight of a molecular substance. The term *molecular weight* is sometimes used for all substances, whether or not they have molecules. Since ionic substances such as NaCl do not occur as molecules, this is not strictly correct. Both molecular and formula weights are calculated the same way, but formula weight is a more general term.

Example 12.1

What is the formula weight of table sugar (sucrose), which has the formula $C_{12}H_{22}O_{11}$?

Solution

The formula identifies the numbers of each atom and the atomic weights are from a periodic table:

Atoms	Atomic Weight		Totals
12 of C	12 × 12.0 u	=	144.0 u
22 of H	22 × 1.0 u	=	22.0 u
11 of O	11 × 16.0 u	=	176.0 u
	Formula weight	=	342.0 u

Example 12.2

What is the molecular weight of ethyl alcohol, C_2H_5OH? (Answer: 46.0 u)

Percent Composition of Compounds

The formula weight of a compound can provide useful information about the elements making up a compound (figure 12.3). For example, suppose you want to know how much calcium is provided by a dietary supplement. The label lists the main ingredient as calcium carbonate, $CaCO_3$. To find how much calcium is supplied by a pill with a certain mass you need to find the *mass percentage* of calcium in the compound.

Percent is simply the fractional part of the whole times 100 percent (meaning "per 100"), or

$$\left(\frac{\text{part}}{\text{whole}}\right)(100\% \text{ of whole}) = \% \text{ of part} \qquad \textbf{equation 12.1}$$

For example, if 13 students in a class of 50 are freshmen, the percentage of freshmen in the class is

$$\left(\frac{13 \text{ freshmen}}{50 \text{ classmates}}\right)(100\% \text{ of classmates})$$

$$= \left(0.26 \frac{\text{freshmen}}{\text{classmates}}\right)(100\% \text{ of classmates})$$

$$= 26\% \text{ freshmen}$$

Note the classmate units cancel, giving the answer in percent freshmen.

Since the formula weight of a compound represents all of its composition, the formula weight is the "whole" in equation 12.1, with all the atoms contributing a part of the whole weight. The "part" in equation 12.1 is the atomic weight times the number of atoms of the element in which you are interested. Thus, the mass percentage of an element in a compound can be found from

$$\frac{\left(\begin{array}{c}\text{atomic weight} \\ \text{of element}\end{array}\right)\left(\begin{array}{c}\text{number of atoms} \\ \text{of element}\end{array}\right)}{\text{formula weight of compound}} \times \begin{array}{c}100\% \text{ of} \\ \text{compound}\end{array} = \begin{array}{c}\% \text{ of} \\ \text{element}\end{array}$$

$$\textbf{equation 12.2}$$

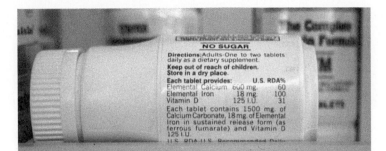

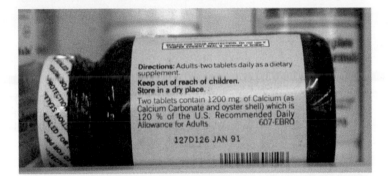

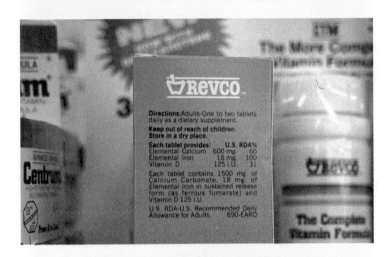

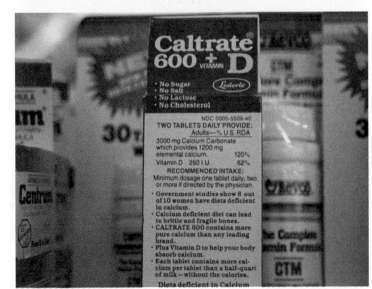

The mass percentage of calcium in $CaCO_3$ can be found in two steps:

Step 1: Determine formula weight:

Atoms	Atomic Weight		Totals
1 of Ca	1×40.1 u	=	40.1 u
1 of C	1×12.0 u	=	12.0 u
3 of O	3×16.0 u	=	48.0 u
	Formula weight	=	100.1 u

Step 2: Determine percentage of Ca:

$$\frac{(40.1 \text{ u Ca})(1)}{100.1 \text{ u } \cancel{CaCO_3}} \times 100\% \text{ } \cancel{CaCO_3} = 40.1\% \text{ Ca}$$

Knowing the percentage of the total mass contributed by the calcium, this fractional part (as a decimal) can be multiplied by the mass of the supplement pill to find the calcium supplied. The mass percentage of the other elements can also be determined with equation 12.2.

Example 12.3

Sodium fluoride is added to water supplies and to some toothpastes for fluoridation. What is the percentage composition of the elements in sodium fluoride?

Solution
Step 1: Write the formula for sodium fluoride, NaF.

Step 2: Determine the formula weight.

Atoms	Atomic Weight		Totals
1 of Na	1×23.0 u	=	23.0 u
1 of F	1×19.0 u	=	19.0 u
	Formula weight	=	42.0 u

Step 3: Determine the percentage of Na and F.

For Na:

$$\frac{(23.0 \text{ u Na})(1)}{42.0 \text{ u NaF}} \times 100\% \text{ NaF} = \boxed{54.7\% \text{ Na}}$$

For F:

$$\frac{(19.0 \text{ u F})(1)}{42.0 \text{ u NaF}} \times 100\% \text{ NaF} = \boxed{45.2\% \text{ F}}$$

The percentage often does not total to exactly 100 percent because of rounding.

Example 12.4

Calculate the percentage composition of carbon in table sugar, sucrose, which has a formula of $C_{12}H_{22}O_{11}$. (Answer: 42.1% C)

◀ **Figure 12.3**

If you know the name of an ingredient, you can write a chemical formula and the percent composition of a particular substance can be calculated from the formula and this can be useful information for consumer decisions.

       Chemical Formulas and Equations

Chemical fertilizers are added to the soil when it does not contain sufficient elements essential for plant growth. The three critical elements are nitrogen, phosphorus, and potassium, and these are the basic ingredients in most chemical fertilizers. In general, lawns require fertilizers high in nitrogen and gardens require fertilizers high in phosphorus.

Read the labels on commercial packages of chemical fertilizers sold in a garden shop. Find the name of the chemical that supplies each of these critical elements, for example, nitrogen is sometimes supplied by ammonium sulfate $(NH_4)_2SO_4$. Calculate the mass percentage of each critical element supplied according to the label information. Compare these percentages to the grade number of the fertilizer, for example, 10–20–10. Determine which fertilizer brand gives you the most nutrients for the money.

Chemical Equations

Chemical reactions occur when bonds between the outermost parts of atoms are formed or broken. Bonds are formed, for example, when a green plant uses sunlight—a form of energy—to create molecules of sugar, starch, and plant fibers. Bonds are broken and energy is released when you digest the sugars and starches or when plant fibers are burned. Chemical reactions thus involve changes in matter, the creation of new materials with new properties, and energy exchanges. So far you have considered chemical symbols as a concise way to represent elements, and formulas as a concise way to describe what a compound is made of. There is also a concise way to describe a chemical reaction, the **chemical equation.**

Balancing Equations

Word equations were introduced in the previous chapter. Word equations are useful in identifying the *reactants,* the substances that existed before the reaction, and the *products,* the new substances formed as a result of the reaction. For example, the charcoal used in a barbecue grill is carbon (figure 12.4). The carbon reacts with oxygen while burning, and the reaction (1) releases energy and (2) forms carbon dioxide. The reactants and products for this reaction can be described as

carbon + oxygen → carbon dioxide

The arrow means *yields* and the word equation is read as, "Carbon reacts with oxygen to yield carbon dioxide." This word equation describes what happens in the reaction but says nothing about the quantities of reactants or products.

Chemical symbols and formulas can be used in the place of words in an equation and the equation will have a whole new meaning. For example, the equation describing carbon reacting with oxygen to yield carbon dioxide becomes

$$C + O_2 \rightarrow CO_2 \qquad \text{(balanced)}$$

The new, added meaning is that one atom of carbon (C) reacts with one molecule of oxygen (O_2) to yield one molecule of carbon dioxide (CO_2). Note that the equation also shows one atom of carbon and two atoms of oxygen (recall that oxygen occurs as

Figure 12.4

The charcoal used in a grill is basically carbon. The carbon reacts with oxygen to yield carbon dioxide. The chemical equation for this reaction, $C + O_2 \rightarrow CO_2$, contains the same information as the English sentence but has quantitative meaning as well.

a diatomic molecule) as reactants on the left side and one atom of carbon and two atoms of oxygen as products on the right side. Since the same number of each kind of atom appears on both sides of the equation, the equation is said to be *balanced*.

You would not want to use a charcoal grill in a closed room because there might not be enough oxygen. An insufficient supply of oxygen produces a completely different product, the poisonous gas, carbon monoxide (CO). An equation for this reaction is

$$C + O_2 \rightarrow CO \qquad \text{(not balanced)}$$

As it stands, this equation describes a reaction that violates the **law of conservation of mass,** that matter is neither created nor destroyed in a chemical reaction. From the point of view of atoms, this law states that *atoms are neither created nor destroyed in a chemical reaction.* A chemical reaction is the making or breaking of chemical bonds between atoms or groups of atoms. Atoms are not lost or destroyed in the process nor are they changed to a different kind. The equation for the formation of carbon monoxide has two oxygen atoms in the reactants (O_2) but only one in the product (in CO). An atom of oxygen has disappeared somewhere and that violates the law of conservation of mass. You cannot fix the equation by changing the CO to a CO_2, because this would change the identity of the compounds. Carbon monoxide is a poisonous gas that is different from carbon dioxide, a relatively harmless product of burning and respiration. *You cannot change the subscript in a formula* because that would change the formula. A different formula means a different composition and thus a different compound.

You cannot change the subscripts of a formula but you can place a number called a *coefficient* in *front* of the formula. Changing a coefficient changes the *amount* of a substance, not

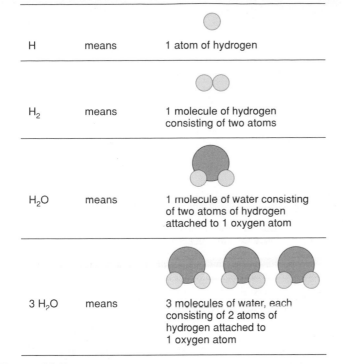

H	means	1 atom of hydrogen
H₂	means	1 molecule of hydrogen consisting of two atoms
H₂O	means	1 molecule of water consisting of two atoms of hydrogen attached to 1 oxygen atom
3 H₂O	means	3 molecules of water, each consisting of 2 atoms of hydrogen attached to 1 oxygen atom

Figure 12.5

The meaning of subscripts and coefficients used with a chemical formula. The subscripts tell you how many atoms of a particular element are in a compound. The coefficient tells you about the quantity, or number, of molecules of the compound.

Figure 12.6

These large tanks collapse as the natural gas is removed, then grow larger as they are filled. Why do you suppose collapsing and inflating tanks are necessary? The answer is that you cannot pump a gas as easily as you can pump a liquid. You can, however, push the gas out of a container just as you squeeze a balloon to get all the air out.

the identity. Thus 2 CO means two molecules of carbon monoxide and 3 CO means three molecules of carbon monoxide. If there is no coefficient, 1 is understood as with subscripts. The meaning of coefficients and subscripts is illustrated in figure 12.5.

Placing a coefficient of 2 in front of the C and a coefficient of 2 in front of the CO in the equation will result in the same numbers of each kind of atom on both sides:

$$2 C + O_2 \rightarrow 2 CO$$

| Reactants: | 2 C | Products: | 2 C |
| | 2 O | | 2 O |

The equation is now balanced.

Suppose your barbecue grill burns natural gas, not charcoal (figure 12.6). Natural gas is mostly methane, CH_4. Methane burns by reacting with oxygen (O_2) to produce carbon dioxide (CO_2) and water vapor (H_2O). A balanced chemical equation for this reaction can be written by following a procedure of four steps.

Step 1: Write the correct formulas for the reactants and products in an unbalanced equation. The reactants and products could have been identified by chemical experiments, or they could have been predicted from what is known about chemical properties. This will be discussed in more detail later. For now, assume that the reactants and products are known and are given in words. For the burning of methane, the unbalanced, but otherwise correct, formula equation would be

$$CH_4 + O_2 \rightarrow CO_2 + H_2O \qquad \textbf{(not balanced)}$$

Step 2: Inventory the number of each kind of atom on both sides of the unbalanced equation. In the example there are

Reactants:	1 C	Products:	1 C
	4 H		2 H
	2 O		3 O

This shows that the H and O are unbalanced.

Step 3: Determine where to place coefficients in front of formulas to balance the equation. It is often best to focus on the simplest thing you can do with whole number ratios. The H and the O are unbalanced, for example, and there are 4 H atoms on the left and 2 H atoms on the right. Placing a coefficient 2 in front of H_2O will balance the H atoms:

$$CH_4 + O_2 \rightarrow CO_2 + 2 H_2O \qquad \textbf{(not balanced)}$$

Now take a second inventory:

Reactants:	1 C	Products:	1 C
	4 H		4 H
	2 O		4 O (O_2 + 2 O)

This shows the O atoms are still unbalanced with 2 on the left and 4 on the right. Placing a coefficient of 2 in front of O_2 will balance the O atoms.

$$CH_4 + 2 O_2 \rightarrow CO_2 + 2 H_2O \qquad \textbf{(balanced)}$$

Step 4: Take another inventory to determine (a) if the number of atoms on both sides are now equal and, if so, (b) if the coefficients are in the lowest possible whole number ratio. The inventory is now

Reactants:	1 C	Products:	1 C
	4 H		4 H
	4 O		4 O

Chemical Formulas and Equations

Reaction:	Methane reacts with oxygen to yield carbon dioxide and water
Balanced equation:	$CH_4 + 2\ O_2 \longrightarrow CO_2 + 2\ H_2O$

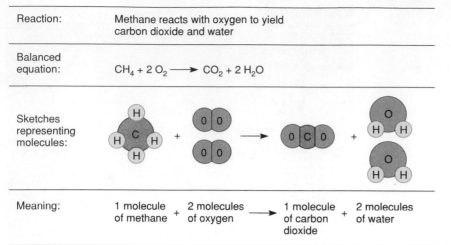

| Meaning: | 1 molecule of methane | + | 2 molecules of oxygen | → | 1 molecule of carbon dioxide | + | 2 molecules of water |

◀ **Figure 12.7**

Compare the numbers of each kind of atom in the balanced equation with the numbers of each kind of atom in the sketched representation. Both the equation and the sketch have the same number of atoms in the reactants and in the products.

(a) The number of each kind of atom on each side of the equation is the same, and (b) the ratio of 1:2 → 1:2 is the lowest possible whole number ratio. The equation is balanced, which is illustrated with sketches of molecules in figure 12.7.

Balancing chemical equations is mostly a trial-and-error procedure. But with practice, you will find there are a few generalized "role models" that can be useful in balancing equations for many simple reactions. The key to success at balancing equations is to think it out step-by-step while remembering the following:

1. Atoms are neither lost nor gained nor do they change their identity in a chemical reaction. The same kind and number of atoms in the reactants must appear in the products, meaning atoms are conserved.
2. A correct formula of a compound cannot be changed by altering the number or placement of subscripts. Changing subscripts changes the identity of a compound and the meaning of the entire equation.
3. A coefficient in front of a formula multiplies everything in the formula by that number.

There are also a few generalizations that can be helpful for success in balancing equations:

1. Look first to formulas of compounds with the most atoms and try to balance the atoms or compounds they were formed from or decomposed to.
2. Polyatomic ions that appear on both sides of the equation should be treated as independent units with a charge. That is, consider the polyatomic ion as a unit while taking an inventory rather than the individual atoms making up the polyatomic ion. This will save time and simplify the procedure.
3. Both the "crossover technique" and the use of "fractional coefficients" can be useful in finding the least common multiple to balance an equation. All of these generalizations are illustrated in examples 12.5, 12.6, and 12.7.

The physical state of reactants and products in a reaction is often identified by the symbols (g) for gas, (l) for liquid, (s) for solid, and (aq) for an aqueous solution (aqueous means water). If a gas escapes, this is identified with an arrow pointing up (↑). A solid formed from a solution is identified with an arrow pointing down (↓). The Greek symbol delta (Δ) is often used to indicate that heat is supplied.

Example 12.5

Propane is a liquified petroleum gas (LPG) that is often used as a bottled substitute for natural gas (figure 12.8). Propane (C_3H_8) reacts with oxygen (O_2) to yield carbon dioxide (CO_2) and water vapor (H_2O). What is the balanced equation for this reaction?

Solution

Step 1: Write the correct formulas of the reactants and products in an unbalanced equation.

$$C_3H_8(g) + O_2(g) \rightarrow CO_2(g) + H_2O(g) \quad \text{(unbalanced)}$$

Step 2: Inventory the numbers of each kind of atom.

Reactants:	3 C	Products:	1 C
	8 H		2 H
	2 O		3 O

Step 3: Determine where to place coefficients to balance the equation. Looking at the compound with the most atoms (generalization 1), you can see that a propane molecule has 3 C and 8 H. Placing a coefficient of 3 in front of CO_2 and a 4 in front of H_2O will balance these atoms (3 of C and 4 × 2 = 8 H atoms on the right has the same number of atoms as C_3H_8 on the left),

$$C_3H_8(g) + O_2(g) \rightarrow 3\ CO_2(g) + 4\ H_2O(g) \quad \text{(not balanced)}$$

A second inventory shows

Reactants:	3 C
	8 H
	2 O

Products:	3 C
	8 H (4 × 2 = 8)
	10 O [(3 × 2) + (4 × 1) = 10]

The O atoms are still unbalanced. Place a 5 in front of O_2 and the equation is balanced (5 × 2 = 10). Remember that you cannot change the subscripts and that oxygen occurs as a diatomic molecule of O_2.

$$C_3H_8(g) + 5\ O_2(g) \rightarrow 3\ CO_2(g) + 4\ H_2O(g) \quad \text{(balanced)}$$

Step 4: Another inventory shows (a) the number of atoms on both sides are now equal and (b) the coefficients are 1:5 → 3:4, the lowest possible whole number ratio. The equation is balanced.

Figure 12.8

One of two burners is operating at the moment as this hot air balloon prepares to ascend. The burners are fueled by propane (C_3H_8), a liquified petroleum gas (LPG). As other forms of petroleum, propane releases large amounts of heat during the chemical reaction of burning.

Example 12.6

One type of water hardness is caused by the presence of calcium bicarbonate in solution, $Ca(HCO_3)_2$. One way to remove the troublesome calcium ions from wash water is to add washing soda, which is sodium carbonate, Na_2CO_3. The reaction yields sodium bicarbonate ($NaHCO_3$) and calcium carbonate ($CaCO_3$), which is insoluble. Since $CaCO_3$ is insoluble, the reaction removes the calcium ions from solution. Write a balanced equation for the reaction.

Solution

Step 1: Write the unbalanced equation

$$Ca(HCO_3)_2(aq) + Na_2CO_3(aq) \rightarrow NaHCO_3(aq) + CaCO_3\downarrow$$
(not balanced)

Step 2: Inventory the numbers of each kind of atom. This reaction has polyatomic ions that appear on both sides, so they should be treated as independent units with a charge (generalization 2). The inventory is

Reactants:		Products:	
1 Ca		1 Ca	
2 $(HCO_3)^{1-}$		1 $(HCO_3)^{1-}$	
2 Na		1 Na	
1 $(CO_3)^{2-}$		1 $(CO_3)^{2-}$	

Step 3: Placing a coefficient of 2 in front of $NaHCO_3$ will balance the equation,

$$Ca(HCO_3)_2(aq) + Na_2CO_3(aq) \rightarrow 2\ NaHCO_3(aq) + CaCO_3\downarrow \quad \textbf{(balanced)}$$

Step 4: An inventory shows

Reactants:		Products:	
1 Ca		1 Ca	
2 $(HCO_3)^{1-}$		2 $(HCO_3)^{1-}$	
2 Na		2 Na	
1 $(CO_3)^{2-}$		1 $(CO_3)^{2-}$	

The coefficient ratio of 1:1 → 2:1 is the lowest whole number ratio. The equation is balanced.

Example 12.7

Gasoline is a mixture of hydrocarbons, including octane (C_8H_{18}). Combustion of octane produces CO_2 and H_2O, with the release of energy. Write a balanced equation for this reaction.

Solution

Step 1: Write the correct formulas in an unbalanced equation,

$$C_8H_{18}(g) + O_2(g) \rightarrow CO_2(g) + H_2O(g) \quad \textbf{(not balanced)}$$

Step 2: Take an inventory,

Reactants:		Products:	
8 C		1 C	
18 H		2 H	
2 O		3 O	

Step 3: Start with the compound with the most atoms (generalization 1) and place coefficients to balance these atoms,

$$C_8H_{18}(g) + O_2(g) \rightarrow 8\ CO_2(g) + 9\ H_2O(g) \quad \textbf{(not balanced)}$$

Redo the inventory,

Reactants:		Products:	
8 C		8 C	
18 H		18 H	
2 O		25 O	

The O atoms are still unbalanced. There are 2 O atoms in the reactants but 25 O atoms in the products. Since the subscript cannot be changed, it will take 12.5 O_2 to produce 25 oxygen atoms (generalization 3).

$$C_8H_{18}(g) + 12.5\ O_2(g) \rightarrow 8\ CO_2(g) + 9\ H_2O(g) \quad \textbf{(balanced)}$$

Step 4: (a) An inventory will show that the atoms balance,

Reactants:		Products:	
8 C		8 C	
18 H		18 H	
25 O		25 O	

(b) The coefficients are not in the lowest whole number ratio (one-half an O_2 does not exist). To make the lowest possible whole number ratio, all coefficients are multiplied by two. This results in a correct balanced equation of

$$2\ C_8H_{18}(g) + 25\ O_2(g) \rightarrow 16\ CO_2(g) + 18\ H_2O(g) \quad \textbf{(balanced)}$$

Generalizing Equations

In the previous chapters you learned that the act of classifying, or grouping, something according to some property makes the study of a large body of information less difficult. Generalizing from groups of chemical reactions also makes it possible to predict what will happen in similar reactions. For example, you have studied equations in the previous section describing the combustion of methane (CH_4), propane (C_3H_8), and octane (C_8H_{18}). Each of these reactions involves a *hydrocarbon,* a compound of the elements hydrogen and carbon. Each hydrocarbon reacted with O_2, yielding CO_2 and releasing the energy of combustion. Generalizing from these reactions, you could predict that the combustion of any hydrocarbon would involve the combination of atoms of the hydrocarbon molecule with O_2

Chemical Formulas and Equations

◀ *Figure 12.9*

Hydrocarbons are composed of the elements hydrogen and carbon. Propane (C_3H_8) and gasoline, which contains octane (C_8H_{18}) are examples of hydrocarbons. *Carbohydrates* are composed of the elements hydrogen, carbon, and oxygen. Table sugar, for example, is the carbohydrate $C_{12}H_{22}O_{11}$. Generalizing, all hydrocarbons and carbohydrates react completely with oxygen to yield CO_2 and H_2O.

to produce CO_2 and H_2O with the release of energy. Such reactions could be analyzed by chemical experiments, and the products could be identified by their physical and chemical properties. You would find your predictions based on similar reactions would be correct, thus justifying predictions from such generalizations. Butane, for example, is a hydrocarbon with the formula C_4H_{10}. The balanced equation for the combustion of butane is

$$2\ C_4H_{10}(g) + 13\ O_2(g) \rightarrow 8\ CO_2(g) + 10\ H_2O(g)$$

You could extend the generalization further, noting that the combustion of compounds containing oxygen as well as carbon and hydrogen also produces CO_2 and H_2O (figure 12.9). These compounds are *carbohydrates,* composed of carbon and water. Glucose, for example, was identified earlier as a compound with the formula $C_6H_{12}O_6$. Glucose combines with oxygen to produce CO_2 and H_2O and the balanced equation is

$$C_6H_{12}O_6(s) + 6\ O_2(g) \rightarrow 6\ CO_2(g) + 6\ H_2O(g)$$

Note that three molecules of oxygen were not needed from the O_2 reactant since the other reactant, glucose, contains six oxygen atoms per molecule. An inventory of atoms will show that the equation is thus balanced.

Combustion is a rapid reaction with O_2 that releases energy, usually with a flame. A very similar, although much slower reaction takes place in plant and animal respiration. In respiration, carbohydrates combine with O_2 and release energy used for biological activities. This reaction is slow compared to combustion and requires enzymes to proceed at body temperature. Nonetheless, CO_2 and H_2O are the products.

Oxidation-Reduction Reactions

The reactions involving hydrocarbons and carbohydrates with oxygen are examples of an important group of chemical reactions called *oxidation-reduction* reactions. When the term "oxidation" was first used, it specifically meant reactions involving the combination of oxygen with other atoms. But fluorine, chlorine, and other nonmetals were soon understood to have similar reactions as oxygen, so the definition was changed to one concerning the shifts of electrons in the reaction.

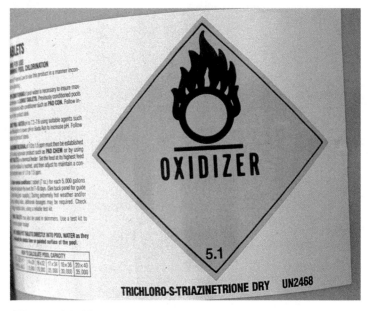

Figure 12.10

Oxidizing agents take electrons from other substances that are being oxidized. Oxygen and chlorine are commonly used strong oxidizing agents.

There are two broad classes of chemical reactions: (1) those in which electrons are transferred as a result of the reaction and (2) those in which electrons are not transferred. Reactions in which electrons are not transferred will be considered in the next section. For now, consider the **oxidation-reduction reaction,** a chemical reaction in which electrons are transferred from one atom to another. From the first letters, it is sometimes called a **redox reaction** for short.

Substances that take electrons from other substances are called **oxidizing agents.** Oxidizing agents take electrons from the substances being oxidized. Oxygen is the most common oxidizing agent, and several examples have already been given about how it oxidizes foods and fuels. Chlorine is another commonly used oxidizing agent, often for the purposes of bleaching or killing bacteria (figure 12.10).

A **reducing agent** supplies electrons to the substance being reduced. Hydrogen and carbon are commonly used reducing agents. Carbon is commonly used as a reducing agent to extract metals from their ores. For example, carbon (from coke, which is coal that has been baked) reduces Fe_2O_3, an iron ore, in the reaction

$$2 \ Fe_2O_3(s) + 3 \ C(s) \rightarrow 4 \ Fe(s) + 3 \ CO_2 \uparrow$$

The Fe in the ore gained electrons from the carbon, the reducing agent in this reaction.

Types of Chemical Reactions

Many chemical reactions can be classified as redox or nonredox reactions. Another way to classify chemical reactions is to consider what is happening to the reactants and products. This type of classification scheme leads to four basic categories of chemical reactions, which are (1) *combination,* (2) *decomposition,* (3) *replacement,* and (4) *ion exchange* reactions. The first three categories are subclasses of redox reactions. It is in the ion exchange reactions that you will find the first example of a reaction that is not a redox reaction.

Combination Reactions

A **combination reaction** is a synthesis reaction in which two or more substances combine to form a single compound. The combining substances can be (1) elements, (2) compounds, or (3) combinations of elements and compounds. In generalized form, a combination reaction is

$$X + Y \rightarrow XY \qquad \textbf{equation 12.3}$$

Many redox reactions are combination reactions. For example, metals are oxidized when they burn in air, forming a metal oxide. Consider magnesium, which gives off a bright white light as it burns:

$$2 \ Mg(s) + O_2(g) \rightarrow 2 \ MgO(s)$$

Note how the magnesium-oxygen reaction follows the generalized form of equation 12.3.

The rusting of metals is oxidation that takes place at a slower pace than burning, but metals are nonetheless oxidized

in the process (figure 12.11). Again noting the generalized form of a combination reaction, consider the rusting of iron:

$$4 \ Fe(s) + 3 \ O_2(g) \rightarrow 2 \ Fe_2O_3(s)$$

Nonmetals are also oxidized by burning in air, for example, when carbon burns with a sufficient supply of O_2:

$$C(s) + O_2(g) \rightarrow CO_2(g)$$

Note that all the combination reactions follow the generalized form of $X + Y \rightarrow XY$.

Decomposition Reactions

A **decomposition reaction,** as the term implies, is the opposite of a combination reaction. In decomposition reactions a compound is broken down (1) into the elements that make up the compound, (2) into simpler compounds, or (3) into elements and simpler compounds. Decomposition reactions have a generalized form of

$$XY \rightarrow X + Y + \ldots \qquad \textbf{equation 12.4}$$

Decomposition reactions generally require some sort of energy, which is usually supplied in the form of heat or electrical energy. An electric current, for example, decomposes water into hydrogen and oxygen:

$$2 \ H_2O(l) \overset{electricity}{\rightarrow} 2 \ H_2(g) + O_2(g)$$

Mercury(II) oxide is decomposed by heat, an observation that led to the discovery of oxygen (figure 12.12):

$$2 \ HgO(s) \overset{\Delta}{\rightarrow} 2 \ Hg(s) + O_2 \uparrow$$

Plaster is a building material made from a mixture of calcium hydroxide, $Ca(OH)_2$, and plaster of Paris, $CaSO_4$. The calcium hydroxide is prepared by adding water to calcium oxide (CaO), which is commonly called quicklime. Calcium oxide is made by heating limestone or chalk ($CaCO_3$), and

$$CaCO_3(s) \overset{\Delta}{\rightarrow} CaO(s) + CO_2 \uparrow$$

Note that all the decomposition reactions follow the generalized form of $XY \rightarrow X + Y + \ldots$

Replacement Reactions

In a **replacement reaction,** an atom or polyatomic ion is replaced in a compound by a different atom or polyatomic ion. The replaced part can be either the negative or positive part of the compound. In generalized form, a replacement reaction is

$$XY + Z \rightarrow XZ + Y \qquad \textbf{equation 12.5}$$
(negative part replaced)

or

$$XY + A \rightarrow AY + X \qquad \textbf{equation 12.6}$$
(positive part replaced)

Replacement reactions occur because some elements have a stronger electron holding ability than other elements. Elements that have the least ability to hold on to their electrons are the most chemically active. Figure 12.13 shows a list of chemical activity of some metals, with the most chemically active at the top. Hydrogen is included because of its role in acids (see

Chemical Formulas and Equations

Figure 12.11

Rusting iron is a common example of a combination reaction, where two or more substances combine to form a new compound. Rust is iron(III) oxide formed from the combination of iron and oxygen. These new girders are made of unprotected iron that has begun to rust during construction of a new building.

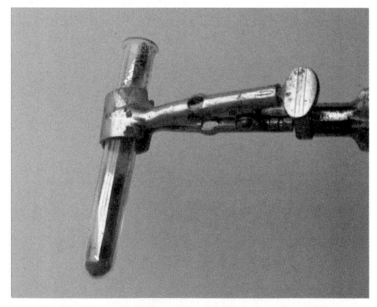

Figure 12.12

Mercury(II) oxide is decomposed by heat, leaving the silver-colored element mercury behind as oxygen is driven off. This is an example of a decomposition reaction, $2 HgO \rightarrow 2 Hg + O_2 \uparrow$. Compare this equation to the general form of a decomposition reaction.

chapter 13). Take a few minutes to look over the generalizations listed in figure 12.13. The generalizations apply to combination, decomposition, and replacement reactions.

Replacement reactions take place as more active metals give up electrons to elements lower on the list with a greater electron holding ability. For example, aluminum is higher on the activity series than copper. When aluminum foil is placed in a solution of copper(II) chloride, aluminum is oxidized, losing electrons to the copper. The loss of electrons from metallic aluminum forms aluminum ions in solution, and the copper comes out of solution as a solid metal (figure 12.14).

$$2 Al(s) + 3 CuCl_2(aq) \rightarrow 2 AlCl_3(aq) + 3 Cu(s)$$

A metal will replace any metal ion in solution that it is above in the activity series. If the metal is listed below the metal ion in solution, no reaction occurs. For example, $Ag(s) + CuCl_2(aq) \rightarrow$ no reaction.

The very active metals (lithium, potassium, calcium, and sodium) react with water to yield metal hydroxides and hydrogen. For example,

$$2 Na(s) + 2 H_2O(l) \rightarrow 2 NaOH(aq) + H_2\uparrow$$

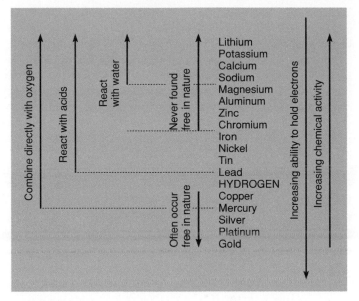

Figure 12.13

The activity series for common metals, together with some generalizations about the chemical activities of the metals. The series is used to predict which replacement reactions will take place and which reactions will not occur. (Note that hydrogen is not a metal and is placed in the series for reference to acid reactions.)

Acids yield hydrogen ions in solution, and metals above hydrogen in the activity series will replace hydrogen to form a metal salt. For example,

$$Zn(s) + H_2SO_4(aq) \rightarrow ZnSO_4(aq) + H_2\uparrow$$

In general, the energy involved in replacement reactions is less than the energy involved in combination or decomposition reactions.

Ion Exchange Reactions

An **ion exchange reaction** is a reaction that takes place when the ions of one compound interact with the ions of another compound, forming (1) a solid that comes out of solution (a precipitate), (2) a gas, or (3) water.

A water solution of dissolved ionic compounds is a solution of ions. For example, solid sodium chloride dissolves in water to become ions in solution,

$$NaCl(s) \rightarrow Na^{1+}Cl^{1-}(aq)$$

If a second ionic compound is dissolved with a solution of another, a mixture of ions results. The formation of a precipitate, a gas, or water, however, removes ions from the solution, and this must occur before you can say that an ionic exchange reaction has taken place. For example, water being treated for domestic use sometimes carries suspended matter that is removed by adding aluminum sulfate and calcium hydroxide to the water. The reaction is

$$3\ Ca(OH)_2(aq) + Al_2(SO_4)_3(aq) \rightarrow 3\ CaSO_4(aq) + 2\ Al(OH)_3\downarrow$$

The aluminum hydroxide is a jellylike solid, which traps the suspended matter for sand filtration. The formation of the in-

Figure 12.14

This shows a reaction between metallic aluminum and the blue solution of copper(II) chloride. Aluminum is above copper in the activity series, and aluminum replaces the copper ions from the solution as copper is deposited as a metal. The aluminum loses electrons to the copper and forms aluminum ions in solution.

soluble aluminum hydroxide removed the aluminum and hydroxide ions from the solution, so an ion exchange reaction took place.

In general, an ion exchange reaction has the form

$$XY + AZ \rightarrow XZ + AY \qquad \text{equation 12.7}$$

where one of the products removes ions from the solution. The calcium hydroxide and aluminum sulfate reaction took place as the aluminum and calcium ions traded places. A solubility table such as the one in appendix B will tell you if an ionic exchange reaction has taken place. Aluminum hydroxide is insoluble, according to the table, so the reaction did take place. No ionic exchange reaction occurred if the new products are both soluble.

Another way for an ion exchange reaction to occur is if a gas or water molecule forms to remove ions from the solution. When an acid reacts with a base (an alkaline compound), a salt and water are formed

$$HCl(aq) + NaOH(aq) \rightarrow NaCl(aq) + H_2O(l)$$

The reactions of acids and bases are discussed in chapter 13.

Example 12.8

Write complete balanced equations for the following, and identify if each reaction is combination, decomposition, replacement, or ion exchange:

(a) silver(s) + sulfur(g) → silver sulfide(s)
(b) aluminum(s) + iron(III) oxide(s) → aluminum oxide(s) + iron
(c) sodium chloride(aq) + silver nitrate(aq) → ?
(d) potassium chlorate(s) $\xrightarrow{\Delta}$ potassium chloride(s) + oxygen(g)

Chemical Formulas and Equations

Solution

(a) The reactants are two elements and the product is a compound, following the general form X + Y → XY of a combination reaction. Table 11.2 gives the charge on silver as Ag^{1+}, and sulfur (as the other nonmetals in family VIA) is S^{2-}. The balanced equation is

$$2 \text{ Ag(s)} + \text{S(g)} \rightarrow \text{Ag}_2\text{S(s)}$$

Silver sulfide is the tarnish that appears on silverware.

(b) The reactants are an element and a compound that react to form a new compound and an element. The general form is XY + Z → XZ + Y, which describes a replacement reaction. The balanced equation is

$$2 \text{ Al(s)} + \text{Fe}_2\text{O}_3\text{(s)} \rightarrow \text{Al}_2\text{O}_3\text{(s)} + 2 \text{ Fe(s)}$$

This is known as a thermite reaction, and in the reaction aluminum reduces the iron oxide to metallic iron with the release of sufficient energy to melt the iron. The thermite reaction is sometimes used to weld large steel pieces, such as railroad rails.

(c) The reactants are water solutions of two compounds with the general form of XY + AZ →, so this must be the reactant part of an ion exchange reaction. Completing the products part of the equation by exchanging parts as shown in the general form and balancing,

$$\text{NaCl(aq)} + \text{AgNO}_3\text{(aq)} \rightarrow \text{NaNO}_3\text{(?)} + \text{AgCl(?)}$$

The solubility chart in appendix B is now consulted to find out if either of the products is insoluble. $NaNO_3$ is soluble and AgCl is insoluble. Since at least one of the products is insoluble, the reaction did take place and the equation is rewritten as

$$\text{NaCl(aq)} + \text{AgNO}_3\text{(aq)} \rightarrow \text{NaNO}_3\text{(aq)} + \text{AgCl}\downarrow$$

(d) The reactant is a compound and the products are a simpler compound and an element, following the generalized form of a decomposition reaction, XY → X + Y. The delta sign (Δ) also means that heat was added, which provides another clue that this is a decomposition reaction. The formula for the chlorate ion is in table 11.4. The balanced equation is

$$2 \text{ KClO}_3\text{(s)} \xrightarrow{\Delta} 2 \text{ KCl(s)} + 3 \text{ O}_2\uparrow$$

Information from Chemical Equations

A balanced chemical equation describes what happens in a chemical reaction in a concise, compact way. The balanced equation also carries information about (1) atoms, (2) molecules, and (3) atomic weights. The balanced equation for the combustion of hydrogen, for example, is

$$2 \text{ H}_2\text{(g)} + \text{O}_2\text{(g)} \rightarrow 2 \text{ H}_2\text{O(l)}$$

An inventory of each kind of atom in the reactants and products shows

Reactants:	4 hydrogen	Products:	4 hydrogen
	2 oxygen		2 oxygen
Total:	6 atoms	Total:	6 atoms

There are six atoms before the reaction and there are six atoms after the reaction, which is in accord with the law of conservation of mass.

In terms of molecules, the equation says that two diatomic molecules of hydrogen react with one (understood) diatomic molecule of oxygen to yield two molecules of water. The number of coefficients in the equation is the number of molecules involved in the reaction. If you are concerned how two molecules plus one molecule could yield two molecules, remember that *atoms* are conserved in a chemical reaction, not molecules.

Since atoms are conserved in a chemical reaction, their atomic weights should be conserved, too. One hydrogen atom has an atomic weight of 1.0 u so the formula weight of a diatomic hydrogen molecule must be 2×1.0 u, or 2.0 u. The formula weight of O_2 is 2×16.0 u, or 32 u. If you consider the equation in terms of atomic weights, then

Equation

$$2 \text{ H}_2 + \text{O}_2 \rightarrow 2 \text{ H}_2\text{O}$$

Atomic weights

$$2 (1.0 \text{ u} + 1.0 \text{ u}) + (16.0 \text{ u} + 16.0 \text{ u}) \rightarrow 2 (2 \times 1.0 \text{ u} + 16.0 \text{ u})$$

$$4 \text{ u} + 32 \text{ u} \rightarrow 36 \text{ u}$$

$$36 \text{ u} \rightarrow 36 \text{ u}$$

The formula weight for H_2O is $(1.0 \text{ u} \times 2) + 16$ u, or 18 u. The coefficient of 2 in front of H_2O means there are two molecules of H_2O, so the mass of the products is 2×18 u, or 36 u. Thus, the reactants had a total mass of 4 u + 32 u, or 36 u, and the products had a total mass of 36 u. Again, this is in accord with the law of conservation of mass.

The equation says that 4 u of hydrogen will combine with 32 u of oxygen. Thus hydrogen and oxygen combine in a mass ratio of 4:32, which reduces to 1:8. So one gram of hydrogen will combine with eight grams of oxygen, and, in fact, they will combine in this ratio no matter what the measurement units are (gram, kilogram, pound, etc.). They always combine in this mass ratio because this is the mass of the individual reactants.

Back in the early 1800s John Dalton attempted to work out a table of atomic weights as he developed his atomic theory (see chapter 10). Dalton made two major errors in determining the atomic weights, including (1) measurement errors about mass ratios of combining elements and (2) incorrect assumptions about the formula of the resulting compound. For water, for example, Dalton incorrectly measured that 5.5 g of oxygen combined with 1.0 g of hydrogen. He assumed that one atom of hydrogen combined with one atom of oxygen, resulting in a formula of HO. Thus, Dalton concluded that the atomic mass of oxygen was 5.5 u and the atomic mass of hydrogen was 1.0 u. Incorrect atomic weights for hydrogen and oxygen led to conflicting formulas for other substances, and no one could show that the atomic theory worked.

The problem was solved during the first decade of the 1800s through the separate work of a French chemistry professor, Joseph Gay-Lussac, and an Italian physics professor, Amadeo Avogadro. In 1808, Gay-Lussac reported that reacting gases combined in small, whole number *volumes* when the temperature and pressure were constant. Two volumes of hydrogen,

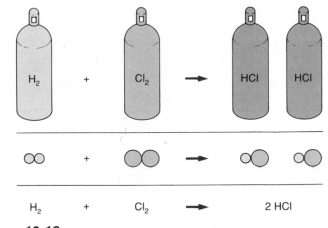

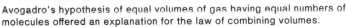

Figure 12.16

Avogadro's hypothesis of equal volumes of gas having equal numbers of molecules offered an explanation for the law of combining volumes.

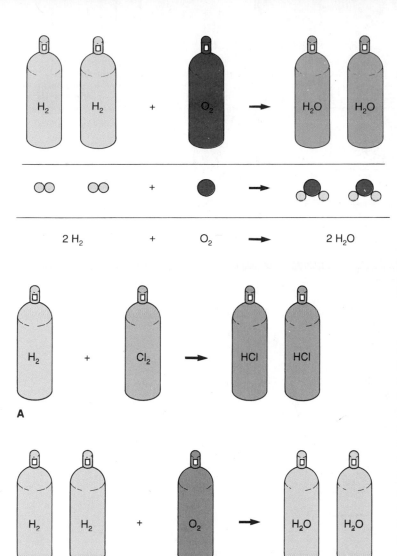

Figure 12.15

Reacting gases combine in ratios of small, whole number volumes when the temperature and pressure are the same for each volume. (a) One volume of hydrogen gas combines with one volume of chlorine gas to yield two volumes of hydrogen chloride gas. (b) Two volumes of hydrogen gas combine with one volume of oxygen gas to yield two volumes of water vapor.

for example, combined with one volume of oxygen to form two volumes of water vapor. The term "volume" means any measurement unit, for example, a liter. Other reactions between gases were also observed to combine in small, whole number ratios and the pattern became known as the *law of combining volumes* (figure 12.15).

Avogadro proposed an explanation for the law of combining volumes in 1811. Avogadro proposed that equal volumes of all gases at the same temperature and pressure *contain the same number of molecules.* Avogadro's hypothesis had two important implications for the example of water. First, since two volumes of hydrogen combine with one volume of oxygen, it means that a molecule of water contains twice as many hydrogen atoms as oxygen atoms. The formula for water must be

H_2O, not HO. Second, since *two* volumes of water vapor were produced, each molecule of hydrogen and each molecule of oxygen must be diatomic. Diatomic molecules of hydrogen and oxygen would double the number of hydrogen and oxygen atoms, thus producing twice as much water vapor. These two implications are illustrated in figure 12.16, along with a balanced equation for the reaction. Note that the coefficients in the equation now have two meanings, (1) the number of molecules of each substance involved in the reaction and (2) the ratios of combining volumes. The coefficient of 2 in front of the H_2, for example, means two molecules of H_2. It also means two volumes of H_2 gas when all volumes are measured at the same temperature and pressure. Recall that the volume of any gas at the same temperature and pressure contains the same number of molecules. Thus, the coefficient in a balanced equation means a ratio of *any number* of molecules, from 2 of H_2 and 1 of O_2, 20 of H_2 and 10 of O_2, 2,000 of H_2 and 1,000 of O_2, or however many are found in 2 L of H_2 and 1 L of O_2.

Example 12.9

Propane is a hydrocarbon with the formula C_3H_8 that is used as a bottled gas. (a) How many liters of oxygen are needed to burn 1 L of propane gas? (b) How many liters of carbon dioxide are produced by the reaction? Assume all volumes to be measured at the same temperature and pressure.

Solution

The balanced equation is

$$C_3H_8(g) + 5\ O_2(g) \rightarrow 3\ CO_2(g) + 4\ H_2O(g)$$

The coefficients tell you the relative number of molecules involved in the reaction, that 1 molecule of propane reacts with 5 molecules of oxygen to produce 3 molecules of carbon dioxide and 4 molecules of water. Since equal volumes of gases at the same temperature and pressure contain equal numbers of molecules, the coefficients also tell you the relative volumes of gases. Thus 1 L of propane (a) requires 5 L of oxygen and (b) yields 3 L of carbon dioxide (and 4 L of water vapor) when reacted completely.

Chemical Formulas and Equations

A. Each of the following represents one mole of an element:

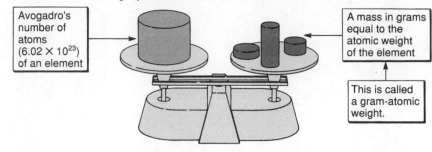

Avogadro's number of atoms (6.02 × 10²³) of an element → A mass in grams equal to the atomic weight of the element

This is called a gram-atomic weight.

B. Each of the following represents one mole of a compound:

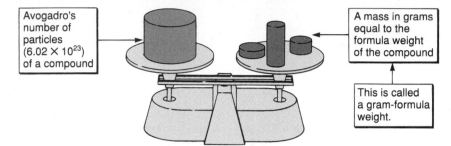

Avogadro's number of particles (6.02 × 10²³) of a compound → A mass in grams equal to the formula weight of the compound

This is called a gram-formula weight.

C. Each of the following represents one mole of a molecular substance:

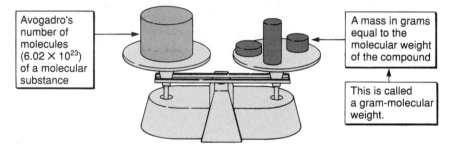

Avogadro's number of molecules (6.02 × 10²³) of a molecular substance → A mass in grams equal to the molecular weight of the compound

This is called a gram-molecular weight.

◀ *Figure 12.17*

The mole concept for (*a*) elements, (*b*) compounds, and (*c*) molecular substances. A mole contains 6.02 × 10²³ particles. Since every mole contains the same number of particles, the ratio of the mass of any two moles is the same as the ratio of the masses of individual particles making up the two moles.

Units of Measurement Used with Equations

The coefficients in a balanced equation represent a ratio of any *number* of molecules involved in a chemical reaction. The equation has meaning about the atomic *weights* and formula *weights* of reactants and products. The counting of numbers and the use of atomic weights are brought together in a very important measurement unit called a *mole* (from the Latin meaning "a mass"). Here are the important ideas in the mole concept:

1. Recall that the atomic weights of elements are average relative masses of the isotopes of an element. The weights are based on a comparison to carbon-12, with an assigned mass of exactly 12.00 (see chapter 10).
2. The *number* of C-12 atoms in exactly 12.00 g of C-12 has been measured experimentally to be 6.02 × 10²³. This number is called **Avogadro's number,** named after the scientist who reasoned that equal volumes of gases contain equal numbers of molecules.
3. An amount of a substance that contains Avogadro's number of atoms, ions, molecules or any other chemical unit is defined as a **mole** of the substance. Thus a mole is 6.02 × 10²³ atoms, ions, etc., just as a dozen is 12 eggs, apples, etc. The mole is the chemist's measure of atoms, molecules, or other chemical units. A mole of Na⁺ ions is 6.02 × 10²³ Na⁺ ions.

4. A mole of C-12 atoms is defined as having a mass of exactly 12.00 g, a mass that is numerically equal to its atomic mass. So the mass of a mole of C-12 atoms is 12.00 g, or

mass of one atom × one mole = mass of a mole of C-12
(12.00 u)   (6.02 × 10²³) =   12.00 g

The masses of all the other isotopes are *based* on a comparison to the C-12 atom. Thus a He-4 atom has one-third the mass of a C-12 atom. An atom of Mg-24 is twice as massive as a C-12 atom. Thus

	1 Atom	×	1 Mole	=	Mass of Mole
C-12:	12.00 u	×	6.02 x 10²³	=	12.00 g
He-4:	4.00 u	×	6.02 x 10²³	=	4.00 g
Mg-24:	24.00 u	×	6.02 x 10²³	=	24.00 g

Therefore, the mass of a mole of any element is numerically equal to its atomic mass. And any mass of an element that is in the same ratio as their atomic masses must contain the same number of atoms (figure 12.17).

This reasoning can be used to generalize about formula weights, molecular weights, and atomic weights since they are all based on atomic mass units relative to C-12. The **gram-atomic weight** is the mass in grams of one mole of an element that is

The Catalytic Converter

The modern automobile produces two troublesome products in the form of (1) nitrogen monoxide (NO) and (2) hydrocarbons from the incomplete combustion of gasoline. These products from the exhaust enter the air to react in sunlight, eventually producing an irritating haze known as *photochemical smog.* To reduce photochemical smog, modern automobiles are fitted with a catalytic converter as part of the automobile exhaust system (box figure 12.1). This feature is about how the catalytic converter combats smog-forming pollutants.

Chemical reactions proceed at a rate that is affected by (1) the concentration of the reactants, (2) the temperature at which a reaction occurs, and (3) the surface area of the reaction. In general, a higher concentration, higher temperatures, and greater surface area mean a faster reaction. The problem with nitrogen monoxide is that it is easily oxidized to nitrogen dioxide (NO_2), a reddish brown, damaging gas that also plays a key role in the formation of photochemical smog. Nitrogen dioxide and the hydrocarbons are oxidized slowly in the air when left to themselves. What is needed is a means to decompose nitrogen monoxide and uncombusted hydrocarbons rapidly before they are released into the air.

The rate at which a chemical reaction proceeds is affected by a *catalyst,* a material that speeds up a chemical reaction without being permanently changed by the reaction. Apparently, molecules require a certain amount of energy to change the chemical bonds that tend to keep them as they are, unreacted. This certain amount of energy is called the *activation energy,* and it represents an energy barrier that must be overcome before a chemical reaction can take place. This explains why chemical reactions proceed at a faster rate at higher temperatures. At higher temperatures, molecules have greater average kinetic energies, thus they already have part of the minimum energy needed for a reaction to take place.

A catalyst appears to speed a chemical reaction by lowering the activation energy. Molecules become temporarily attached to the surface of the catalyst, which weakens the chemical bonds holding the molecule together. Thus, the weakened molecule is easier to break apart and the activation energy is

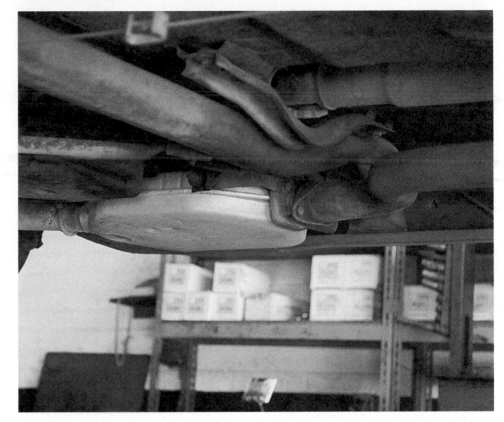

Box Figure 12.1

This silver-colored canister is the catalytic converter. The catalytic converter is located between the engine and the muffler, which is farther back toward the rear of the car.

lowered. Some catalysts do this better with some specific compounds than others, and extensive chemical research programs are devoted to finding new and more effective catalysts.

Automobile catalytic converters use unreactive metals, such as platinum, and transition metal oxides such as copper(II) oxide and chromium(III) oxide. Catalytic reactions that occur in the converter include the following:

$$H_2O + CO \xrightarrow{\text{catalyst}} H_2 + CO_2$$

$$2\,NO + 2\,CO \xrightarrow{\text{catalyst}} N_2 + 2\,CO_2$$

$$2\,NO + 2\,H_2 \xrightarrow{\text{catalyst}} N_2 + 2\,H_2O$$

$$O_2 + CO^* \xrightarrow{\text{catalyst}} CO_2 + H_2O$$

*or other hydrocarbons

Thus nitrogen monoxide is reduced to nitrogen gas and hydrocarbons are oxidized to CO_2 and H_2O. Sometimes small quantities of sulfur in gasoline are burned, producing SO_2 in the exhaust, which is converted to H_2S by the catalyst. H_2S has a rotten egg odor. Leaded fuels such as those containing tetraethyl lead, $Pb(C_2H_5)_4$, will poison a catalytic converter, rendering it useless. In spite of these problems, a catalytic converter can reduce or oxidize about 90 percent of the hydrocarbons, 85 percent of the carbon monoxide, and 40 percent of the nitrogen monoxide from exhaust gases. Other controls, such as exhaust gas recirculation (EGR) are used to further reduce NO formation.

numerically equal to its atomic weight. The atomic weight of carbon is 12.01 u, the gram-atomic weight of carbon is 12.01 g. The atomic weight of magnesium is 24.3 u; the gram-atomic weight of magnesium is 24.3 g. Any gram-atomic weight contains Avogadro's number of atoms. Therefore the gram-atomic weights of the elements all contain the same number of atoms.

Similarly, the **gram-formula weight** of a compound is the mass in grams of one mole of the compound that is numerically equal to its formula weight. The **gram-molecular weight** is the gram-formula weight of a molecular compound. Note that one mole of O atoms (6.02×10^{23} oxygen atoms) has a gram-atomic weight of 16.0 g, but one mole of O_2 molecules (6.02×10^{23} oxygen molecules) has a gram-molecular weight of 32.0 g. Stated the other way around, 32.0 g of O_2 and 16.0 g of O both contain the same Avogadro's number of particles.

Example 12.10

(a) A 100 percent silver chain has a mass of 107.9 g. How many silver atoms are in the chain? (b) What is the mass of one mole of sodium chloride, NaCl?

Solution

The mole concept and Avogadro's number provide a relationship between numbers and masses. (a) The atomic weight of silver is 107.9 u, so the gram-atomic weight of silver is 107.9 g. A gram-atomic weight is one mole of an element, so the silver chain contains 6.02×10^{23} silver atoms. (b) The formula weight of NaCl is 58.5 u, so the gram-formula weight is 58.5 g. One mole of NaCl has a mass of 58.5 g.

Quantitative Uses of Equations

A balanced chemical equation can be interpreted in terms of (1) a *molecular ratio* of the reactants and products, (2) a *mole ratio* of the reactants and products, or (3) a *mass ratio* of the reactants and products. Consider, for example, the balanced equation for reacting hydrogen with nitrogen to produce ammonia,

$$3 H_2(g) + N_2(g) \rightarrow 2 NH_3(g)$$

From a *molecular* point of view, the equation says that three molecules of hydrogen combine with one molecule of N_2 to form two molecules of NH_3. The coefficients of $3:1 \rightarrow 2$ thus express a molecular ratio of the reactants and the products.

The molecular ratio leads to the concept of a *mole ratio* since any number of molecules can react as long as they are in the ratio of $3:1 \rightarrow 2$. The number could be Avogadro's number, so (3) $\times$ (6.02×10^{23}) molecules of H_2 will combine with (1) $\times$ (6.02×10^{23}) molecules of N_2 to form (2) $\times$ (6.02×10^{23}) molecules of ammonia. Since 6.02×10^{23} molecules is the number of particles in a mole, the coefficients therefore represent the *numbers of moles* involved in the reaction. Thus, three moles of H_2 react with one mole of N_2 to produce two moles of NH_3.

The mole ratio of a balanced chemical equation leads to the concept of a *mass ratio* interpretation of a chemical equation. The gram-formula weight of a compound is the mass in grams of *one mole* that is numerically equal to its formula weight. Therefore, the equation also describes the mass ratios

Table 12.1

Three interpretations of a chemical equation

Equation: $3 H_2 + N_2 \rightarrow 2 NH_3$

Molecular Ratio:
     3 molecules H_2 + 1 molecule $N_2 \rightarrow$ 2 molecules NH_3
 Mole Ratio:
     3 moles H_2 + 1 mole $N_2 \rightarrow$ 2 moles NH_3
 Mass Ratio:
     6.0 g H_2 + 28.0 g $N_2 \rightarrow$ 34.0 g NH_3

of the reactants and the products. The mass ratio can be calculated from the mole relationship described in the equation. The three interpretations are summarized in table 12.1.

Thus, the coefficients in a balanced equation can be interpreted in terms of molecules, which leads to an interpretation of moles, mass, or any formula unit. The mole concept thus provides the basis for calculations about the quantities of reactants and products in a chemical reaction.

Summary

A chemical formula is a shorthand way of describing the composition of a compound. An *empirical formula* identifies the simplest whole number ratio of element. A *molecular formula* identifies the actual number of atoms in a molecule.

The sum of the atomic weights of all the atoms in any formula is called the *formula weight*. The *molecular weight* is the formula weight of a molecular substance. The formula weight of a compound can be used to determine the *mass percentage* of elements making up a compound.

A concise way to describe a chemical reaction is to use formulas in a *chemical equation*. A chemical equation with the same number of each kind of atom on both sides is called a *balanced equation*. A balanced equation is in accord with the *law of conservation of mass*, which states that atoms are neither created nor destroyed in a chemical reaction. To balance a chemical equation, *coefficients* are placed in front of chemical formulas. Subscripts of formulas may not be changed since this would change the formula, meaning a different compound.

One important group of chemical reactions is called *oxidation-reduction reactions*, or *redox* reactions for short. Redox reactions are reactions where shifts of electrons occur. The process of losing electrons is called *oxidation*, and the substance doing the losing is said to be *oxidized*. The process of gaining electrons is called *reduction*, and the substance doing the gaining is said to be *reduced*. The process is called reduction because gaining electrons reduces the *oxidation number*, the charge an atom appears to have while in a compound. Substances that take electrons from other substances are called *oxidizing agents*. Substances that supply electrons are called *reducing agents*.

Chemical reactions can also be classified as (1) *combination*, (2) *decomposition*, (3) *replacement*, or (4) *ion exchange*. The first three of these are redox reactions, but ion exchange is not.

A balanced chemical equation describes chemical reactions and has quantitative meaning about numbers of atoms, numbers of molecules, and conservation of atomic weights. The coefficients also describe the *volumes* of combining gases. At a constant temperature and pressure gases combine in small, whole number ratios that are given by the coefficients. Each volume at the same temperature and pressure contains the *same number of molecules*.

The number of atoms in exactly 12.00 g of C-12 is called *Avogadro's number,* which has a value of 6.02×10^{23}. Any substance that contains Avogadro's number of atoms, ions, molecules, or any chemical unit is called a *mole* of that substance. The mole is a measure of a number of atoms, molecules, or other chemical units. The mass of a mole of any substance is equal to the atomic mass of that substance.

The mass, number of atoms, and mole concepts are generalized to other units. The *gram-atomic weight* of an element is the mass in grams that is numerically equal to its atomic weight. The *gram-formula weight* of a compound is the mass in grams that is numerically equal to the formula weight of the compound. The *gram-molecular weight* is the gram-formula weight of a molecular compound. The relationships between the mole concept and the mass ratios can be used with a chemical equation for calculations about the quantities of reactants and products in a chemical reaction.

Summary of Equations

12.1

$$\left(\frac{\text{part}}{\text{whole}}\right)(100\% \text{ of whole}) = \% \text{ of part}$$

12.2

$$\frac{\left(\begin{array}{c}\text{atomic weight} \\ \text{of element}\end{array}\right)\left(\begin{array}{c}\text{number of atoms} \\ \text{of element}\end{array}\right)}{\text{formula weight of compound}} \times \begin{array}{c}100\% \text{ of} \\ \text{compound}\end{array} = \begin{array}{c}\% \text{ of} \\ \text{element}\end{array}$$

12.3

Combination reaction, general form

$$X + Y \rightarrow XY$$

12.4

Decomposition reaction, general form

$$XY \rightarrow X + Y + \ldots$$

12.5

Replacement reaction, general form for negative part replaced

$$XY + Z \rightarrow XZ + Y$$

12.6

Replacement reaction, general form for positive part replaced

$$XY + A \rightarrow AY + X$$

12.7

Ion exchange reaction, general form

$$XY + AZ \rightarrow XZ + AY$$

(Note: One of the products must remove ions by forming an insoluble product, water, or a gas.)

Key Terms

Avogadro's number (p. 258)
chemical equation (p. 248)
combination reaction (p. 253)
decomposition reaction (p. 253)
empirical formula (p. 245)
formula weight (p. 246)
gram-atomic weight (p. 258)

gram-formula weight (p. 260)
gram-molecular weight (p. 260)
ion exchange reaction (p. 255)
law of conservation of mass (p. 248)
mole (p. 258)
molecular formula (p. 245)

molecular weight (p. 246)
oxidation-reduction reaction (p. 252)
oxidizing agents (p. 252)

redox reaction (p. 252)
reducing agent (p. 253)
replacement reaction (p. 253)

Applying the Concepts

1. A formula for a compound is given as KCl. This is a (an)
   a. empirical formula.
   b. molecular formula.
   c. structural formula.
   d. formula, but type unknown without further information.

2. A formula for a compound is given as $C_{12}H_{22}O_{11}$. This is a (an)
   a. empirical formula.
   b. molecular formula.
   c. structural formula.
   d. formula, but type unknown without further information.

3. The formula weight of sulfuric acid, H_2SO_4 is
   a. 49 u.
   b. 50 u.
   c. 98 u.
   d. 194 u.

4. A balanced chemical equation has
   a. the same number of molecules on both sides of the equation.
   b. the same kinds of molecules on both sides of the equation.
   c. the same number of each kind of atom on both sides of the equation.
   d. all of the above.

5. The law of conservation of mass means that
   a. atoms are not lost or destroyed in a chemical reaction.
   b. the mass of a newly formed compound cannot be changed.
   c. in burning, part of the mass must be converted into fire in order for mass to be conserved.
   d. molecules cannot be broken apart because this would result in less mass.

6. A chemical equation is balanced by changing (the)
   a. subscripts.
   b. superscripts.
   c. coefficients.
   d. any of the above as necessary to achieve a balance.

7. Since wood is composed of carbohydrates, you should expect what gases to exhaust from a fireplace when complete combustion takes place?
   a. carbon dioxide, carbon monoxide, and pollutants
   b. carbon dioxide and water vapor
   c. carbon monoxide and smoke
   d. It depends on the type of wood being burned.

8. When carbon burns with an insufficient supply of oxygen, carbon monoxide is formed according to the following equation: $2 C + O_2 \rightarrow 2 CO$. What category of chemical reaction is this?
   a. combination
   b. ion exchange
   c. replacement
   d. None of the above, because the reaction is incomplete.

9. According to the activity series for metals, adding metallic iron to a solution of aluminum chloride should result in
   a. a solution of iron chloride and metallic aluminum.
   b. a mixed solution of iron and aluminum chloride.
   c. the formation of iron hydroxide with hydrogen given off.
   d. no reaction.

Chemical Formulas and Equations

10. In a replacement reaction, elements that have the most ability to hold onto their electrons are
   a. the most chemically active.
   b. the least chemically active.
   c. not generally involved in replacement reactions.
   d. none of the above.

11. Of the elements listed below, the one with the greatest electron holding ability is
   a. sodium.
   b. zinc.
   c. copper.
   d. platinum.

12. Of the elements listed below, the one with the greatest chemical activity is
   a. aluminum.
   b. zinc.
   c. iron.
   d. mercury.

13. You know that an ion exchange reaction has taken place if the reactants include
   a. a precipitate.
   b. a gas.
   c. water.
   d. any of the above.

14. The incomplete equation of $2 \: KClO_3(s) \xrightarrow{\Delta}$ probably represents which type of chemical reaction?
   a. combination
   b. decomposition
   c. replacement
   d. ion exchange

15. In the equation of $2 \: H_2 + O_2 \rightarrow 2 \: H_2O$,
   a. the total mass of the gaseous reactants is less than the total mass of the liquid product.
   b. the total number of molecules in the reactants is equal to the total number of molecules in the products.
   c. one volume of oxygen combines with two volumes of hydrogen to produce 2 volumes of water.
   d. All of the above are correct.

16. An amount of a substance that contains Avogadro's number of atoms, ions, or molecules is (a)
   a. mole.
   b. gram-atomic weight.
   c. gram-formula weight.
   d. any of the above.

17. If you have 6.02×10^{23} atoms of metallic iron you will have how many grams of iron?
   a. 26
   b. 55.8
   c. 334.8
   d. 3.4×10^{25}

Answers

1. a   2. b   3. c   4. c   5. a   6. c   7. b   8. a   9. d
10. b   11. d   12. a   13. d   14. b   15. c   16. d   17. b

Questions for Thought

1. How is an empirical formula like and unlike a molecular formula?

2. Describe the basic parts of a chemical equation. Identify how the physical state of elements and compounds is identified in an equation.

3. What is the law of conservation of mass? How do you know if a chemical equation is in accord with this law?

4. Describe in your own words how a chemical equation is balanced.

5. What is a hydrocarbon? What is a carbohydrate? In general, what are the products of complete combustion of hydrocarbons and carbohydrates?

6. Define and give an example in the form of a balanced equation of (a) a combination reaction, (b) a decomposition reaction, (c) a replacement reaction, and (d) an ion exchange reaction.

7. What must occur in order for an ion exchange reaction to take place? What is the result if this does not happen?

8. Predict the products for the following reactions: (a) The combustion of ethyl alcohol, C_2H_5OH, (b) the rusting of aluminum, and (c) the reaction between iron and sodium chloride.

9. The formula for butane is C_4H_{10}. Is this an empirical formula or a molecular formula? Explain the reason(s) for your answer.

10. How is the activity series for metals used to predict if a replacement reaction will occur or not?

11. What is a gram-formula weight? How is it calculated?

12. What is the meaning and the value of Avogadro's number? What is a mole?

Exercises

Group A—Solutions Provided in Appendix

1. Identify the following as empirical formulas or molecular formulas and indicate any uncertainty with (?):
   (a) $MgCl_2$
   (b) C_2H_2
   (c) BaF_2
   (d) C_8H_{18}
   (e) CH_4
   (f) S_8

Group B—Solutions Not Given

1. Identify the following as empirical formulas or molecular formulas and indicate any uncertainty with (?):
   (a) CH_2O
   (b) $C_6H_{12}O_6$
   (c) $NaCl$
   (d) CH_4
   (e) F_6
   (f) CaF_2

2. What is the formula weight for each of the following compounds?
   (a) Copper(II) sulfate
   (b) Carbon disulfide
   (c) Calcium sulfate
   (d) Sodium carbonate

3. What is the mass percentage composition of the elements in the following compounds?
   (a) Fool's gold, FeS_2
   (b) Boric acid, H_3BO_3
   (c) Baking soda, $NaHCO_3$
   (d) Aspirin, $C_9H_8O_4$

4. Write balanced chemical equations for each of the following unbalanced reactions:
   (a) $SO_2 + O_2 \rightarrow SO_3$
   (b) $P + O_2 \rightarrow P_2O_5$
   (c) $Al + HCl \rightarrow AlCl_3 + H_2$
   (d) $NaOH + H_2SO_4 \rightarrow Na_2SO_4 + H_2O$
   (e) $Fe_2O_3 + CO \rightarrow Fe + CO_2$
   (f) $Mg(OH)_2 + H_3PO_4 \rightarrow Mg_3(PO_4)_2 + H_2O$

5. Identify the following as combination, decomposition, replacement, or ion exchange reaction:
   (a) $NaCl_{(aq)} + AgNO_{3(aq)} \rightarrow NaNO_{3(aq)} + AgCl\downarrow$
   (b) $H_2O_{(l)} + CO_{2(g)} \rightarrow H_2CO_{3(l)}$
   (c) $2\ NaHCO_{3(s)} \rightarrow Na_2CO_{3(s)} + H_2O_{(g)} + CO_{2(g)}$
   (d) $2\ Na_{(s)} + Cl_{2(g)} \rightarrow 2\ NaCl_{(s)}$
   (e) $Cu_{(s)} + 2\ AgNO_{3(aq)} \rightarrow Cu(NO_3)_{2(aq)} + 2\ Ag_{(s)}$
   (f) $CaO_{(s)} + H_2O_{(l)} \rightarrow Ca(OH)_{2(aq)}$

6. Write complete, balanced equations for each of the following reactions:
   (a) $C_5H_{12(g)} + O_{2(g)} \rightarrow$
   (b) $HCl_{(aq)} + NaOH_{(aq)} \rightarrow$
   (c) $Al_{(s)} + Fe_2O_{3(s)} \rightarrow$
   (d) $Fe_{(s)} + CuSO_{4(aq)} \rightarrow$
   (e) $MgCl_{(aq)} + Fe(NO_3)_{2(aq)} \rightarrow$
   (f) $C_6H_{10}O_{5(s)} + O_{2(g)} \rightarrow$

7. Write complete, balanced equations for each of the following decomposition reactions. Include symbols for physical states, heating, and others as needed:
   (a) Solid potassium chloride and oxygen gas are formed when solid potassium chlorate is heated.
   (b) Upon electrolysis, molten bauxite (aluminum oxide) yields solid aluminum metal and oxygen gas.
   (c) Upon heating, solid calcium carbonate yields solid calcium oxide and carbon dioxide gas.

8. Write complete, balanced equations for each of the following replacement reactions. If no reaction is predicted, write "no reaction" as the product:
   (a) $Na_{(s)} + H_2O_{(l)} \rightarrow$
   (b) $Au_{(s)} + HCl_{(aq)} \rightarrow$
   (c) $Al_{(s)} + FeCl_{3(aq)} \rightarrow$
   (d) $Zn_{(s)} + CuCl_{2(aq)} \rightarrow$

2. Calculate the formula weight for each of the following compounds:
   (a) Dinitrogen monoxide
   (b) Lead(II) sulfide
   (c) Magnesium sulfate
   (d) Mercury(II) chloride

3. What is the mass percentage composition of the elements in the following compounds?
   (a) Potash, K_2CO_3
   (b) Gypsum, $CaSO_4$
   (c) Saltpeter, KNO_3
   (d) Caffeine, $C_8H_{10}N_4O_2$

4. Write balanced chemical equations for each of the following unbalanced reactions:
   (a) $NO + O_2 \rightarrow NO_2$
   (b) $KClO_3 \rightarrow KCl + O_2$
   (c) $NH_4Cl + Ca(OH)_2 \rightarrow CaCl_2 + NH_3 + H_2O$
   (d) $NaNO_3 + H_2SO_4 \rightarrow Na_2SO_4 + HNO_3$
   (e) $PbS + H_2O_2 \rightarrow PbSO_4 + H_2O$
   (f) $Al(SO_4)_3 + BaCl_2 \rightarrow AlCl_3 + BaSO_4$

5. Identify the following as combination, decomposition, replacement, or ion exchange reaction:
   (a) $ZnCO_{3(s)} \rightarrow ZnO_{(s)} + CO_2\uparrow$
   (b) $2\ NaBr_{(aq)} + Cl_{2(g)} \rightarrow 2\ NaCl_{(aq)} + Br_{2(g)}$
   (c) $2\ Al_{(s)} + 3\ Cl_{2(g)} \rightarrow 2\ AlCl_{3(s)}$
   (d) $Ca(OH)_{2(aq)} + H_2SO_{4(aq)} \rightarrow CaSO_{4(aq)} + 2\ H_2O_{(l)}$
   (e) $Pb(NO_3)_{2(aq)} + H_2S_{(g)} \rightarrow 2\ HNO_{3(aq)} + PbS\downarrow$
   (f) $C_{(s)} + ZnO_{(s)} \rightarrow Zn_{(s)} + CO\uparrow$

6. Write complete, balanced equations for each of the following reactions:
   (a) $C_3H_{6(g)} + O_{2(g)} \rightarrow$
   (b) $H_2SO_{4(aq)} + KOH_{(aq)} \rightarrow$
   (c) $C_6H_{12}O_{6(s)} + O_{2(g)} \rightarrow$
   (d) $Na_3PO_{4(aq)} + AgNO_{3(aq)} \rightarrow$
   (e) $NaOH_{(aq)} + Al(NO_3)_{3(aq)} \rightarrow$
   (f) $Mg(OH)_{2(aq)} + H_3PO_{4(aq)} \rightarrow$

7. Write complete, balanced equations for each of the following decomposition reactions. Include symbols for physical states, heating, and others as needed:
   (a) When solid zinc carbonate is heated, solid zinc oxide and carbon dioxide gas are formed.
   (b) Liquid hydrogen peroxide decomposes to liquid water and oxygen gas.
   (c) Solid ammonium nitrite decomposes to liquid water and nitrogen gas.

8. Write complete, balanced equations for each of the following replacement reactions. If no reaction is predicted, write "no reaction" as the product:
   (a) $Zn_{(s)} + FeCl_{2(aq)} \rightarrow$
   (b) $Zn_{(s)} + AlCl_{3(aq)} \rightarrow$
   (c) $Cu_{(s)} + HgCl_{2(aq)} \rightarrow$
   (d) $Al_{(s)} + HCl_{(aq)} \rightarrow$

Chemical Formulas and Equations

9. Write complete, balanced equations for each of the following ion exchange reactions. If no reaction is predicted, write "no reaction" as the product:
   (a) $NaOH_{(aq)} + HNO_{3(aq)} \rightarrow$
   (b) $CaCl_{2(aq)} + KNO_{3(aq)} \rightarrow$
   (c) $Ba(NO_3)_{2(aq)} + Na_3PO_{4(aq)} \rightarrow$
   (d) $KOH_{(aq)} + ZnSO_{4(aq)} \rightarrow$

10. The gas welding torch is fueled by two tanks, one containing acetylene (C_2H_2) and the other pure oxygen (O_2). The very hot flame of the torch is produced as acetylene burns,

$$2\ C_2H_{2(g)} + O_{2(g)} \rightarrow 4\ CO_{2(g)} + H_2O_{(g)}$$

According to this equation, how many liters of oxygen are required to burn one liter of acetylene?

9. Write complete, balanced equations for each of the following ion exchange reactions. If no reaction is predicted, write "no reaction" as the product:
   (a) $Ca(OH)_{2(aq)} + H_2SO_{4(aq)} \rightarrow$
   (b) $NaCl_{(aq)} + AgNO_{3(aq)} \rightarrow$
   (c) $NH_4NO_{3(aq)} + MgPO_{4(aq)} \rightarrow$
   (d) $Na_3PO_{4(aq)} + AgNO_{3(aq)} \rightarrow$

10. Iron(III) oxide, or hematite, is one mineral used as an iron ore. Other iron ores are magnetite (Fe_3O_4) and siderite ($FeCO_3$). Assume that you have pure samples of all three ores that will be reduced by reaction with carbon monoxide. Which of the three ores will have the highest yield of metallic iron?

Chapter

13

Water and Solutions

Water is often referred to as the universal
solvent because it makes so many different
kinds of solutions. Eventually, moving water
can dissolve solid rock, carrying it away in
solution.

T_{HE} previous three chapters were concerned with elements, compounds, and their chemical reactions. Elements and compounds are pure substances, materials with a definite, fixed composition that is the same throughout. Mixtures, on the other hand, have a composition that may vary from one sample to the next. A mixture contains the particles of one substance physically dispersed throughout the particles of another substance. These particles can be any size, from the size of rock particles in a gravel mixture down to the smallest size possible—particles the size of ions or molecules. The size of the particles gives a mixture its appearance. When the particles are relatively large and visible to the eye, a mixture appears to be heterogeneous. Thus, a pile of gravel appears to be a heterogeneous mixture. At the other end of the size scale, a uniform mixture of ion- or molecule-sized particles appears to be homogeneous, or the same throughout. Such homogeneous mixtures are called *solutions.* This chapter is concerned with solutions, those involving water in particular.

Many common liquids are solutions. Solutions are commonly used in everyday household activities, as well as in the chemistry laboratory. The household activities of cooking, cleaning, and painting all involve solutions that will be considered in this chapter. Detergents, cleaners, and drain openers all function because a solution is a medium for rapid chemical change. Solids react slowly, if at all, because they have limited contact only at their immediate surfaces. Thus, solutions are commonly used to speed reactions. The water you drink is also a solution (figure 13.1). Hard water results in certain reactions that occur between soap and the water solution. This chapter considers hard water and how it is softened, in addition to acids, bases, the pH scale, and many common solutions used in everyday activities.

Solutions

A **solution** is defined as a homogeneous mixture of ions or molecules of two or more substances. The process of making a solution is called *dissolving,* and during dissolving the different components that make up the solution become mixed. The components could be sugar and water, for example, and when sugar dissolves in water, molecules of sugar are uniformly dispersed throughout the molecules of water. The uniform taste of sweetness of any part of the sugar solution is a result of this uniform mixing. In a salt and water solution, however, the salt dissolves into sodium and chlorine ions. The components of a salt and water solution are sodium and chlorine ions dissolved in water.

Solutions are not limited to solids, such as sugar or salt, dissolved in liquids, such as water. There are three states of matter, and a solution can involve any combination of the ions or molecules of gases, liquids, or solids (figure 13.2). Thus, it is possible to have nine kinds of solutions. Table 13.1 gives an example of each of these nine kinds of solutions.

The amounts of the components of a solution are identified by the general terms of *solvent* and *solute.* The **solvent** is the component present in the larger amount. The **solute** is the component that dissolves in the solvent. Atmospheric air, for example, is about 78 percent nitrogen, so nitrogen is considered the solvent. Oxygen (about 21 percent), argon (about 0.9 percent), and other gases make up the solutes.

Figure 13.1

Water is the most abundant liquid on the earth and is necessary for all life. Because of water's great dissolving properties, any sample is a solution containing solids, other liquids, and gases from the environment. The stream also carries suspended ground-up rocks, called rock flour, from a nearby glacier.

Figure 13.2

Above this city there are at least three kinds of solutions. These are (1) gas in gas—oxygen dissolved in nitrogen, (2) liquid in gas—water vapor dissolved in air, and (3) solid in gas—tiny particles of smoke dissolved in the air.

Table 13.1

Examples of each kind of solution

Kind of Solution	Example
Gas in gas	*Air* is O_2, CO_2, and other gases dissolved in nitrogen gas.
Liquid in gas	*Water vapor* is water molecules dissolved in air.
Solid in gas	*Smoke* is solid particles dissolved in air (many smoke particles are larger than molecules, so they are not part of the solution).
Gas in liquid	*Soda water* is CO_2 dissolved in water.
Liquid in liquid	*Alcohol* is alcohol molecules dissolved in water (unless 200 proof, which is pure alcohol).
Solid in liquid	*Seawater* is the ions of salts dissolved in water (mostly sodium chloride).
Liquid in solid	*Dental fillings* are prepared from a solution of mercury in silver.
Solid in solid	*Brass* is zinc dissolved in copper.

Figure 13.3

There are different ways to express the concentration of a solution. How many different ways can you identify in this photograph?

If one of the components of a solution is a liquid, it is usually identified as the solvent. An *aqueous solution* is a solution of a solid, a liquid, or a gas in water. Water is the solvent in an aqueous solution. A *tincture* is a solution of something dissolved in alcohol. Tincture of iodine, for example, is iodine dissolved in the solvent alcohol.

Concentration of Solutions

The relative amounts of solute and solvent are described by the **concentration** of a solution. In general, a solution with a large amount of solute is *concentrated* and a solution with much less solute is *dilute*. The terms "dilute" and "concentrated" are somewhat arbitrary, and it is sometimes difficult to know the difference between a solution that is "weakly concentrated" and one that is "not very diluted." More meaningful information is provided by measurement of the *amount of solute in a solution*. There are different ways to express concentration measurements, each lending itself to a particular kind of solution or to how the information will be used. For example, you read about concentrations of parts per million in an article about pollution, but most of the concentration of solutions sold in stores are reported in percent by volume or percent by weight (figure 13.3). Each of these concentrations is concerned with the amount of *solute* in the *solution*.

Concentration ratios that describe small concentrations of solute are sometimes reported as a ratio of **parts per million** (ppm) or **parts per billion** (ppb). This ratio could mean ppm by volume or ppm by weight, depending if the solution is a gas or a liquid. For example, a drinking water sample with 1 ppm Na^+ by weight has 1 weight measure of solute, sodium ions, *in* every 1,000,000 weight measures of the total solution. By way of analogy, 1 ppm expressed in money means 1 cent in every $10,000 (which is one million cents). A concentration of 1 ppb means 1 cent in $10,000,000. Thus, the concentrations of very dilute solutions, such as certain salts in seawater, minerals in drinking water, and pollutants in water or in the atmosphere are often reported in ppm or ppb.

Sometimes it is useful to know the conversion factors between ppm or ppb and the more familiar percent concentration by weight. These factors are ppm $\div (1 \times 10^4)$ = percent concentration and ppb $\div (1 \times 10^7)$ = percent concentration. For example, very hard water (water containing Ca^{2+} or Mg^{2+} ions), by definition, contains more than 300 ppm of the ions. This is a percent concentration of $300 \div 1 \times 10^4$, or 0.03 percent. To be suitable for agricultural purposes, irrigation water must not contain more than 700 ppm of total dissolved salts, which means a concentration no greater than 0.07 percent salts.

The concentration term of **percent by volume** is defined as the *volume of solute in 100 volumes of solution*. This concentration term is just like any other percentage ratio, that is, "part" divided by the "whole" times 100 percent. The distinction is that the part and the whole are concerned with a volume of solute and a volume of solution. Knowing the meaning of percent by volume can be useful in consumer decisions. Rubbing alcohol, for example, can be purchased at a wide range of prices. The various brands range from a concentration, according to the labels, of "12% by volume" to "70% by volume." If the volume unit is mL, a "12% by volume" concentration contains 12 mL of pure isopropyl (rubbing) alcohol in every 100 mL of solution. The "70% by volume" contains 70 mL of isopropyl alcohol in every 100 mL of solution. The relationship for % by volume is

$$\frac{\text{volume solute}}{\text{volume solution}} \times 100\% \text{ solution} = \% \text{ solute}$$

or

$$\frac{V_{solute}}{V_{solution}} \times 100\% \text{ solution} = \% \text{ solute} \qquad \text{equation 13.1}$$

The concentration term of **percent by weight** is defined as the *weight of solute in 100 weight units of solution*. This concentration term is just like any other percentage composition, the difference being that it is concerned with the weight of solute (the part) in a weight of solution (the whole) (figure 13.4). Hydrogen peroxide, for example, is usually sold in a concentration of "3% by weight." This means that 3 oz (or other weight units) of pure hydrogen peroxide are in 100 oz of solution. Since weight is proportional to mass in a given location, mass units such as grams are sometimes used to calculate a percent by weight. The relationship for percent by weight (using mass units) is

$$\frac{\text{mass of solute}}{\text{mass of solution}} \times 100\% \text{ solution} = \% \text{ solute}$$

or

$$\frac{m_{solute}}{m_{solution}} \times 100\% \text{ solution} = \% \text{ solute} \qquad \text{equation 13.2}$$

Example 13.1

Vinegar that is prepared for table use is a mixture of acetic acid in water, usually 5.00% by weight. How many grams of pure acetic acid are in 25.0 g of vinegar?

Solution

The percent by weight is given (5.00%), the mass of the solution is given (25.0 g), and the mass of the solute ($H_2C_2H_3O_2$) is the un-

Water and Solutions

known. The relationship between these quantities is found in equation 13.2, which can be solved for the mass of the solute:

% solute = 5.00%

$m_{solution}$ = 25.0 g

m_{solute} = ?

$$\frac{m_{solute}}{m_{solution}} \times 100\% \text{ solution} = \% \text{ solute}$$

$$\therefore$$

$$m_{solute} = \frac{(m_{solution})(\% \text{ solute})}{100\% \text{ solution}}$$

$$= \frac{(m_{solution})(\% \text{ solute})}{100\% \text{ solution}}$$

$$= \frac{(25.0 \text{ g})(5.00)}{100} \text{ solute}$$

$$= \boxed{1.25 \text{ g solute}}$$

Example 13.2

A solution used to clean contact lenses contains 0.002% by volume of thimerosal as a preservative. How many mL of this preservative are needed to make 100,000 L of the cleaning solution? (Answer: 2.0 mL)

Both percent by volume and percent by weight are defined as the volume or weight per 100 units of solution because percent *means* parts per hundred. The measure of dissolved salts in seawater is called *salinity*. **Salinity** is defined as the mass of salts dissolved in 1,000 g of solution. As illustrated in figure 13.5, evaporation of 965 g of water from 1,000 g of seawater will leave an average of 35 g salts. Thus, the average salinity of the seawater is 35‰. Note the ‰, which means parts per thousand just as % means parts per hundred. Thus, the average salinity of seawater is 35‰, which means there are 35 g of salts dissolved in every 1,000 g of seawater. The equivalent percent measure for salinity is 3.5%, which equals 35‰.

Solubility

Gases and liquids appear to be soluble in all proportions, but there is an obvious limit to how much solid can be dissolved in a liquid. You may have noticed that a cup of hot tea will dissolve several teaspoons of sugar, but the limit of solubility is reached quickly in a glass of iced tea. The limit of how much sugar will dissolve seems to depend on the temperature of the tea. More sugar added to the cold tea after the limit is reached will not dissolve, and solid sugar granules begin to accumulate at the bottom of the glass. At this limit the sugar and tea solution is said to be *saturated*. Dissolving does not actually stop when a solution becomes saturated and undissolved sugar continues to enter the solution. However, dissolved sugar is now returning to the undissolved state at the same rate. The overall equilibrium condition of sugar dissolving and sugar coming out of solution

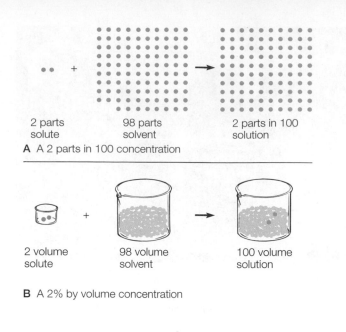

2 parts solute + 98 parts solvent → 2 parts in 100 solution

A A 2 parts in 100 concentration

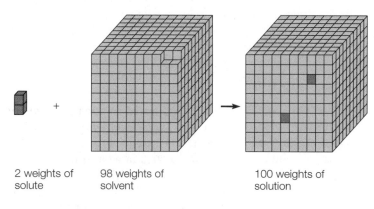

2 volume solute + 98 volume solvent → 100 volume solution

B A 2% by volume concentration

2 weights of solute + 98 weights of solvent → 100 weights of solution

C A 2% by weight concentration

Figure 13.4

Three ways of expressing the amount of solute in a solution include (a) expressed as parts, as in parts per million, (b) expressed as percent by volume, and (c) expressed as percent by weight. Note that all three expressions are concerned with amounts of solute and amounts of solution.

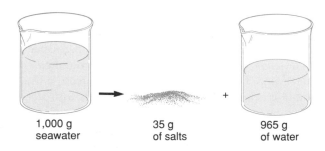

1,000 g seawater → 35 g of salts + 965 g of water

Figure 13.5

Salinity is a measure of the amount of salts dissolved in 1 kg of solution. If 1,000 g of seawater were evaporated, 35.0 g of salts would remain as 965.0 g of water leave.

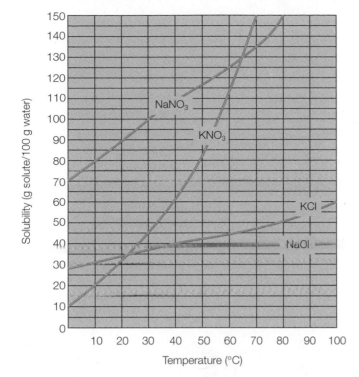

Figure 13.6
Approximate solubility curves for sodium nitrate, potassium nitrate, potassium chloride, and sodium chloride.

is called a **saturated solution.** A saturated solution is a *state of equilibrium that exists between dissolving solute and solute coming out of solution.* You actually cannot see the dissolving and coming out of solution that occurs in a saturated solution because the exchanges are taking place with particles the size of molecules or ions.

Not all compounds dissolve as sugar does, and more or less of a given compound may be required to produce a saturated solution at a particular temperature. In general, the difficulty of dissolving a given compound is referred to as solubility. More specifically, the **solubility** of a solute is defined as the *concentration that is reached in a saturated solution at a particular temperature.* Solubility varies with the temperature as the sodium and potassium salt examples show in figure 13.6. These solubility curves describe the amount of solute required to reach the saturation equilibrium at a particular temperature. In general, the solubilities of most ionic solids increase with temperature, but there are exceptions. In addition, some salts release heat when dissolved in water and other salts absorb heat when dissolved. The "instant cold pack" used for first aid is a bag of water containing a second bag of ammonium nitrate (NH_4NO_3). When the bag of ammonium nitrate is broken, the compound dissolves and absorbs heat.

You can usually dissolve more of a solid, such as salt or sugar, as the temperature of the water is increased. Contrary to what you might expect, gases usually become *less* soluble in water as the temperature increases. As a glass of water warms, small bubbles collect on the sides of the glass as dissolved air

comes out of solution. The first bubbles that appear when warming a pot of water to boiling are also bubbles of dissolved air coming out of solution. This is why water that has been boiled usually tastes "flat." The dissolved air has been removed by the heating. The "normal" taste of water can be restored by pouring the boiled water back and forth between two glasses. The water dissolves more air during this process, restoring the usual taste.

Changes in pressure have no effect on the solubility of solids in liquids but greatly affect the solubility of gases. The fizz of an opened bottle or can of soda occurs because pressure is reduced on the beverage and dissolved carbon dioxide comes out of solution. In general, *gas solubility decreases with temperature and increases with pressure.* As usual, there are exceptions to this generalization.

Water Solutions

Water is the one chemical that is absolutely essential for all organisms, both plant and animal. Organisms use water to transport food and essential elements and to carry biological molecules in a solution. In fact, living organisms are mostly cells filled with water solutions. Your foods are mostly water, with fruits and vegetables consisting of up to 95 percent water and meat consisting of 50 percent water. Your body consists of over 70 percent water by weight. It is the specific properties of water that make it so important for life, in particular the unusual ability of water to act as a solvent. The properties of a water molecule must be considered in order to account for the solvent abilities of water.

Properties of Water Molecules

A water molecule is composed of two atoms of hydrogen and one atom of oxygen joined by a polar covalent bond. The electron dot formula for a water molecule is

$$\ddot{\underset{H\quad H}{O}}:$$

Notice the four pairs of electrons around the oxygen atom, consisting of two lone pairs and two bonding pairs. These four electron pairs are arranged in the direction of a tetrahedral arrangement (figure 13.7). Since there are only two bonding pairs, however, a water molecule has a bent molecular arrangement, represented with a structural formula as

$$\underset{H\quad\quad H}{\overset{O}{\diagup\,\diagdown}}$$

The angle between the two bonds is not 90° but has been experimentally measured to be about 105°, which is very close to the tetrahedral angle (figure 13.7).

Oxygen has a stronger electronegativity (3.5) than hydrogen (2.1), and thus it has a greater ability to attract shared electrons. This results in the bonding electrons spending more

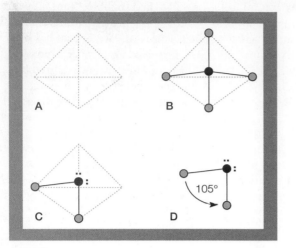

Figure 13.7

(a) A tetrahedron is a four-sided pyramid. (b) Molecules with four bonding electron pairs have a tetrahedral shape like the CCl_4 molecule illustrated here. Note the carbon atom is in the center of the pyramid, with chlorine atoms at each of the four corners. (c) A water molecule has two bonding pairs and two lone pairs in a tetrahedral arrangement. (d) The water molecule has an angular arrangement called bent. If something were attached to the two lone pairs, it would be a tetrahedral arrangement (see also figure 13.9).

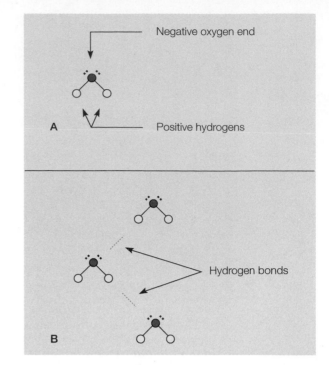

Figure 13.8

(a) The water molecule is polar, with centers of positive and negative charges. (b) Attractions between these positive and negative centers establish hydrogen bonds between adjacent molecules.

time near the oxygen atom than the hydrogen atoms, essentially leaving the hydrogen as exposed protons on one end of the molecule. Thus, the water molecule has polar covalent bonds and is a *polar molecule* with centers of negative or positive charges.

The polar water molecule, with its negative oxygen end and positive hydrogen end, sets the stage for **intermolecular forces,** forces of interaction between molecules. The positive end of the water molecule can attract the negative end of another molecule, including the hydrogen end of another water molecule. The general term for weak attractive intermolecular forces is a **van der Waals force,** named after the Dutch physicist who first proposed the concept. Specifically, the intermolecular force of attraction between a hydrogen atom in a polar molecule and electrons in another molecule is called a **hydrogen bond.** A hydrogen bond is a weak to moderate bond between the hydrogen end (+) of a polar molecule and the negative end (−) of a second polar molecule (figure 13.8). In general, hydrogen bonding occurs between the hydrogen atom of one molecule and the oxygen, fluorine, or nitrogen atom of another molecule.

Hydrogen bonding accounts for the physical properties of water, including its unusually high heat of fusion and heat of vaporization as well as its unusual density changes. Figure 13.9 shows the hydrogen bonded structure of ice. Each oxygen atom is associated with four hydrogen atoms, two in the H_2O molecule and two from other water molecules, held by hydrogen bonds. The arrangement is tetrahedral, forming a six-sided hexagonal structure. The open space of the hexagonal channel in this structure results in ice being less dense than water. The shape of the channel also suggests why snowflakes always have six sides.

When ice is warmed, the increased vibrations of the molecules begin to expand and stretch the hydrogen bond structure. When ice melts, about 15 percent of the hydrogen bonds break

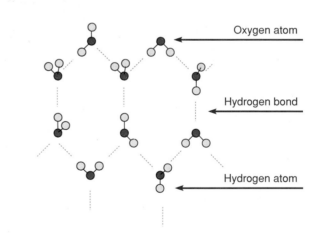

Figure 13.9

The hexagonal structure of ice. Hydrogen bonding between the oxygen atom and two hydrogen atoms of other water molecules results in a tetrahedral arrangement, which forms the open, hexagonal structure of ice.

and the open structure collapses into the more compact arrangement of liquid water. As the liquid water is warmed from 0° C still more hydrogen bonds break down and the density of the water steadily increases. At 4° C the expansion of water from the increased molecular vibrations begins to predominate and the density decreases steadily with further warming (figure 13.10). Thus water has its greatest density at a temperature of 4° C.

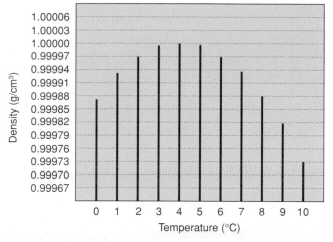

Figure 13.10

The density of water from 0° C to 10° C. The density of water is at a maximum at 4° C, becoming less dense as it is cooled or warmed from this temperature. Hydrogen bonding explains this unusual behavior.

The heat of fusion, specific heat, and heat of vaporization of water are unusually high when compared to other, but chemically similar, substances. These high values are accounted for by the additional energy needed to break hydrogen bonds.

The Dissolving Process

A solution is formed when the molecules or ions of two or more substances become homogeneously mixed. But the process of dissolving must be more complicated than the simple mixing together of particles because (1) solutions become saturated, meaning there is a limit on solubility, and (2) some substances are *insoluble,* not dissolving at all or at least not noticeably. In general, the forces of attraction between molecules or ions of the solvent and solute determine if something will dissolve and any limits on the solubility. These forces of attraction and their role in the dissolving process will be considered in the following examples.

First, consider the dissolving process in gaseous and liquid solutions. In a gas, the intermolecular forces are small so gases can mix in any proportion. Fluids that can mix in any proportion like this are called **miscible fluids.** Fluids that do not mix are called *immiscible fluids.* Air is a mixture of gases and vapors, so gases and vapors are miscible.

Liquid solutions can have a gas, another liquid, or a solid as a solute. Gases are miscible in liquids, and a carbonated beverage (your favorite cola) is the common example, consisting of carbon dioxide dissolved in water. Whether or not two given liquids form solutions depends on some similarities in their molecular structures. The water molecule, for example, is a polar molecule with a negative end and a positive end. On the other hand, carbon tetrachloride (CCl_4) is a molecule with polar bonds that are symmetrically arranged. Because of the symmetry, CCl_4 has no negative or positive ends so it is nonpolar. So some liquids have polar molecules and some liquids have nonpolar molecules. The general rule for forming solutions is *like dissolves like* (figure 13.11). A nonpolar compound, such as carbon tetrachloride, will dissolve oils and greases because they are non-

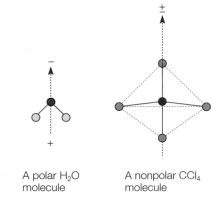

A polar H_2O molecule   A nonpolar CCl_4 molecule

Figure 13.11

Water is polar and carbon tetrachloride is nonpolar. Since like dissolves like, water and carbon tetrachloride are immiscible.

polar compounds. Water, a polar compound, will not dissolve the nonpolar oils and greases. Carbon tetrachloride was at one time used as a cleaning solvent because of its oil and grease dissolving abilities. Its use is no longer recommended because it is also a possible health hazard (liver damage).

Some molecules, such as ethyl alcohol and soap, have a part of the molecule that is polar and a part that is nonpolar. Washing with water alone will not dissolve oils because water and oil are immiscible. When soap is added to the water, however, the polar end of the soap molecule is attracted to the polar water molecules and the nonpolar end is absorbed into the oil. A particle (larger than a molecule) is formed and the oil is washed away with the water (figure 13.12).

The "like dissolves like" rule applies to solids and liquid solvents as well as liquids and liquid solvents. Polar solids, such as salt, will readily dissolve in water, which has polar molecules, but do not dissolve readily in oil, grease, or other nonpolar solvents. Polar water readily dissolves salt because the charged polar water molecules are able to exert an attraction force on the ions, pulling them away from the crystal structure. Thus, ionic compounds dissolve in water.

As noted in figure 13.6, ionic compounds vary in their solubilities in water. This difference is explained by the existence of two different forces involved in an ongoing "tug of war." One force is the attraction between an ion on the surface of the crystal and a water molecule, an *ion-polar molecule force.* When solid sodium chloride and water are mixed together, the negative ends of the water molecules (the oxygen end) become oriented toward the positive sodium ions on the crystal. Likewise, the positive ends of water molecules (the hydrogen ends) become oriented toward the negative chlorine ions. The attraction of water molecules for ions is called **hydration.** If the force of hydration is greater than the attraction between the ions in the solid, they are pulled away from the solid and dissolving occurs (figure 13.13). Not considering the role of water in this dissolving process, the equation is

$$Na^+Cl^-(s) \rightarrow Na^+(aq) + Cl^-(aq)$$

which shows that the ions were separated from the solid to become a solution of ions. In other compounds the attraction

Water and Solutions

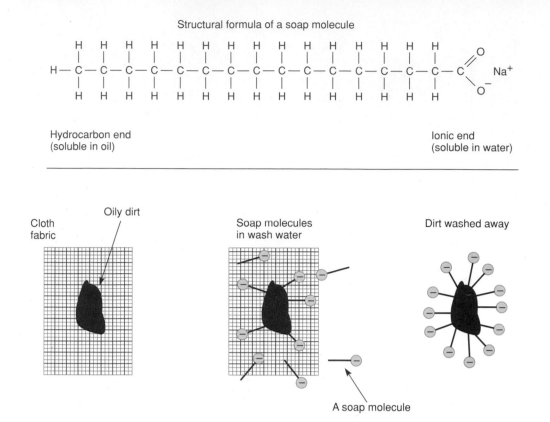

Structural formula of a soap molecule

Hydrocarbon end
(soluble in oil)

Ionic end
(soluble in water)

Cloth fabric

Oily dirt

Soap molecules
in wash water

Dirt washed away

A soap molecule

Figure 13.12

Soap cleans oil and grease because one end of the soap molecule is
soluble in water and the other end is soluble in oil and grease. Thus, the
soap molecule provides a link between two substances that would
otherwise be immiscible.

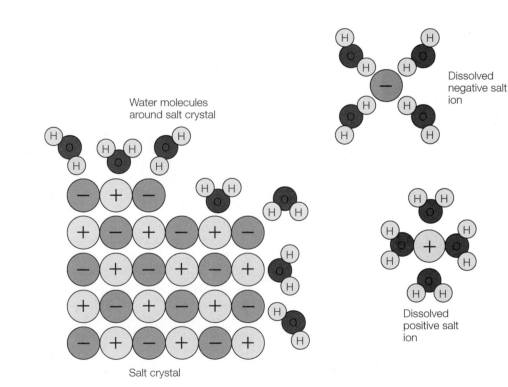

Dissolved
negative salt
ion

Water molecules
around salt crystal

Dissolved
positive salt
ion

Salt crystal

Figure 13.13

An ionic solid dissolves in water because the number of water molecules
around the surface is greater than the number of other ions of the solid.
This attraction between polar water molecules and a charged ion enables
the water molecules to pull ions away from the crystal, a process called
dissolving.

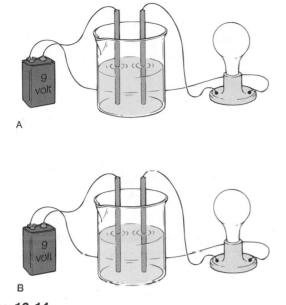

Figure 13.14

(a) Water solutions that conduct an electric current are called electrolytes. (b) Water solutions that do not conduct electricity are called nonelectrolytes.

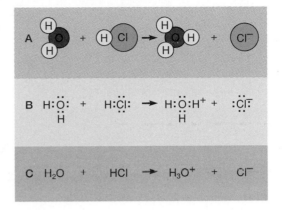

Figure 13.15

Three representations of water and hydrogen chloride in an ionizing reaction. (a) Sketches of molecules involved in the reaction. (b) Electron dot equation of the reaction. (c) The chemical equation for the reaction. Each of these representations show the hydrogen being pulled away from the chlorine atom to form H_3O^+, the hydronium ion.

between the ions in the solid might be greater than the energy of hydration. In this case, the ions of the solid would win the "tug of war" and the ionic solid is insoluble.

The saturation of soluble compounds is explained in terms of hydration eventually occupying a large number of the polar water molecules. Fewer available water molecules means less attraction on the ionic solid, with more solute ions being pulled back to the surface of the solid. The tug of war continues back and forth as an equilibrium condition is established.

Properties of Solutions

Pure solvents have characteristic physical and chemical properties that are changed by the presence of the solute. Following are some of the more interesting changes.

Electrolytes

Water solutions of ionic substances will conduct an electric current, so they are called **electrolytes.** Ions must be present and free to move in a solution to carry the charge, so electrolytes are solutions containing ions. Pure water will not conduct an electric current as it is a covalent compound, which ionizes only very slightly. Water solutions of sugar, alcohol, and most other covalent compounds are nonconductors, so are called **nonelectrolytes.** Nonelectrolytes are covalent compounds that form molecular solutions, so they cannot conduct an electric current (figure 13.14).

Some covalent compounds are nonelectrolytes as a pure liquid but become electrolytes when dissolved in water. Pure hydrogen chloride (HCl), for example, does not conduct an electric current, so you can assume that it is a molecular substance. When dissolved in water, hydrogen chloride does conduct a current, so it must now contain ions. Evidently, the hydrogen chloride has become **ionized** by the water. The process of forming ions from molecules is called **ionization.** Hydrogen chloride, just

as water, has polar molecules. The positive hydrogen atom on the HCl molecule is attracted to the negative oxygen end of a water molecule, and the force of attraction is strong enough to break the hydrogen-chlorine bond, forming charged particles (figure 13.15). The reaction is

$$HCl(l) + H_2O(l) \rightarrow H_3O^+(aq) + Cl^-(aq)$$

The H_3O^+ ion is called a **hydronium ion.** A hydronium ion is basically a molecule of water with an attached hydrogen ion. The presence of the hydronium ion gives the solution new chemical properties, and the solution is no longer hydrogen chloride but is *hydrochloric acid.* Hydrochloric acid, and other acids, will be discussed shortly.

Boiling Point

Boiling occurs when the pressure of the vapor escaping from a liquid is equal to the atmospheric pressure on the liquid. The *normal* boiling point is defined as the temperature at which the vapor pressure is equal to the average atmospheric pressure at sea level. For pure water, this temperature is 100° C (212° F).

During the later 1880s, a French chemist named Francois Raoult observed that the vapor pressure over a solution is *less* than the vapor pressure over the pure solvent at the same temperature. Molecules of a liquid can escape into the air only at the surface of the liquid, and the presence of molecules of a solute means that fewer solvent molecules can be at the surface to escape. Thus, the vapor pressure over a solution is less than the vapor pressure over a pure solvent (figure 13.16).

Because the vapor pressure of a solution is less than over the pure solvent, the solution boils at a higher temperature. A higher temperature is required to increase the vapor pressure to that of the atmospheric pressure. Some cooks have been observed to add a "pinch" of salt to a pot of water before boiling. Is this to increase the boiling point, and therefore cook the food more quickly? How much does a pinch of salt increase the boiling temperature? The answers are found in the relationship between the concentration of a solute and the boiling point of the solution.

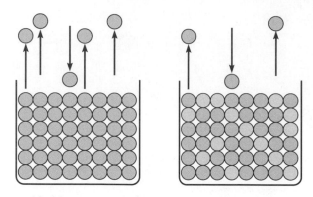

Figure 13.16

The rate of evaporation, and thus the vapor pressure, is less for a solution than for a solvent in the pure state. The greater the solute concentration, the less the vapor pressure.

It is the number of solute particles (ions or molecules) at the surface of a solution that increases the boiling point. Recall from chapter 12 that a mole is a measure that can be defined as a number of particles called Avogadro's number. Since the number of particles at the surface is proportional to the ratio of particles in the solution, the concentration of the solute will directly influence the increase in the boiling point. In other words, the boiling point of any dilute solution is increased proportional to the concentration of the solute. For water, the boiling point is increased 0.521° C for every mole of solute dissolved in 1,000 g of water. Thus, any water solution will boil at a higher temperature than pure water. Since it boils at a higher temperature, it also takes a longer time to reach the boiling point.

It makes no difference what the substance is that is dissolved in the water, one mole of solute in 1,000 g of water will elevate the boiling point by 0.521° C. A mole contains Avogadro's number of particles, so a mole of any solute will lower the vapor pressure by the same amount. Sucrose, or table sugar, for example, is $C_{12}H_{22}O_{11}$ and has a gram-formula weight of 342 g. So 342 g of sugar in 1,000 g of water (about a liter) will increase the boiling point by 0.521° C. Thus, if you measure the boiling point of a sugar solution you can determine the concentration of sugar in the solution. For example, pancake syrup that boils at 100.261° C (sea level pressure) must contain 171 g of sugar dissolved in 1,000 g of water. You know this because the increase of 0.261° C over 100° C is one-half of 0.521° C. If the boiling point were increased by 0.521° C over 100° C, the syrup would have the full gram-formula weight (342 g) dissolved in a kg of water.

Since it is the number of particles of solute in a specific sample of water that elevates the boiling point, there is a different effect by dissolved covalent and dissolved ionic compounds (figure 13.17). Sugar is a covalent compound, and the solute is molecules of sugar moving between the water molecules. Sodium chloride, on the other hand, is an ionic compound and dissolves by the separation of ions, or

$$Na^+Cl^-(s) \rightarrow Na^+(aq) + Cl^-(aq)$$

This equation tells you that one mole of NaCl separates into one mole of sodium ions and one mole of chlorine ions for a total

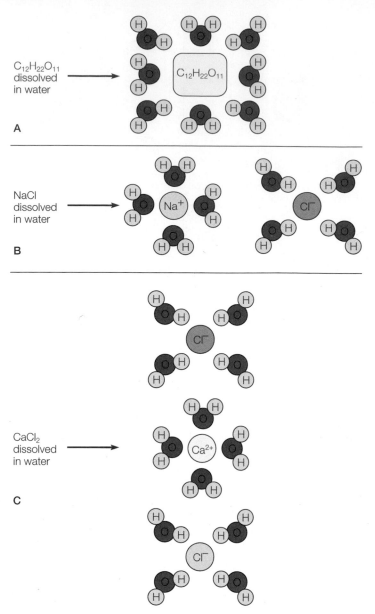

Figure 13.17

Since ionic compounds dissolve by separation of ions, they provide more particles in solution than molecular compounds. (a) A mole of sugar provides Avogadro's number of particles. (b) A mole of NaCl provides two times Avogadro's number of particles. (c) A mole of $CaCl_2$ provides three times Avogadro's number of particles.

of *two* moles of solute. The boiling point elevation of a solution made from one mole of NaCl (58.5 g) is therefore multiplied by two, or $2 \times 0.521°$ C = 1.04° C. The boiling point of a solution made by adding 58.5 g of NaCl to 1,000 g of water is therefore 101.04° C at normal sea level pressure.

Now back to the question of how much a pinch of salt increases the boiling point of a pot of water. Assuming the pot contains about a liter of water (about a quart), and assuming that a "pinch" of salt has a mass of about 0.2 gram, the boiling point will be increased by 0.0037° C. So, there must be some reason other than increasing the boiling point that a cook adds a pinch of salt to a pot of boiling water.

Table 13.2

Some common acids

Name	Formula	Comment
Acetic acid	CH_3COOH	A weak acid found in vinegar
Boric acid	H_3BO_3	A weak acid used in eyedrops
Carbonic acid	H_2CO_3	The weak acid of carbonated beverages
Formic acid	$HCOOH$	Makes the sting of insects and certain plants
Hydrochloric acid	HCl	Also called muriatic acid; used in swimming pools, soil acidifiers, and stain removers
Lactic acid	$CH_3CHOHCOOH$	Found in sour milk, sauerkraut, and pickles; gives tart taste to yogurt
Nitric acid	HNO_3	A strong acid
Phosphoric acid	H_3PO_4	Used in cleaning solutions; added to carbonated beverages for tartness
Sulfuric acid	H_2SO_4	Also called oil of vitriol; used as battery acid and in swimming pools

Table 13.3

Some common bases

Name	Formula	Comment
Sodium hydroxide	$NaOH$	Also called lye or caustic soda; a strong base used in oven cleaners and drain cleaners
Potassium hydroxide	KOH	Also called caustic potash; a strong base used in drain cleaners
Ammonia	NH_3	A weak base used in household cleaning solutions
Calcium hydroxide	$Ca(OH)_2$	Also called slaked lime; used to make brick mortar
Magnesium hydroxide	$Mg(OH)_2$	Solution is called milk of magnesia; used as antacid and laxative

Freezing Point

Freezing occurs when the kinetic energy of molecules has been reduced sufficiently so the molecules can come together, forming the crystal structure of the solid. Reduced kinetic energy of the molecules, that is, reduced temperature, results in a specific freezing point for each pure liquid. The *normal* freezing point for pure water, for example, is 0° C (32° F) under normal pressure. The presence of solute particles in a solution interferes with the water molecules as they attempt to form the six-sided hexagonal structure. The water molecules cannot get by the solute particles until the kinetic energy of the solute particles is reduced, that is, until the temperature is below the normal freezing point. Thus, the presence of solute particles lowers the freezing point, and solutions freeze at a lower temperature than the pure solvent.

The freezing-point depression of a solution has a number of interesting implications for solutions such as seawater. When seawater freezes, the water molecules must work their way around the salt particles as was described earlier. Thus, the solute particles are *not* normally included in the hexagonal structure of ice. Ice formed in seawater is practically pure water. Since the solute was *excluded* when the ice formed, the freezing of seawater increases the salinity. Increased salinity means increased concentration, so the freezing point of seawater is further depressed and more ice forms only at a lower temperature. When this additional ice forms more pure water is removed, and the process goes on. Thus, seawater does not have a fixed freezing point but has a lower and lower freezing point as more and more ice freezes.

The depression of the freezing point by a solute has a number of interesting applications in colder climates. Salt, for example, is spread on icy roads to lower the freezing point (and thus the melting point) of the ice. Calcium chloride, $CaCl_2$, is a salt that is often used for this purpose. Water in a car radiator would also freeze in colder climates if a solute, called antifreeze, were not added to the radiator water. Methyl alcohol has been used as an antifreeze because it is soluble in water and does not damage the cooling system. Methyl alcohol, however, has a low boiling point and tends to boil away. Ethylene glycol has a higher boiling point, so it is called a "permanent" antifreeze. As other solutes, ethylene glycol also raises the boiling point, which is an added benefit for summer driving.

Acids, Bases, and Salts

The electrolytes known as acids, bases, and salts are evident in environmental quality, foods, and everyday living. Environmental quality includes the hardness of water, which is determined by the presence of certain salts, the acidity of soils, which determines how well plants grow, and acid rain, which is a by-product of industry and automobiles. Many concerns about air and water pollution are often related to the chemistry concepts of acids, bases, and salts. These concepts, and uses of acids, bases, and salts, will be considered in this section.

Properties of Acids and Bases

Acids and bases are classes of chemical compounds that have certain characteristic properties. These properties can be used to identify if a substance is an acid or a base (tables 13.2 and 13.3). The following are the properties of *acids* dissolved in water:

1. Acids have a sour taste such as the taste of citrus fruits.
2. Acids change the color of certain substances; for example, litmus changes from blue to red when placed in an acid solution (figure 13.18).

A

B

Figure 13.18

(a) Acid solutions will change the color of blue litmus to red. (b) Solutions of bases will change the color of red litmus to blue.

3. Acids react with active metals, such as magnesium or zinc, releasing hydrogen gas.
4. Acids *neutralize* bases, forming water and salts from the reaction.

Likewise, *bases* have their own characteristic properties. Bases are also called alkaline substances, and the following are the properties of bases dissolved in water:

1. Bases have a bitter taste, for example, the taste of caffeine.
2. Bases reverse the color changes that were caused by acids. Red litmus is changed back to blue when placed in a solution containing a base (figure 13.18).
3. Basic solutions feel slippery on the skin. They have a *caustic* action on plant and animal tissue, converting tissue into soluble materials. A strong base, for example, reacts with fat to make soap and glycerine. This accounts for the slippery feeling on the skin.
4. Bases *neutralize* acids, forming water and salts from the reaction.

Tasting an acid or base to see if it is sour or bitter can be hazardous, since some are highly corrosive or caustic. Many organic acids are not as corrosive and occur naturally in foods. Citrus fruit, for example, contains citric acid, vinegar is a solution of acetic acid, and sour milk contains lactic acid. The stings or bites of some insects (bees, wasps, and ants) and some plants (stinging nettles) are painful because an organic acid, formic acid, is injected by the insect or plant. Your stomach contains a solution of hydrochloric acid. In terms of relative strength, the hydrochloric acid in your stomach is about ten times stronger than the carbonic acid (H_2CO_3) of carbonated beverages.

Examples of bases include solutions of sodium hydroxide (NaOH), which has a common name of lye or caustic soda, and potassium hydroxide (KOH), which has a common name of caustic potash. These two bases are used in products known as drain cleaners. They open plugged drains because of their caustic action, turning grease, hair, and other organic "plugs"

into soap and other soluble substances that are washed away. A weaker base is a solution of ammonia (NH_3), which is often used as a household cleaner. A solution of magnesium hydroxide, $Mg(OH)_2$, has a common name of milk of magnesia and is sold as an antacid and laxative.

Many natural substances change color when mixed with acids or bases. You may have noticed that tea changes color slightly, becoming lighter, when lemon juice (which contains citric acid) is added. Some plants have flowers of one color when grown in acidic soil and flowers of another color when grown in basic soil. Vegetable dyes that change color in the presence of acids or bases can be used as an **acid-base indicator.** An indicator is simply a vegetable dye that is used to distinguish between acid and base solutions by a color change. Litmus, for example, is an acid-base indicator made from a dye extracted from certain species of lichens. The dye is applied to paper strips, which turn red in acidic solutions and blue in basic solutions.

| Activities |

To see how acids and bases change the color of certain vegetable dyes, consider the dye that gives red cabbage its color. Shred several leaves of red cabbage and boil them in a pan of water to extract the dye. After you have a purple solution, squeeze the juice from the cabbage into the pan and allow the solution to cool. Add vinegar in small amounts as you stir the solution, continuing until the color changes. Add ammonia in small amounts, again stirring until the color changes again. Reverse the color change again by adding vinegar in small amounts. Will this purple cabbage acid-base indicator tell you if other substances are acids or bases?

Explaining Acid-Base Properties

Comparing the lists in tables 13.2 and 13.3, you can see that acids and bases appear to be chemical opposites. Notice in table 13.2 that the acids all have an H, or hydrogen atom, in their formulas. In table 13.3, most of the bases have a hydroxide ion, OH^-, in their formulas. Could this be the key to acid-base properties?

Lavoisier was one of the first to attempt an explanation for the properties of acids. Lavoisier thought that all acids contained oxygen, so in 1777 he proposed the name *oxygen,* which means "acid former" in Greek. The formulas of acids listed in table 13.2 show that most acids do contain oxygen, with the exception of hydrochloric acid, which is HCl. During Lavoisier's time chlorine was believed to be an oxygen compound. Not until 1810 was chlorine discovered as an element. The modern understanding of acids and bases originated with the introduction of a *theory of ionization,* developed by Svante Arrhenius in 1883. Arrhenius proposed that electrolytes such as acids, bases, and salts produced equal numbers of positive and negative ions when dissolved in dilute solutions. In concentrated solutions, these ions were in equilibrium with solute molecules that were not ionized.

Chemical equilibrium occurs when two opposing reactions happen at the same time and at the same rate. Pure water, for example, ionizes very slightly to produce a hydronium ion, H_3O^+, and a hydroxide ion, OH^-, from the dissociation of a water molecule:

$$H_2O(l) + H_2O(l) \rightarrow H_3O^+(aq) + OH^-(aq)$$

This is termed the *forward reaction,* which occurs to about 1 molecule in 500 million molecules of pure water. The H_3O^+ and OH^- now react in a *reverse reaction* to produce a molecule of water. Chemical equilibrium occurs when the forward reaction occurs at the same rate as the reverse reaction. This is shown by a double arrow:

$$H_2O(l) + H_2O(l) \rightleftharpoons H_3O^+(aq) + OH^-(aq)$$

In a condition of chemical equilibrium the concentration of molecules and ions remains constant, even though both reactions are always occurring.

The modern concept of an acid considers the properties of acids in terms of the hydronium ion, H_3O^+. As was mentioned earlier, the hydronium ion is a water molecule to which a H^+ ion is attached. Since a hydrogen ion is a hydrogen atom without its single electron, it could be considered as an ion consisting of a single proton. Thus the H^+ ion can be called a *proton.* An **acid** is defined as any substance that is a *proton donor* when dissolved in water, increasing the hydronium ion concentration. For example, hydrogen chloride dissolved in water has the following reaction:

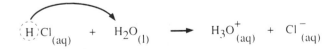

The dotted circle and arrow were added to show that the hydrogen chloride donated a proton to a water molecule. The resulting solution contains H_3O^+ ions and has acid properties, so the solution is called hydrochloric acid.

The bases listed in table 13.3 all appear to have a hydroxide ion, OH^-. Water solutions of these bases do contain OH^-

ions, but the definition of a base is much broader. A **base** is defined as any substance that is a *proton acceptor* when dissolved in water, increasing the hydroxide ion concentration. For example, ammonia dissolved in water has the following reaction:

$$NH_3(g) + H_2O(l) \rightleftharpoons (NH_4)^+ + OH^-$$

The dotted circle and arrow show that the ammonia molecule accepted a proton from a water molecule, providing a hydroxide ion. The resulting solution contains OH^- ions and has basic properties, so a solution of ammonium hydroxide is a base.

Carbonates, such as sodium carbonate (Na_2CO_3), form basic solutions because the carbonate ion reacts with water to produce hydroxide ions.

$$(CO_3)^{2-}(aq) + H_2O(l) \rightarrow (HCO_3)^-(aq) + OH^-(aq)$$

Thus, sodium carbonate produces a basic solution.

Acids could be thought of as simply solutions of hydronium ions in water, and bases could be considered solutions of hydroxide ions in water. The proton donor and proton acceptor definition is much broader, and it does include the definition of acids and bases as hydronium and hydroxide compounds. The broader, more general definition covers a wider variety of reactions and is therefore more useful.

The modern concept of acids and bases explains why the properties of acids and bases are **neutralized,** or lost, when acids and bases are mixed together. For example, consider the hydronium ion produced in the hydrochloric acid solution and the hydroxide ion produced in the ammonia solution. When these solutions are mixed together, the hydronium ion reacts with the hydroxide ion,

$$H_3O^+(aq) + OH^+(aq) \rightarrow H_2O(l) + H_2O(l)$$

Thus, a proton is transferred from the hydronium ion (an acid), and the proton is accepted by the hydroxide ion (a base). A molecule of water is produced and both the acid and base properties disappear or are neutralized.

Strong and Weak Acids and Bases

Acids and bases are classified according to their degree of ionization when placed in water. **Strong acids** ionize completely in water, with all molecules dissociating into ions. Nitric acid, for example, reacts completely in the following equation:

$$HNO_3(aq) + H_2O(l) \rightarrow H_3O^+(aq) + (NO_3)^-(aq)$$

Nitric acid, hydrochloric acid (figure 13.19), and sulfuric acid are common strong acids.

Acids that react only partially produce fewer hydronium ions, so they are weaker acids. **Weak acids** are only partially ionized because of an equilibrium reaction with water. Vinegar, for example, contains acetic acid that reacts with water in the following reaction:

$$HC_2H_3O_2 + H_2O \rightleftharpoons H_3O^+ + (C_2H_3O_2)^-$$

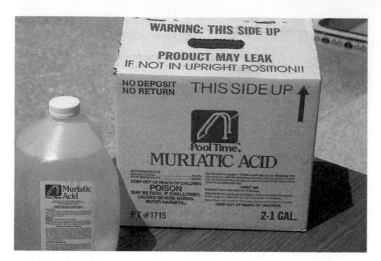

Figure 13.19
Hydrochloric acid (HCl) has the common name of muriatic acid. Hydrochloric acid is a strong acid used in swimming pools, soil acidifiers, and stain removers.

The double yield arrows indicate an equilibrium reaction, which means that at any given time, not many of the hydronium ions are in solution. In fact, only about 1 percent or less of the acetic acid molecules ionize, depending on the concentration.

Bases are also classified as strong or weak. A **strong base** is completely ionic in solution and has hydroxide ions. Sodium hydroxide, or lye, is the most common example of a strong base. It dissolves in water to form sodium and hydroxide ions:

$$Na^+OH^-(s) \rightarrow Na^+(aq) + OH^-(aq)$$

A **weak base** is only partially ionized because of an equilibrium reaction with water. Ammonia, magnesium hydroxide, and calcium hydroxide are examples of weak bases. Magnesium and calcium hydroxide are only slightly soluble in water, and this reduces the *concentration* of hydroxide ions in a solution. It would appear that $Ca(OH)_2$ would produce two moles of hydroxide ions. It would, if it were completely soluble and reacted completely. It is the concentration of hydroxide ions in solution that determines if a base is weak or strong, not the number of ions per mole.

The pH Scale
The strength of an acid or a base is usually expressed in terms of a range of values called a **pH scale.** The pH scale is based on the concentration of the hydronium ion (in moles/L) in an acidic or a basic solution. To understand how the scale is able to express both acid and base strength in terms of the hydronium ion, first recall that pure water is very slightly ionized in the equilibrium reaction:

$$H_2O(l) + H_2O(l) \rightleftharpoons H_3O^+(aq) + OH^-(aq)$$

The amount of self-ionization by water has been determined through measurements. In pure water at 25° C or any neutral water solution at that temperature, the H_3O^+ concentration is 1×10^{-7} moles/L and the OH^- concentration is also 1×10^{-7} moles/L. Since both ions are produced in equal numbers, then

Table 13.4
The pH and hydronium ion concentration (moles/L)

Hydronium Ion Concentration (moles/L)	Reciprocal of Hydronium Ion Concentration	pH
1×10^{0}	1×10^{0}	0
1×10^{-1}	1×10^{1}	1
1×10^{-2}	1×10^{2}	2
1×10^{-3}	1×10^{3}	3
1×10^{-4}	1×10^{4}	4
1×10^{-5}	1×10^{5}	5
1×10^{-6}	1×10^{6}	6
1×10^{-7}	1×10^{7}	7
1×10^{-8}	1×10^{8}	8
1×10^{-9}	1×10^{9}	9
1×10^{-10}	1×10^{10}	10
1×10^{-11}	1×10^{11}	11
1×10^{-12}	1×10^{12}	12
1×10^{-13}	1×10^{13}	13
1×10^{-14}	1×10^{14}	14

the H_3O^+ concentration equals the OH^- concentration and pure water is neutral, neither acidic nor basic.

In general, adding an acid substance to pure water increases the H_3O^+ concentration. Adding a base substance to pure water increases the OH^- concentration. Adding a base also *reduces* the H_3O^+ concentration as the additional OH^- ions are able to combine with more of the hydronium ions to produce unionized water. Thus, at a given temperature, an increase in OH^- concentration is matched by a *decrease* in H_3O^+ concentration. The concentration of the hydronium ion can be used as a measure of acidic, neutral, and basic solutions. In general, (1) acidic solutions have H_3O^+ concentrations above 1×10^{-7} moles/L, (2) neutral solutions have H_3O^+ concentrations equal to 1×10^{-7} moles/L, and (3) basic solutions have H_3O^+ concentrations less than 1×10^{-7} moles/L. These three statements lead directly to the pH scale, which is named from the French *pouvoir hydrogene,* meaning "hydrogen power." Power refers to the exponent of the hydronium ion concentration, and the pH is a *power of ten notation that expresses the H_3O^+ concentration* (table 13.4).

A neutral solution has a pH of 7.0. Acidic solutions have pH values below 7, and smaller numbers mean greater acidic properties. Increasing the OH^- concentration decreases the H_3O^+ concentration, so the strength of a base is indicated on the same scale with values greater than 7. Note that the pH scale is logarithmic, so a pH of 2 is ten times more acidic than a pH of 3. Likewise, a pH of 10 is one hundred times more basic than a pH of 8. Figure 13.20 is a diagram of the pH scale, and table 13.5 compares the pH of some common substances (figure 13.21).

H_3O^+ concentration (moles/liters)	pH	Meaning
$1 \times 10^{-0} (=1)$	0	
1×10^{-1}	1	
1×10^{-2}	2	
1×10^{-3}	3	Increasing acidity
1×10^{-4}	4	
1×10^{-5}	5	
1×10^{-6}	6	
1×10^{-7}	7	Neutral
1×10^{-8}	8	
1×10^{-9}	9	
1×10^{-10}	10	Increasing basicity
1×10^{-11}	11	
1×10^{-12}	12	
1×10^{-13}	13	
1×10^{-14}	14	

Figure 13.20
The pH scale.

Figure 13.21
The pH decreases as the acidic strength of these substances increases from right to left. Did you know that lemon juice is more acidic than vinegar? That a soft drink is more acidic than orange juice or grapefruit juice?

Properties of Salts

Salt is produced by a neutralization reaction between an acid and a base. A **salt** is defined as any ionic compound except those with hydroxide or oxide ions. Table salt, NaCl, is but one example of this large group of ionic compounds. As an example of a salt produced by a neutralization reaction, consider the reaction of HCl (an acid in solution) with $Ca(OH)_2$ (a base in solution). The reaction is

$$2\ HCl(aq) + Ca(OH)_2(aq) \rightarrow CaCl_2(aq) + 2\ H_2O(l)$$

This is an ionic exchange reaction that forms molecular water, leaving Ca^{2+} and Cl^- in solution. As the water is evaporated,

Table 13.5

The approximate pH of some common substances

Substance	pH (or pH Range)
Hydrochloric acid (4%)	0
Gastric (stomach) solution	1.6–1.8
Lemon juice	2.2–2.4
Vinegar	2.4–3.4
Carbonated soft drinks	2.0–4.0
Grapefruit	3.0–3.2
Oranges	3.2–3.6
Acid rain	4.0–5.5
Tomatoes	4.2–4.4
Potatoes	5.7–5.0
Natural rain water	5.6–6.2
Milk	6.3–6.7
Pure water	7.0
Seawater	7.0–8.3
Blood	7.4
Sodium bicarbonate solution	8.4
Milk of magnesia	10.5
Ammonia cleaning solution	11.9
Sodium hydroxide solution	13.0

these ions begin forming ionic crystal structures as the solution concentration increases. When the water is all evaporated, the white crystalline salt of $CaCl_2$ remains.

If sodium hydroxide had been used as the base instead of calcium hydroxide, a different salt would be produced:

$$HCl(aq) + NaOH(aq) \rightarrow NaCl(aq) + H_2O(l)$$

Salts are also produced when elements combine directly, when an acid reacts with a metal, and by other reactions. Salts are usually prepared commercially by a neutralization reaction between an acid and a base that furnishes the desired ions.

Salts are essential in the diet both as electrolytes and as a source of certain elements, usually called *minerals* in this context. Plants must have certain elements that are derived from water-soluble salts. Potassium, nitrates, and phosphate salts are often used to supply the needed elements. There is no scientific evidence that plants prefer to obtain these elements from natural sources, as compost, or from chemical fertilizers. After all, a nitrate ion is a nitrate ion, no matter what its source. Table 13.6 lists some common salts and their uses.

Hard and Soft Water

Salts vary in their solubility in water, and a solubility chart is in appendix B. Table 13.7 lists some generalizations concerning the various common salts. Some of the salts are dissolved by water that will eventually be used for domestic supply. When the salts are soluble calcium or magnesium compounds, the water will contain calcium or magnesium ions in solution. A

Table 13.6

Some common salts and their uses

Common Name	Formula	Use
Alum	$KAl(SO_4)_2$	Medicine, canning, baking powder
Baking soda	$NaHCO_3$	Fire extinguisher, antacid, deodorizer, baking powder
Bleaching powder (chlorine tablets)	$CaOCl_2$	Bleaching, deodorizer, disinfectant in swimming pools
Borax	$Na_2B_4O_7$	Water softener
Chalk	$CaCO_3$	Antacid tablets, scouring powder
Cobalt chloride	$CoCl_2$	Hygrometer (pink in damp weather, blue in dry weather)
Chile saltpeter	$NaNO_3$	Fertilizer
Epsom salt	$MgSO_4 \cdot 7\ H_2O$	Laxative
Fluorspar	CaF_2	Metallurgy flux
Gypsum	$CaSO_4 \cdot 2\ H_2O$	Plaster of Paris, soil conditioner
Lunar caustic	$AgNO_3$	Germicide and cauterizing agent
Niter (or saltpeter)	KNO_3	Meat preservative, makes black gunpowder (75 parts KNO_3, 15 of carbon, 10 of sulfur)
Potash	K_2CO_3	Makes soap, glass
Rochelle salt	$KNaC_4H_4O_6$	Baking powder ingredient
TSP	Na_3PO_4	Water softener, fertilizer

Table 13.7

Generalizations about salt solubilities

Salts	Solubility	Exceptions
Sodium Potassium Ammonium	Soluble	None
Nitrate Acetate Chlorate	Soluble	None
Chlorides	Soluble	Ag and Hg (I) are insoluble
Sulfates	Soluble	Ba, Sr, and Pb are insoluble
Carbonates Phosphates Silicates	Insoluble	Na, D, and NH_4 are soluble
Sulfides	Insoluble	Na, K, and NH_4 are soluble: Mg, Ca, Sr, and Ba decompose

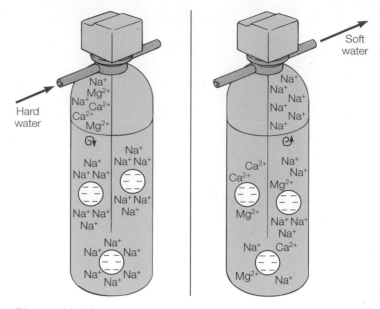

Figure 13.22

A water softener exchanges sodium ions for the calcium and magnesium ions of hard water. Thus, the water is now soft, but it contains the same number of ions as before.

solution of Ca^{2+} of Mg^{2+} ions is said to be **hard water** because it is hard to make soap lather in the water. "Soft" water, on the other hand, makes a soap lather easily. The difficulty occurs because soap is a sodium or potassium compound that is soluble in water. The calcium or magnesium ions, when present, replace the sodium or potassium ions in the soap compound, forming an insoluble compound. It is this insoluble compound that forms a "bathtub ring" and also collects on clothes being washed, preventing cleansing.

The key to "softening" hard water is to remove the troublesome calcium and magnesium ions (figure 13.22). If the hardness is caused by magnesium or calcium bicarbonates, the removal is accomplished by simply heating the water. Upon heating, they decompose, forming an insoluble compound that effectively removes the ions from solution. The decomposition reaction for calcium bicarbonate is

$$Ca^{2+}(HCO_3)_2(aq) \rightarrow CaCO_3(s) + H_2O(l) + CO_2\uparrow$$

The reaction is the same for magnesium bicarbonate. As the solubility chart in appendix B shows, magnesium and calcium carbonates are insoluble, so the ions are removed from solution in the solid that is formed. Perhaps you have noticed such a white compound forming around faucets if you live where bicarbonates are a problem. Commercial products to remove such deposits usually contain an acid, which reacts with the carbonate to make a new, soluble salt that can be washed away.

Water hardness is also caused by magnesium or calcium sulfate, which requires a different removal method. Certain chemicals such as sodium carbonate (washing soda), trisodium phosphate (TSP), and borax will react with the troublesome ions, forming an insoluble solid that removes them from solution. For example, washing soda and calcium sulfate react as follows:

$$Na_2CO_3(aq) + CaSO_4(aq) \rightarrow Na_2SO_4(aq) + CaCO_3\downarrow$$

Acid Rain

Acid rain is a general term used to describe any acidic substances, wet or dry, that fall from the atmosphere. Wet acidic deposition could be in the form of rain, but snow, sleet, and fog could also be involved. Dry acidic deposition could include gases, dust, or any solid particles that settle out of the atmosphere to produce an acid condition.

Pure, unpolluted rain is naturally acidic. Carbon dioxide in the atmosphere is absorbed by rainfall, forming carbonic acid (H_2CO_3). Carbonic acid lowers the pH of pure rainfall to a range of 5.6 to 6.2. Decaying vegetation in local areas can provide more CO_2, making the pH even lower. A pH range of 4.5 to 5.0, for example, has been measured in remote areas of the Amazon jungle. Human-produced exhaust emissions of sulfur and nitrogen oxides can lower the pH of rainfall even more, to a 4.0 to 5.5 range. This is the pH range of acid rain.

The sulfur and nitrogen oxides that produce acid rain come from exhaust emissions of industries and electric utilities that burn coal and from the exhaust of cars, trucks, and buses (Box figure 13.1). The emissions are sometimes called "SO_x" and "NO_x," which is read "socks" and "knox." The "x" subscript implies the variable presence of any or all of the oxides, for example, nitrogen monoxide (NO), nitrogen dioxide (NO_2), and dinitrogen tetroxide (N_2O_4) for NO_x.

SO_x and NO_x are the raw materials of acid rain and are not themselves acidic. They react with other atmospheric chemicals to form sulfates and nitrates, which combine with water vapor to form sulfuric acid (H_2SO_4) and nitric acid (HNO_3). These are the chemicals of concern in acid rain.

Many variables influence how much and how far SO_x and NO_x are carried in the atmosphere and if they are converted to acid rain or simply return to the surface as a dry gas or particles. During the 1960s and 1970s, concerns about local levels of pollution led to the replacement of short smokestacks of about 60 m (about 200 ft) with taller smokestacks of about 200 m (about 650 ft). This did reduce the local levels of pollution by dumping the exhaust higher in the atmosphere where winds could carry it away. It also set the stage for longer-range transport of SO_x and NO_x and their eventual conversion into acids.

Box Figure 13.1

Natural rainwater has a pH of 5.6 to 6.2. Exhaust emissions of sulfur and nitrogen oxides can lower the pH of rainfall to a range of 4.0 to 5.5. The exhaust emissions come from industries, electric utilities, and automobiles. Not all emissions are as visible as those pictured in this illustration.

There are two main reaction pathways by which SO_x and NO_x are converted to acids: (1) reactions in the gas phase and (2) reactions in the liquid phase, such as in water droplets in clouds and fog. In the gas phase, SO_x and NO_x are oxidized to acids, mainly by hydroxyl ions and ozone, and the acid is absorbed by cloud droplets and precipitated as rain or snow. Most of the nitric acid in acid rain and about one-fourth of the sulfuric acid is formed in gas phase reactions. Most of the liquid phase reactions that produce sulfuric acid involve the absorbed SO_x and hydrogen peroxide (H_2O_2), ozone, oxygen, and particles of carbon, iron oxide, and manganese oxide particles. These particles also come from the exhaust of fossil fuel combustion.

Acid rain falls on the land, bodies of water, forests, crops, buildings, and people. The concerns about acid rain center on its environmental impact on lakes, forests, crops, materials, and human health. Lakes in different parts of the world, for example, have been increasing in acidity over the past fifty years. Lakes in northern New England, the Adirondacks, and parts of Canada now have a pH of less than 5.0, and correlations have been established between lake acidity and decreased fish populations. Trees, mostly conifers, are dying at unusually rapid rates in the northeastern United States. Red spruce in Vermont's Green Mountains and the mountains of New York and New Hampshire have been affected by acid rain as have pines in New Jersey's Pine Barrens. It is believed that acid rain leaches essential nutrients, such as calcium, from the soil and also mobilizes aluminum ions. The aluminum ions disrupt the water equilibrium of fine root hairs, and when the root hairs die, so do the trees.

Human-produced emissions of sulfur and nitrogen oxides from burning fossil fuels are the cause of acid rain. The heavily industrialized northeastern part of the United States, from the Midwest through New England, release sulfur and nitrogen emissions that result in a precipitation pH of 4.0 to 4.5. This region is the geographic center of the nation's acid rain problem. The solution to the problem is found in (1) using fuels other than fossil fuels and (2) reducing the thousands of tons of SO_x and NO_x that are dumped into the atmosphere per day when fossil fuels are used.

Calcium carbonate is insoluble, thus the calcium ions are removed from solution before they can react with the soap. Many laundry detergents have Na_2CO_3, TSP, or borax ($Na_2B_4O_7$) added to soften the water. TSP causes other problems, however, as the additional phosphates in the waste water can act as a fertilizer, stimulating the growth of algae to such an extent that other organisms in the water die.

A water softener unit is an ion exchanger. The unit contains a mineral that exchanges sodium ions for calcium and magnesium ions as water is run through it. The softener is regenerated periodically by flushing with a concentrated sodium chloride solution. The sodium ions replace the calcium and magnesium ions, which are carried away in the rinse water. The softener is then ready for use again. The frequency of renewal cycles depends on the water hardness, and each cycle can consume from four to twenty pounds of sodium chloride per renewal cycle. In general, water with less than 75 ppm calcium and magnesium ions is called soft water, and with greater concentrations is called hard water. The greater the concentration above 75 ppm, the harder the water.

Buffers

A **buffer solution** consists of a weak acid together with a salt and has the same negative ion as the acid. A buffer has the ability to resist changes in the pH when small amounts of an acid or a base are added. Acetic acid, for example, is a weak acid that forms hydronium ions and acetate ions in equilibrium:

$$HC_2H_3O_2(aq) + H_2O(l) \rightarrow H_3O^+(aq) + (C_2H_3O_2)^-(aq)$$

When sodium acetate, $NaC_2H_3O_2$, is added to the solution, it becomes a buffer solution. If a small amount of an acid is added, the hydronium ions are neutralized by reacting with the acetate ions in solution:

$$(C_2H_3O_2)^-(aq) + H_3O^+(aq) \rightarrow HC_2H_3O_2(aq) + H_2O(l)$$

If a small amount of a base is added, the hydroxide ions are neutralized by reacting with acetic acid:

$$HC_2H_3O_2(aq) + OH^-(aq) \rightarrow (C_2H_3O_2)^-(aq) + H_2O(l)$$

Thus, the addition of an acid or a base does not change the pH, but it changes the ratio of $HC_2H_3O_2$ and $(C_2H_3O_2)^-$ instead. The solution will continue its buffering action as long as the number of H_3O^+ or OH^- added does not exceed the number of $HC_2H_3O_2$ molecules or acetate ions in the solution. Your blood contains buffer solutions that maintain the pH at about 7.4. Seawater is a buffer solution that maintains a pH of about 8.2. Buffers are also added to medicines and to foods. Many lemon-lime carbonated beverages, for example, contain citric acid and sodium citrate (check the label), which forms a buffer in the acid range. Sometimes the label says that these chemicals are to impart and regulate "tartness." Any acid will produce a tart taste. In this case, the tart taste comes from the citric acid and the addition of sodium citrate makes it a buffered solution.

Summary

A *solution* is a homogeneous mixture of ions or molecules of two or more substances. The substance present in the large amount is the *solvent,* and the *solute* is dissolved in the solvent. If one of the components is a liquid, however, it is called the solvent. The relative amount of solute in a solvent is called the *concentration* of a solution. Concentrations are measured (1) in *parts per million* (ppm) or *parts per billion* (ppb), (2) *percent by volume,* the volume of a solute per 100 volumes of solution, (3) *percent by weight,* the weight of solute per 100 weight units of solution, and (4) *salinity,* the mass of salts in 1 kg of solution.

A limit to dissolving solids in a liquid occurs when the solution is *saturated.* A *saturated solution* is one with equilibrium between solute dissolving and solute coming out of solution. The *solubility* of a solid is the concentration of a saturated solution at a particular temperature.

A water molecule consists of two hydrogen atoms and an oxygen atom with bonding and electron pairs in a tetrahedral arrangement. This results in a *bent molecular arrangement,* with 105° between the hydrogen atoms. Oxygen is more electronegative than hydrogen, so electrons spend more time around the oxygen, producing a *polar molecule,* with centers of negative and positive charge. Polar water molecules interact with an *intermolecular force,* or *van der Waals force,* between the negative center of one molecule and the positive center of another. The force of attraction is called a *hydrogen bond.* The hydrogen bond accounts for the decreased density of ice, the high heat of fusion, and the high heat of vaporization of water. The hydrogen bond is also involved in the *dissolving* process.

Fluids that mix in any proportion are called *miscible fluids,* and *immiscible fluids* do not mix. Polar substances dissolve in polar solvents, but not nonpolar solvents, and the general rule is *like dissolves like.* Thus oil, a nonpolar substance, is immiscible in water, a polar substance. When a polar substance is added to a polar solvent, the substance dissolves if the *ion-polar molecular force* is greater than the *ion-ion force.* If the ion-ion force is greater, the substance is *insoluble.*

Water solutions that carry an electric current are called *electrolytes,* and nonconductors are called *nonelectrolytes.* In general, ionic substances make electrolyte solutions and molecular substances make nonelectrolyte solutions. Polar molecular substances may be *ionized* by polar water molecules, however, making an electrolyte from a molecular solution.

The *boiling point of a solution* is greater than the boiling point of the pure solvent, and the increase depends only on the concentration of the solute (at a constant pressure). For water, the boiling point is increased 0.521° C for each mole of solute in each kg of water. The *freezing point of a solution* is lower than the freezing point of the pure solvent, and the depression also depends on the concentration of the solute.

Acids, bases, and salts are chemicals that form ionic solutions in water, and each can be identified by simple properties. These properties are accounted for by the modern concepts of each. *Acids* are *proton donors* that form *hydronium ions* (H_3O^+) in water solutions. *Bases* are *proton acceptors* that form *hydroxide ions* (OH^-) in water solutions. *Strong acids* and *strong bases* ionize completely in water, and *weak acids* and *weak bases* are only partially ionized because of an *equilibrium reaction* with the solvent. The strength of an acid or base is measured on the *pH scale,* a power of ten notation of the hydronium ion concentration. On the scale, numbers from 0 up to 7 are acids, 7 is neutral, and numbers above 7 and up to 14 are bases. Each unit represents a tenfold increase or decrease in acid or base properties.

A *salt* is any ionic compound except those with hydroxide or oxide ions. Salts provide plants and animals with essential elements. The solubility of salts varies with the ions that make up the compound. Solutions of magnesium or calcium produce *hard water,* water that is hard to make soap lather in. Hard water is softened by removing the magnesium and calcium ions. A *buffer* solution is a solution of a weak acid and one of its salts. The solution resists changes in pH by reacting with acids or bases that are added.

Summary of Equations

13.1

Percent by volume

$$\frac{V_{solute}}{V_{solution}} \times 100\% \text{ solution} = \% \text{ solute}$$

13.2

Percent by weight (mass)

$$\frac{m_{solute}}{m_{solution}} \times 100\% \text{ solution} = \% \text{ solute}$$

Key Terms

acid (p. 277)
acid-base indicator (p. 276)
base (p. 277)
buffer solution (p. 282)
chemical equilibrium (p. 277)
concentration (p. 267)
electrolytes (p. 273)
hard water (p. 280)
hydration (p. 271)
hydrogen bond (p. 270)
hydronium ion (p. 273)
intermolecular forces (p. 270)
ionization (p. 273)
ionized (p. 273)
miscible fluids (p. 271)
neutralized (p. 277)
nonelectrolytes (p. 273)

parts per billion (p. 267)
parts per million (p. 267)
percent by volume (p. 267)
percent by weight (p. 267)
pH scale (p. 278)
salinity (p. 268)
salt (p. 279)
saturated solution (p. 269)
solubility (p. 269)
solute (p. 266)
solution (p. 266)
solvent (p. 266)
strong acids (p. 277)
strong base (p. 278)
van der Waals force (p. 270)
weak acids (p. 277)
weak base (p. 278)

Applying the Concepts

1. Which of the following is *not* a solution?
   a. seawater
   b. carbonated water
   c. sand
   d. brass

2. Atmospheric air is a homogeneous mixture of gases that is mostly nitrogen gas. The nitrogen is therefore (the)
   a. solvent.
   b. solution.
   c. solute.
   d. none of the above.

3. A homogeneous mixture is made up of 95 percent alcohol and 5 percent water. In this case the water is (the)
   a. solvent.
   b. solution.
   c. solute.
   d. none of the above.

4. The solution concentration terms of parts per million, percent by volume, and percent by weight are concerned with the amount of
   a. solvent in the solution.
   b. solute in the solution.
   c. solute compared to solvent.
   d. solvent compared to solute.

5. A concentration of 500 ppm is reported in a news article. This is the same concentration as
   a. 0.005%.
   b. 0.05%.
   c. 5%.
   d. 50%.

6. According to the label, a bottle of vodka has a 40% by volume concentration. This means the vodka contains 40 mL of pure alcohol
   a. in each 140 mL of vodka.
   b. to every 100 mL of water.
   c. to every 60 mL of vodka.
   d. mixed with 60 mL of water.

7. A bottle of vinegar is 4% by weight, so you know that the solution contains 4 weight units of pure vinegar with
   a. 96 weight units of water.
   b. 100 weight units of water.
   c. 104 weight units of water.

8. If a salt solution has a salinity of 40‰, what is the equivalent percentage measure?
   a. 400%
   b. 40%
   c. 4%
   d. 0.4%

9. A salt solution has solid salt on the bottom of the container and salt is dissolving at the same rate that it is coming out of solution. You know the solution is
   a. an electrolyte.
   b. a nonelectrolyte.
   c. a buffered solution.
   d. a saturated solution.

10. As the temperature of water *decreases* the solubility of carbon dioxide gas in the water
    a. increases.
    b. decreases.
    c. remains the same.
    d. increases or decreases, depending on the specific temperature.

11. Water has the greatest density at what temperature?
    a. 100° C
    b. 20° C
    c. 4° C
    d. 0° C

12. A hydrogen bond is a weak-to-moderate bond between
    a. any two hydrogen atoms.
    b. a hydrogen of one polar molecule and another polar molecule.
    c. two hydrogen atoms on two nonpolar molecules.
    d. a hydrogen atom and any nonmetal atom.

13. If two given liquids form solutions or not depends on some similarities in their
    a. electronegativities.
    b. polarities.
    c. molecular structures.
    d. hydrogen bonds.

14. A solid salt is insoluble in water so the strongest force must be the
    a. ion-water molecule force.
    b. ion-ion force.
    c. force of hydration.
    d. polar molecule force.

15. Which of the following will conduct an electric current?
    a. pure water
    b. a water solution of a covalent compound
    c. a water solution of an ionic compound
    d. All of the above are correct.

16. Ionization occurs with
    a. ionic compounds.
    b. polar molecules.
    c. nonpolar molecules.
    d. none of the above.

17. Adding sodium chloride to water raises the boiling point of water because
    a. sodium chloride has a higher boiling point.
    b. sodium chloride ions occupy space at the water surface.
    c. sodium chloride ions have stronger ion-ion bonds than water.
    d. the energy of hydration is higher.

18. The ice that forms in freezing seawater is
    a. pure water.
    b. the same salinity as liquid seawater.
    c. more salty than liquid seawater.
    d. more dense than liquid seawater.

19. Salt solutions freeze at a lower temperature than pure water because
    a. more ionic bonds are present.
    b. salt solutions have a higher vapor pressure.
    c. ions get in the way of water molecules trying to form ice.
    d. salt naturally has a lower freezing point than water.

20. Which of the following would have a pH of *less* than 7?
    a. a solution of ammonia
    b. a solution of sodium chloride
    c. pure water
    d. carbonic acid

21. Which of the following would have a pH of *more* than 7?
    a. a solution of ammonia
    b. a solution of sodium chloride
    c. pure water
    d. carbonic acid

22. The condition of two opposing reactions happening at the same time and at the same rate is called
    a. neutralization.
    b. chemical equilibrium.
    c. a buffering reaction.
    d. cancellation.

23. Solutions of acids, bases, and salts have what in common? All have
    a. proton acceptors.
    b. proton donors.
    c. ions.
    d. polar molecules.

24. When a solution of an acid and a base are mixed together,
    a. a salt and water are formed.
    b. they lose their acid and base properties.
    c. both are neutralized.
    d. All of the above are correct.

25. A substance that ionizes completely into hydronium ions is known as a
    a. strong acid.
    b. weak acid.
    c. strong base.
    d. weak base.

26. A scale of values that expresses the hydronium ion concentration of a solution is known as
    a. an acid-base indicator.
    b. the pH scale.
    c. the solubility scale.
    d. the electrolyte scale.

27. Substance "A" has a pH of 2 and substance "B" has a pH of 3. This means that
    a. substance A has more basic properties than substance B.
    b. substance B has more acidic properties than substance A.
    c. substance A is ten times more acidic than substance B.
    d. substance B is ten times more acidic than substance A.

28. A solution that is able to resist changes in the pH when small amounts of an acid or base are added is called a
    a. neutral solution.
    b. saturated solution.
    c. balanced solution.
    d. buffer solution.

Answers

1. c   2. a   3. c   4. b   5. b   6. d   7. a   8. c   9. d
10. a   11. c   12. b   13. c   14. b   15. c   16. b   17. b
18. a   19. c   20. d   21. a   22. b   23. c   24. d   25. a
26. b   27. c   28. d

Questions for Thought

1. How is a solution different from other mixtures?

2. Explain why some ionic compounds are soluble while others are insoluble in water.

3. Explain why adding salt to water increases the boiling point.

4. A deep lake in Minnesota is covered with ice. What is the water temperature at the bottom of the lake? Explain your reasoning.

5. Explain why water has a greater density at 4° C than at 0° C.

6. What is hard water? How is it softened?

7. According to the definition of an acid and the definition of a base, would the pH increase, decrease, or remain the same when NaCl is added to pure water? Explain.

8. What is a hydrogen bond? Explain how a hydrogen bonds form.

9. What feature of a soap molecule gives it cleaning ability?

10. What ion is responsible for (a) acidic properties? (b) for basic properties?

11. Explain why a pH of 7 indicates a neutral solution—why not some other number?

12. What is a buffer solution?

Exercises

Group A—Solutions Provided in Appendix

1. A 50.0 g sample of a saline solution contains 1.75 g NaCl. What is the percentage by weight concentration?

2. A student attempts to prepare a 3.50 percent by weight saline solution by dissolving 3.50 g NaCl in 100 g of water. Since equation 13.2 calls for 100 g of solution, the correct amount of solvent should have been 96.5 g water ($100 - 3.5 = 96.5$). What percent by weight solution did the student actually prepare?

3. Seawater contains 30,113 ppm by weight dissolved sodium and chlorine ions. What is the percent by weight concentration of sodium chloride in seawater?

4. What is the mass of hydrogen peroxide, H_2O_2, in 250 grams of a 3.0% by weight solution?

5. How many mL of pure alcohol are in a 200 mL glass of wine that is 12 percent alcohol by volume?

6. How many mL of pure alcohol are in a single cocktail made with 50 mL of 40% vodka? (Note: "Proof" is twice the percent, so 80 proof is 40%.)

7. If fish in a certain lake are reported to contain 5 ppm by weight DDT, (a) what percentage of the fish meat is DDT? (b) How much of this fish would have to be consumed to reach a poisoning accumulation of 17.0 grams of DDT?

8. For each of the following reactants, draw a circle around the proton donor and a box around the proton acceptor. Label which acts as an acid and which acts as a base.
   (a) $HC_2H_3O_{2(aq)} + H_2O_{(l)} \rightleftharpoons H_3O^+_{(aq)} + C_2H_3O_2^-_{(aq)}$
   (b) $C_6H_6NH_{2(l)} + H_2O_{(l)} (\rightleftharpoons) C_6H_6NH_3^+_{(aq)} + OH^-_{(aq)}$
   (c) $HClO_{4(aq)} + HC_2H_3O_{2(aq)} \rightleftharpoons H_2C_2H_3O_2^+_{(aq)} + ClO_4^-_{(aq)}$
   (d) $H_2O_{(l)} + H_2O_{(l)} (\rightleftharpoons) H_3O^+_{(aq)} + OH^-_{(aq)}$

Group B—Solutions Not Given

1. What is the percent by weight of a solution containing 2.19 g NaCl in 75 g of the solution?

2. What is the percent by weight of a solution prepared by dissolving 10 g of NaCl in 100 g of H_2O?

3. A concentration of 0.5 ppm SO_2 is harmful to plant life. What is the percent by volume of this concentration?

4. What is the volume of water in a 500 mL bottle of rubbing alcohol that has a concentration of 70% by volume?

5. If a definition of intoxication is an alcohol concentration of 0.05 percent by volume in blood, how much alcohol would be present in the average (155 lb) person's 6,300 mL of blood?

6. How much pure alcohol is in a 355 mL bottle of a "wine cooler" that is 5.0% alcohol by volume?

7. In the 1970s, when lead was widely used in "ethyl" gasoline, the blood level of the average American contained 0.25 ppm lead. The danger level of lead poisoning is 0.80 ppm. (a) What percent of the average person was lead? (b) How much lead would be in an average 80 kg person? (c) How much more lead would the average person need to accumulate to reach the danger level?

8. Draw a circle around the proton donor and a box around the proton acceptor for each of the reactants and label which acts as an acid and which acts as a base.
   (a) $H_3PO_{4(aq)} + H_2O_{(l)} \rightleftharpoons H_3O^+_{(aq)} + H_2PO_4^-_{(aq)}$
   (b) $N_2H_{4(l)} + H_2O_{(l)} (\rightleftharpoons) N_2H_5^+_{(aq)} + OH^-_{(aq)}$
   (c) $HNO_{3(aq)} + HC_2H_3O_{2(aq)} \rightleftharpoons H_2C_2H_3O_2^+_{(aq)} + NO_3^-_{(aq)}$
   (d) $2\ NH_4^+_{(aq)} + Mg_{(s)} (\rightleftharpoons) Mg^{2+}_{(aq)} + 2\ NH_3^+_{(aq)} + H_{2(g)}$

Water and Solutions

Chapter 14

Organic Chemistry

An oil droplet, such as the one shown here on a pin head, is a mixture of certain kinds of hydrocarbons. Other mixtures of hydrocarbons are known as gasoline, kerosene, diesel fuel, asphalt, lubricating oil, wax, and petroleum jelly.

T_{HE} impact of ancient Aristotelian ideas on the development of understandings of motion, elements, and matter was discussed in earlier chapters. Historians also trace the "vitalist theory" back to Aristotle. According to Aristotle's idea, all living organisms are composed of the four elements (earth, air, fire, and water) and have in addition an *actuating force,* the life or soul that makes the organism different from nonliving things made of the same four elements. Plants, as well as animals, were considered to have this actuating, or vital, force in the Aristotelian scheme of things.

There were strong proponents of the vitalist theory as recent as the early 1800s. Their basic argument was that organic matter, the materials and chemical compounds recognized as being associated with life, could not be produced in the laboratory. Organic matter could only be produced in a living organism, they argued, because the organism had a vital force that is not present in laboratory chemicals. Then, in 1828, a German chemist named Fredrich Wohler reacted two chemicals that were *not organic* to produce urea (N_2H_4CO), a known *organic* compound that occurs in urine. Wohler's synthesis of an organic compound was soon followed by the synthesis of other organic substances by other chemists. The vitalist theory gradually disappeared with each new synthesis, and a new field of study, organic chemistry, emerged.

This chapter is an introductory survey of the field of study known as organic chemistry. Organic chemistry is concerned with compounds and reactions of compounds that contain carbon. You will find this an interesting, informative introduction, particularly if you have ever wondered about synthetic materials, natural foods and food products, or any of the thousands of carbon-based chemicals you use every day. The survey begins with the simplest of organic compounds, those consisting of only carbon and hydrogen atoms, compounds known as hydrocarbons. Hydrocarbons are the compounds of crude oil, which is the source of hundreds of petroleum products (figure 14.1). In this section you will find information about things you may have wondered about, for example, what an octane rating is and how petroleum products differ.

Most common organic compounds can be considered derivatives of the hydrocarbons, such as alcohols, ethers, fatty acids, and esters. Some of these are the organic compounds that give flavors to foods, and others are used to make hundreds of commercial products, from face cream to oleo. The main groups, or classes, of derivatives will be briefly introduced, along with some interesting examples of each group. Some of the important organic compounds of life, including proteins, carbohydrates, and fats, are discussed next. The chapter concludes with an introduction to synthetic polymers, what they are, and how they are related to the fossil fuel supply.

Organic Compounds

Organic compounds are sensitive to increases in temperature, decomposing or burning when heated to 400°C (about 750°F) or greater. When sugar is decomposed by heating, for example, it often leaves a black residue of carbon. When burned completely, sugar and other organic materials produce carbon dioxide (and other products). Carbon is the essential element of organic matter, and today, **organic chemistry** is defined as the study of compounds in which carbon is the principal element, whether the compound was formed by living things or not. The

Figure 14.1

Refinery and tank storage facilities, like this one in Texas, are needed to change the hydrocarbons of crude oil to many different petroleum products. The classes and properties of hydrocarbons is one topic of study in organic chemistry.

study of all the other elements and compounds is called **inorganic chemistry.** An *organic compound* is thus a compound in which carbon is the principal element, and an *inorganic compound* is any other compound.

Organic compounds, by definition, must contain carbon while all the other compounds can contain all the other elements. Yet, there are *millions* of different organic compounds but fewer than fifty thousand inorganic compounds. It is the unique properties of carbon that allow it to form so many different compounds. A carbon atom has a simple $1s^2 2s^2 2p^2$ electron structure, and there is room for four more electrons in the outer shell. Carbon has a valence of four and can form four electron pairs, with no lone pairs. The molecular shape of a carbon compound such as CH_4 is therefore tetrahedral. The carbon atom has a valence of four, and can combine with one, two, three, or four *other carbon atoms* in addition to a wide range of other kinds of atoms (figure 14.2). The number of possible molecular combinations is almost limitless, which explains why there are so many organic compounds. Fortunately, there are patterns of groups of carbon atoms and groups of other atoms that lead to similar chemical characteristics, making the study of organic chemistry less difficult. The key to success in studying organic chemistry is to recognize patterns and to understand the code and meaning of organic chemical names. The first patterns to be discussed will be those of the simplest organic compounds, consisting of only two elements.

Hydrocarbons

A **hydrocarbon** is an organic compound consisting of only two elements. As the name implies, these elements are hydrogen and carbon. The simplest hydrocarbon has one carbon atom and four

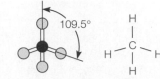

A Three-dimensional model

B An unbranched chain

C—C—C—C—C

C Simplified unbranched chain

Figure 14.2

(*a*) The carbon atom forms bonds in a tetrahedral structure with a bond angle of 109.5°. (*b*) Carbon-to-carbon bond angles are 109.5°, so a chain of carbon atoms makes a zigzag pattern. (*c*) The unbranched chain of carbon atoms is usually simplified in a way that looks like a straight chain, but it is actually a zigzag as shown in (*b*).

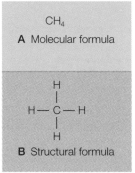

CH$_4$

A Molecular formula

B Structural formula

Figure 14.3

Recall that a molecular formula (*a*) describes the numbers of different kinds of atoms in a molecule, and a structural formula (*b*) represents a two-dimensional model of how the atoms are bonded to each other. Each dash represents a bonding pair of electrons.

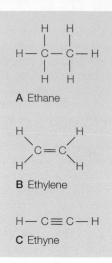

A Ethane

B Ethylene

C Ethyne

Figure 14.4

Carbon-to-carbon bonds can be single (*a*), double (*b*), or triple (*c*). Note that in each example, each carbon atom has four dashes, which represent four bonding pairs of electrons, satisfying the octet rule.

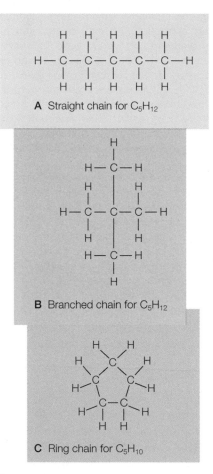

A Straight chain for C$_5$H$_{12}$

B Branched chain for C$_5$H$_{12}$

C Ring chain for C$_5$H$_{10}$

Figure 14.5

Carbon-to-carbon chains can be (*a*) straight, (*b*) branched, or (*c*) in a closed ring.

hydrogen atoms (see figure 14.3), but since carbon atoms can combine with one another, there are thousands of possible structures and arrangements. The carbon-to-carbon bonds are nonpolar covalent and can be single, double, or triple (figure 14.4). Recall that the dash in a structural formula means one shared electron pair. To satisfy the octet rule, this means that each carbon atom must have a total of four dashes around it, no more and no less. Note that when the carbon atom has double or triple bonds, fewer hydrogen atoms can be attached as the octet rule is satisfied. There are four groups of hydrocarbons that are classified according to how the carbon atoms are put together, the (1) *alkanes,* (2) *alkenes,* (3) *alkynes,* and (4) *aromatic hydrocarbons.*

The **alkanes** are *hydrocarbons with single covalent bonds* between the carbon atoms. Alkanes that are large enough to form chains of carbon atoms occur with a straight structure, a branched structure, or a ring structure as shown in figure 14.5.

Table 14.1

The first ten straight-chained alkanes

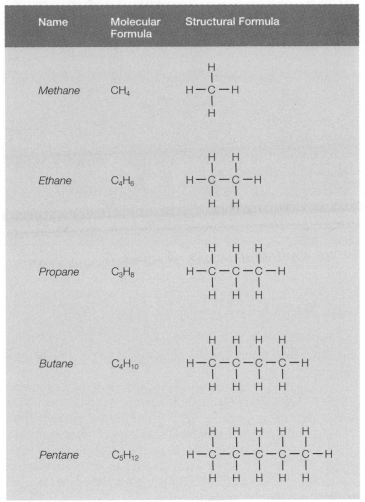

Name	Molecular Formula	Structural Formula
Methane	CH_4	
Ethane	C_4H_6	
Propane	C_3H_8	
Butane	C_4H_{10}	
Pentane	C_5H_{12}	

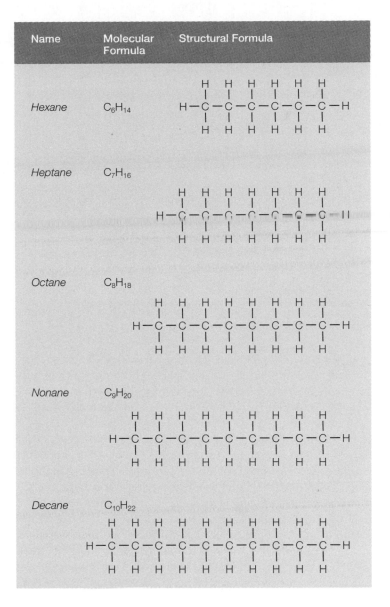

Name	Molecular Formula	Structural Formula
Hexane	C_6H_{14}	
Heptane	C_7H_{16}	
Octane	C_8H_{18}	
Nonane	C_9H_{20}	
Decane	$C_{10}H_{22}$	

(The "straight" structure is actually a zigzag as shown in figure 14.2.) You are familiar with many alkanes, for they make up the bulk of petroleum and petroleum products, which will be discussed shortly. The clues and codes in the names of the alkanes will be considered first.

The alkanes are also called the *paraffin series*. The alkanes are not as chemically reactive as the other hydrocarbons, and the term *paraffin* means "little affinity." They are called a series because *each higher molecular weight alkane has an additional CH2*. The simplest alkane is methane, CH_4, and the next highest molecular weight alkane is propane, C_2H_6. As you can see, C_2H_6 is CH_4 with an additional CH_2. If you compare the first ten alkanes in table 14.1, you will find that each successive compound in the series always has an additional CH_2.

Note the names of the alkanes listed in table 14.1. After pentane the names have a consistent prefix and suffix pattern. The prefix and suffix pattern is a code that provides a clue about the compound. The Greek prefix tells you the *number of carbon atoms* in the molecule, for example, "oct-" means eight, so *oct*ane has eight carbon atoms. The suffix "-ane" tells you this hydrocarbon is a member of the alk*ane* series, so it has single bonds only. With the general alkane formula of C_nH_{2n+2}, you can now write the formula when you hear the name. Octane has eight carbon atoms with single bonds and n = 8. Two times 8 plus 2 (2n + 2) is 18, so the formula for octane is C_8H_{18}. Most organic chemical names provide clues like this, as you will see.

The alkanes in table 14.1 all have straight chains. A straight, continuous chain is identified with the term *normal*, which is abbreviated *n*. Figure 14.6a shows *n*-butane with a straight chain and a molecular formula of C_4H_{10}. Figure 14.6b shows a different branched structural formula that has the same C_4H_{10} molecular formula. Compounds with the same molecular formulas with different structures are called **isomers.** Since the straight-chained isomer is called *n*-butane, the branched isomer

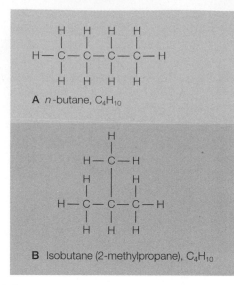

A *n*-butane, C_4H_{10}

B Isobutane (2-methylpropane), C_4H_{10}

Figure 14.6
(*a*) A straight-chain alkane is identified by the prefix "*n*-" for "normal" in the common naming system. (*b*) A branched-chain alkane isomer is identified by the prefix "*iso*-" for "isomer" in the common naming system. In the IUPAC name, isobutane is 2-methylpropane.

Table 14.2

Alkane hydrocarbons and corresponding hydrocarbon groups

Alkane Name	Molecular Formula	Hydrocarbon Group	Molecular Formula
Methane	CH_4	Methyl	$—CH_3$
Ethane	C_2H_6	Ethyl	$—C_2H_5$
Propane	C_3H_8	Propyl	$—C_3H_7$
Butane	C_4H_{10}	Butyl	$—C_4H_9$
Pentane	C_5H_{12}	Amyl	$—C_5H_{11}$
Hexane	C_6H_{14}	Hexyl	$—C_6H_{13}$
Heptane	C_7H_{16}	Heptyl	$—C_7H_{15}$
Octane	C_8H_{18}	Octyl	$—C_8H_{17}$
Nonane	C_9H_{20}	Nonyl	$—C_9H_{19}$
Decane	$C_{10}H_{22}$	Decyl	$—C_{10}H_{21}$

Note: $—CH_3$ means $\cdot—\overset{\displaystyle H}{\underset{\displaystyle H}{C}}—H$ where $\cdot$ denotes unattached. The attachment takes place on a base chain or functional group.

is called *isobutane*. The isomers of a particular alkane, such as butane, have different physical properties because they have different structures. Isobutane, for example, has a boiling point of $-10°C$. The boiling point of *n*-butane, on the other hand, is $-0.5°C$.

Methane, ethane, and propane can have only one structure each, and butane has two isomers. The number of possible isomers for a particular molecular formula increases rapidly as the number of carbon atoms increase. After butane, hexane has five isomers, octane eighteen isomers, and decane seventy-five isomers. Because they have different structures, each isomer has different physical properties. A different naming system is needed because there are just too many isomers to keep track of. The system of naming the branched-chain alkanes is described by rules agreed upon by the International Union of Pure and Applied Chemistry, or IUPAC. Here are the steps in naming the alkane isomers.

Step 1: The longest continuous chain of carbon atoms determines the *base name* of the molecule. The longest continuous chain is not necessarily straight and can take any number of right-angle turns as long as the continuity is not broken. The base name corresponds to the number of carbon atoms in this chain as in table 14.1. For example, the structure

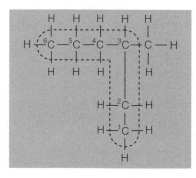

has six carbon atoms in the longest chain, so the base name is *hexane*.

Step 2: The locations of other groups of atoms attached to the base chain are identified by counting carbon atoms from either the left or from the right. The direction selected is the one that results in the *smallest* numbers for attachment locations. For example, the hexane chain has a CH_3 attached to the third or the fourth carbon atom, depending on which way you count. The third atom direction is chosen since it results in a smaller number.

Step 3: The hydrocarbon groups attached to the base chain are named from the number of carbons in the group by changing the alkane suffix "-ane" to "-yl." Thus, a hydrocarbon group attached to a base chain that has one carbon atom is called meth*yl*. Note that the "-yl" hydrocarbon groups have one less hydrogen than the corresponding alkane. Therefore, methane is CH_4 and a *methyl group* is CH_3. The first ten alkanes and their corresponding hydrocarbon group names are listed in table 14.2. In the example, a methyl group is attached to the third carbon atom of the base hexane chain. The name and address of this hydrocarbon group is 3-methyl. The compound is named 3-methylhexane.

Step 4: The prefixes "di-," "tri-," and so on are used to indicate if a particular hydrocarbon group appears on the main chain more than once. For example,

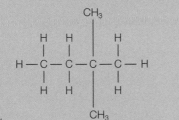

(or)

is 2,2-dimethylbutane and

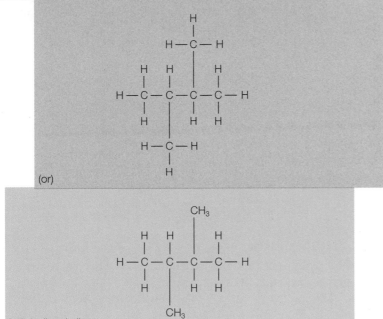

(or)

is 2,3-dimethylbutane.

If hydrocarbon groups with different numbers of carbon atoms are on a main chain they are listed by order of increasing size, for example,

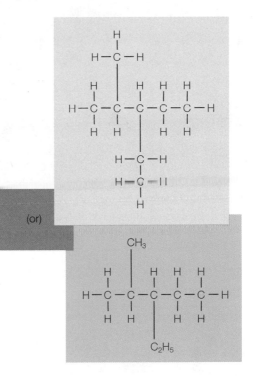

(or)

is named 2-methyl-3-ethylpentane.

Example 14.1

What is the name of an alkane with the following formula?

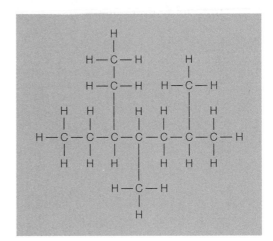

Solution

The longest continuous chain has seven carbon atoms, so the base name is heptane. The smallest numbers are obtained by counting from right to left and counting the carbons on this chain; there is a methyl group in carbon atom 2, a second methyl group on atom 4, and an ethyl group on atom 5. There are two methyl groups, so the prefix "di-" is needed, and the methyl group contains fewer carbon atoms, so it is listed first. The name of the compound is 2,4-dimethyl-5-ethylheptane.

·

Figure 14.7
Ethylene is the gas that ripens fruit and a ripe fruit emits the gas, which will act on unripe fruit. Thus, a ripe tomato placed in a sealed bag with green tomatoes will help ripen them.

Example 14.2

Write the structural formula for 2,2-dichloro-3-methyloctane.

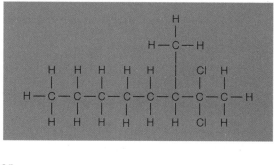

Answer

Alkenes and Alkynes

The alkanes are hydrocarbons with single carbon-to-carbon bonds. The **alkenes** are *hydrocarbons with a double covalent carbon-to-carbon bond.* To denote the presence of a double bond the "-ane" suffix of the alkanes is changed to "-ene" as in alk*ene.* Figure 14.4 shows the structural formula for (a) ethane, C_2H_6, and (b) ethylene, C_2H_4. Alkenes have room for two less hydrogen atoms because of the double bond, so the general alkene formula is C_nH_{2n}.

Ethylene is an important raw material in the chemical industry. Obtained from the processing of petroleum, about half of the commercial ethylene is used to produce the familiar polyethylene plastic. It is also produced by plants to ripen fruit, which explains why unripe fruit enclosed in a sealed plastic bag with ripe fruit will ripen more quickly (figure 14.7). The ethylene produced by the ripe fruit acts on the unripe fruit. Commercial

fruit packers sometimes use small quantities of ethylene gas to ripen fruit quickly that was picked while green.

Perhaps you have heard the terms "saturated" and "unsaturated" in advertisements for cooking oil and oleomargarine. The meaning of these terms with reference to foods will be discussed shortly. First, you need to understand the meaning of the terms. An organic molecule, such as a hydrocarbon, that does not contain the maximum number of hydrogen atoms is an **unsaturated** hydrocarbon. For example, ethylene can add more hydrogen atoms by reacting with hydrogen gas to form ethane:

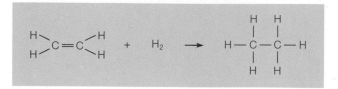

The ethane molecule has all the hydrogen atoms possible, so ethane is a **saturated** hydrocarbon. Unsaturated molecules are less stable, which means that they are more chemically reactive than saturated molecules. Again, the role of saturated and unsaturated fats in foods will be discussed later.

Alkenes are named just as the alkanes except (1) the longest chain of carbon atoms must contain the double bond, (2) the base name now ends in "-ene," (3) the carbon atoms are numbered from the end nearest the double bond, and (4) the base name is given a number of its own, which identifies the address of the double bond. For example,

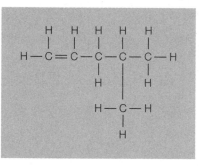

is named 4-methyl-1-pentene. The 1-pentene tells you there is a double bond (-ene) and the 1 tells you the double bond is after the first carbon atom in the longest chain containing the double bond. The methyl group is on the fourth carbon atom in this chain.

An **alkyne** is a *hydrocarbon with a carbon-to-carbon triple bond* and the general formula of C_nH_{2n-2}. The alkynes are highly reactive and the simplest one, ethyne, has a common name of acetylene. Acetylene is commonly burned with oxygen gas in a welding torch because the flame reaches a temperature of about 3,000°C. Acetylene is also an important raw material in the production of plastics. The alkynes are named as the alkenes, except the longest chain must contain the triple bond and the base name suffix is changed to "-yne."

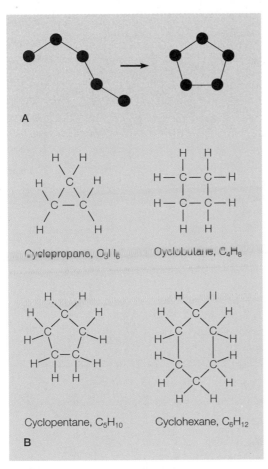

Figure 14.8

(a) The "straight" chain has carbon atoms that are able to rotate freely around their single bonds, sometimes linking up in a closed ring. (b) Ring compounds of the first four cycloalkanes.

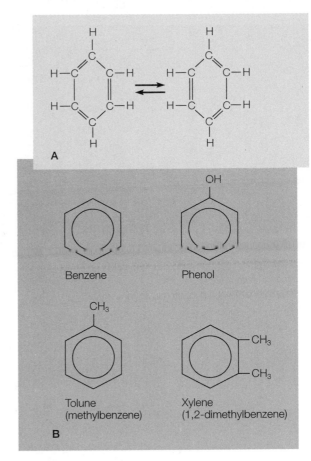

Figure 14.9

(a) The double bonds in C_6H_6 are continually shifting, which gives it different chemical properties than other double-bonded hydrocarbons. (b) The C_6H_6 ring compound with shifting double bonds is benzene. The six-sided symbol with a circle represents the benzene ring. Organic compounds based on the benzene ring are called aromatic hydrocarbons because of their aromatic character.

Cyclo-Alkanes and Aromatic Hydrocarbons

The hydrocarbons discussed up until now have been straight or branched open-ended chains of carbon atoms. Carbon atoms can also bond to each other to form a ring, or cycle, structure. Figure 14.8 shows the structural formulas for some of these cycle structures.

The six-carbon ring structure shown in Figure 14.9a has three double bonds but does not behave like the double bonds in the alkenes. In this six-carbon ring the double bonds are not localized in one place but have many different possible electron locations. This gives the C_6H_6 molecule increased stability. As a result, the molecule does not act unsaturated, that is, it does not readily react to add hydrogen to the ring. The C_6H_6 molecule is the organic compound named *benzene*. Organic compounds that are based on the benzene ring structure are called **aromatic hydrocarbons.** To denote the six-carbon ring with delocalized electrons, benzene is represented by the symbol shown in figure 14.9b.

The circle in the six-sided benzene symbol represents the delocalized electrons. Figure 14.9b illustrates how this benzene ring symbol is used to show the structural formula of some aromatic hydrocarbons. You may have noticed some of the names on labels of paints, paint thinners, and lacquers. Toluene and the xylenes are commonly used in these products as a solvent. A benzene ring attached to another molecule or functional group is given the name *phenyl*.

Petroleum

Petroleum is a mixture of alkanes, cyclo-alkanes, and some aromatic hydrocarbons. The origin of petroleum is uncertain, but it is believed to have formed from the slow anaerobic decomposition of buried marine life, primarily plankton and algae. Time, temperature, pressure, and perhaps bacteria are considered important in the formation of petroleum. As the petroleum formed, it was forced through porous rock until it reached a rock type or rock structure that stopped it. Here, it accumulated to saturate the porous rock, forming an accumulation called an

Organic Chemistry

Figure 14.10

Crude oil from the ground is separated into usable groups of hydrocarbons at this Louisiana refinery. Each petroleum product has a boiling point range, or "cut" of distilled vapors that collect in condensing towers.

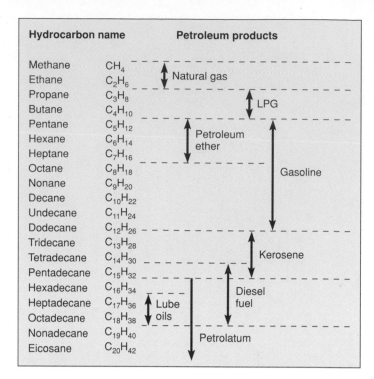

Figure 14.11

Petroleum products and the ranges of hydrocarbons in each product.

Table 14.3

Petroleum products

Name	Boiling Range (°C)	Carbon Atoms per Molecule
Natural Gas	Less than 0	C_1 to C_4
Petroleum Ether	35–100	C_5 to C_7
Gasoline	35–215	C_5 to C_{12}
Kerosene	35–300	C_{12} to C_{15}
Diesel Fuel	300–400	C_{15} to C_{18}
Motor Oil, Grease	350–400	C_{16} to C_{18}
Paraffin	Solid, melts at about 55	C_{20}
Asphalt	Boiler residue	C_{40} or more

oil field. The composition of petroleum varies from one oil field to the next. The oil from a given field might be dark or light in color, and it might have an asphalt base or a paraffin base. Some oil fields contain oil with a high quantity of sulfur, referred to as "sour crude." Because of such variations, some fields have oil with more desirable qualities than oil from other fields.

In some locations an oil field occurs close to the surface and petroleum seeps to the surface, often floating on water from a spring. Such seepage is the source of petroleum that has been collected and used since about 3000 B.C. Ancient Babylonians, Egyptians, and Roman civilizations used this oil for medicinal purposes, for paving roads, and when thickened by drying, as a caulking compound in early wooden ships.

Early settlers found oil seeps in the eastern United States and collected the oil for medicinal purposes. One enterprising oil peddler tried to improve the taste by running the petroleum through a whiskey still. He obtained a clear liquid by distilling the petroleum and, by accident, found that the liquid made an excellent lamp oil. This was fortunate timing, for the lamp oil used at that time was whale oil, and whale oil production was declining. This clear liquid obtained by distilling petroleum is today known as kerosene.

The first oil well was drilled in Titusville, Pennsylvania, in 1859. The well struck oil at a depth of seventy feet and produced two thousand barrels a year. This is not much compared to the billions of barrels produced per year today, but it had an economic impact in 1859. Before the well was drilled, oil was selling for $40 a barrel. Two years later the price was 10¢ a barrel. A "barrel of oil" is an accounting measure of forty-two United States gallons. Such a barrel size does not really exist. When or if oil is shipped in barrels, each drum holds fifty-five United States gallons.

Wells were drilled and crude oil refineries were built to produce the newly discovered lamp oil. Gasoline was a by-product of the distillation process and was used primarily as a spot remover. With Henry Ford's automobile production and Edison's electric light invention, the demand for gasoline increased and the demand for kerosene decreased. The refineries were converted to produce gasoline, and the petroleum industry grew to become one of the world's largest industries.

Crude oil is petroleum that is pumped from the ground, a complex and variable mixture of hydrocarbons with one or more carbon atoms, with an upper limit of about fifty atoms. This thick, smelly black mixture is not usable until it is refined, that is, separated into usable groups of hydrocarbons called petroleum products. The petroleum products are separated by distillation, and any particular product has a boiling point range, or "cut" of the distilled vapors (figure 14.10). Thus, each product, such as gasoline, heating oil, and so forth is made up of hydrocarbons within a range of carbon atoms per molecule (figure 14.11). The products, their boiling ranges, and ranges of carbon atoms per molecule are listed in table 14.3.

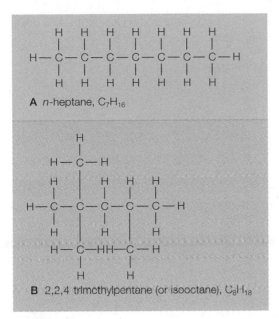

Figure 14.12

The octane rating scale is a description of how rapidly gasoline burns. It is based on (a) n-heptane, with an assigned octane number of 0 and (b) 2,2,4-trimethylpentane, with an assigned number of 100.

The hydrocarbons that have one to four carbon atoms (CH_4 to C_4H_{10}) are gases at room temperature. They can be pumped from certain wells as a gas, but they also occur dissolved in crude oil. *Natural gas* is a mixture of hydrocarbon gases, but it is about 95 percent methane (CH_4). Propane (C_3H_8) and butane (C_4H_{10}) are liquified by compression and cooling and are sold as liquified petroleum gas, or *LPG*. LPG is used where natural gas is not available for cooking or heating and is widely used as a fuel in barbecue grills and camp stoves.

Hydrocarbons with five to seven carbon atoms per molecule are volatile liquids at room temperature. Different groups of these closely related volatile hydrocarbons are used for various commercial purposes under the general heading of *petroleum ether,* also called "petroleum distillates" and naphtha. Petroleum ether is used as a cleaning fluid. It is also used as a solvent. Naphtha is also present in gasoline.

Gasoline is a mixture of volatile hydrocarbons with five to twelve carbon atoms per molecule. Gasoline distilled from crude oil consists mostly of straight-chain molecules not suitable for use as an automotive fuel. Straight-chain molecules burn too rapidly in an automobile engine, producing more of an explosion than a smooth burn. You hear these explosions as a knocking or pinging in the engine, and they mean poor efficiency and they could damage the engine. On the other hand, branched chain molecules burn comparatively slower, without the pinging or knocking explosions. The burning rate of gasoline is described by the *octane number* scale. The scale is based on pure *n*-heptane, straight-chain molecules that are assigned an octane number of 0, and a multiple branched isomer of octane, 2,2,4-trimethylpentane, which is assigned an octane number of

100 (figure 14.12). Most unleaded gasolines have an octane rating of 87, which could be obtained with a mixture that is 87 percent 2,2,4-trimethylpentane and 13 percent *n*-heptane. Gasoline, however, is a much more complex mixture.

The octane rating of gasoline can be improved one of two ways: (1) by adding a substance that slows the burning rate, such as tetraethyl lead, $(C_2H_5)_4Pb$ ("ethyl"), or (2) by converting some of the straight-chain hydrocarbons into branched-chain ones. The use of tetraethyl lead is less expensive but is being phased out because of increased concerns over lead pollution and because lead poisons the antipollution catalytic converter (see the Feature at the end of this chapter). It is more expensive to produce unleaded gasoline because some of the straight-chain hydrocarbon molecules must be converted into branched molecules. The process is one of "cracking and reforming" some of the straight-chain molecules. First, the gasoline is passed through metal tubes heated to 500°C to 800°C (932°F to 1,470°F). At this high temperature, and in the absence of oxygen, the hydrocarbon molecules decompose by breaking into smaller carbon-chain units. These smaller hydrocarbons are then passed through tubes containing a catalyst, which causes them to reform into branched-chain molecules. Unleaded gasoline is produced by the process. Without the reforming that produces unleaded gasoline, low-numbered hydrocarbons (such as ethylene) can be produced. Ethylene is used as a raw material for many plastic materials, antifreeze, and other products. Cracking is also used to convert higher-numbered hydrocarbons, such as heating oil, into gasoline.

Kerosene is a mixture of hydrocarbons that have from twelve to fifteen carbon atoms. The petroleum product called kerosene is also known by other names, depending on its use. Some of these names are lamp oil (with coloring and odorants added), jet fuel (with a flash flame retardant added), heating oil, #1 fuel oil, and in some parts of the country, "coal oil."

Diesel fuel is a mixture of a group of hydrocarbons that have from fifteen to eighteen carbon atoms per molecule. Diesel fuel also goes by other names, again depending on its use. This group of hydrocarbons is called diesel fuel, distillate fuel oil, heating oil, or #2 fuel oil. During the summer season there is a greater demand for gasoline than for heating oil, so some of the supply is converted to gasoline by the cracking process.

Motor oil and *lubricating oils* have sixteen to eighteen carbon atoms per molecule. Lubricating grease is heavy oil that is thickened with soap. *Petroleum jelly,* also called petrolatum (or Vaseline), is a mixture of hydrocarbons with sixteen to thirty-two carbon atoms per molecule. *Mineral oil* is a light lubricating oil that has been decolorized and purified.

Depending on the source of the crude oil, varying amounts of *paraffin* wax (C_{20} or greater) or *asphalt* (C_{36} or more) may be present. Paraffin is used for candles, waxed paper, and home canning. Asphalt is mixed with gravel and used to surface roads. It is also mixed with refinery residues and lighter oils to make a fuel called #6 fuel oil or residual fuel oil. Industries and utilities often use this semisolid material that must be heated before it will pour. Number 6 fuel oil is used as a boiler fuel, costing about half as much as #2 fuel oil.

Organic Chemistry

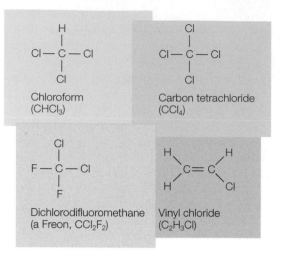

Figure 14.13
Common examples of organic halides.

Hydrocarbon Derivatives

The hydrocarbons account for only about 5 percent of the known organic compounds, but the other 95 percent can be considered as hydrocarbon derivatives. **Hydrocarbon derivatives** are formed when *one or more hydrogen atoms on a hydrocarbon have been replaced by some element or group of elements other than hydrogen.* For example, the halogens (F_2, Cl_2, Br_2) react with an alkane in sunlight or when heated, replacing a hydrogen:

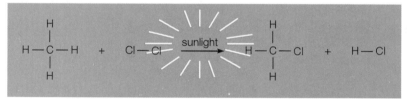

In this particular *substitution reaction* a hydrogen atom on methane is replaced by a chlorine atom to form methyl chloride. Replacement of any number of hydrogen atoms is possible, and a few *organic halides* are illustrated in figure 14.13.

If a hydrocarbon molecule is unsaturated (has a multiple bond), a hydrocarbon derivative can be formed by an *addition reaction:*

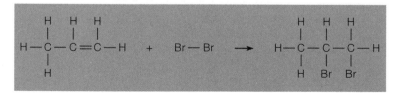

The bromine atoms add to the double bond on propene, forming 1,2-dibromopropane.

Alkene molecules can also add to each other in an addition reaction to form a very long chain consisting of hundreds of molecules. A long chain of repeating units is called a **polymer,** and the reaction is called *addition polymerization.* Ethylene,

for example, is heated under pressure with a catalyst to form *polyethylene*. Heating breaks the double bond,

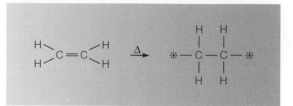

which provides sites for single covalent bonds to join the ethylene units together,

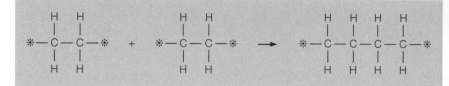

which continues the addition polymerization until the chain is hundreds of units long. Synthetic polymers such as polyethylene are discussed in a later section.

The addition reaction and the addition polymerization reaction can take place because of the double bond of the alkenes, and, in fact, the double bond is the site of most alkene reactions. The atom or group of atoms in an organic molecule that is the site of a chemical reaction is identified as a **functional group.** *It is the functional group that is responsible for the chemical properties of an organic compound.* Functional groups usually have (1) multiple bonds or (2) lone pairs of electrons that cause them to be sites of reactions. Table 14.4 lists some of the common hydrocarbon functional groups. Look over this list, comparing the structure of the functional group with the group name. Some of the more interesting examples from a few of these groups will be considered next. Note that the R and R′ stand for one or more of the hydrocarbon groups from table 14.2. For example, in the reaction between methane and chlorine, the product is methyl chloride. In this case the R in RCl stands for methyl, but it could represent any hydrocarbon group.

Alcohols

An **alcohol** is an organic compound formed by replacing one or more hydrogens on an alkane with a hydroxyl functional group (−OH). The hydroxyl group should not be confused with the hydroxide ion, OH^-. The hydroxyl group is attached to an organic compound and does not form ions in solution as the hydroxide ion does. It remains attached to a hydrocarbon group (R), giving the compound its set of properties that are associated with alcohols.

The name of the hydrocarbon group (table 14.2) determines the name of the alcohol. If the hydrocarbon group in ROH is methyl, for example, the alcohol is called *methyl alcohol.* Using the IUPAC naming rules, the name of an alcohol has the suffix "-ol." Thus, the IUPAC name of methyl alcohol is *methanol.* If the molecule has a sufficient number of carbon atoms that further definition is needed, the base name is determined

Table 14.4

Selected organic functional groups

Name of Functional Group	General Formula	General Structure
Organic Halide	RCl	R — C̈l :
Alcohol	ROH	R — Ö — H
Ether	ROR'	R — Ö — R'
Aldehyde	RCHO	R — C — H, ‖ :O:
Ketone	RCOR'	R — C — R', ‖ :O:
Organic Acid	RCOOH	R — C — Ö — H, ‖ :O:
Ester	RCOOR'	R — C — Ö — R', ‖ :O:
Amine	RNH₂	R — N̈ — H, │ H

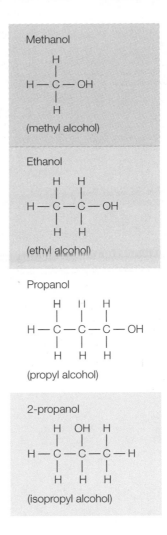

Methanol

$$H-C-OH$$

(methyl alcohol)

Ethanol

(ethyl alcohol)

Propanol

(propyl alcohol)

2-propanol

(isopropyl alcohol)

Figure 14.14

Four different alcohols. The IUPAC name is given above each structural formula and the common name is given below.

from the longest continuous chain of carbon atoms that has the —OH. The location of the hydroxyl group is identified with a number (figure 14.14).

All alcohols have the hydroxyl functional group and all are chemically similar. Alcohols are toxic to humans, for example, except that ethanol can be consumed in limited quantities. Consumption of other alcohols such as methanol or 2-propanol (isopropyl alcohol) can result in blindness and death. Ethanol, C_2H_5OH, is produced by the action of yeast or by a chemical reaction of ethylene derived from petroleum refining. Yeast acts on sugars to produce ethanol and CO_2. When beer, wine, and other such beverages are the desired product, the CO_2 escapes during fermentation and the alcohol remains in solution. In baking, the same reaction utilizes the CO_2 to make the dough rise and the alcohol is evaporated during baking. Most alcoholic beverages are produced by the yeast fermentation reaction, but some are made from ethanol derived from petroleum refining.

The hydroxyl group is strongly polar, and alcohols are soluble in both alkanes and water. A solution of ethanol and gas-

oline is called **gasohol** (figure 14.15). Alcoholic beverages are a solution of ethanol and water. The **proof** of such a beverage is double the ethanol concentration by volume. Therefore, a solution of 40 percent ethanol by volume in water is 80 proof, and wine that is 12 percent alcohol by volume is 24 proof. Distillation alone will produce a 190 proof concentration, but other techniques are necessary to obtain 200 proof absolute alcohol. *Denatured alcohol* is ethanol with acetone, formaldehyde, and other chemicals in solution that are difficult to separate by distillation. Since these denaturants make consumption impossible, denatured alcohol is sold without the consumption tax.

Methanol, ethanol, and isopropyl alcohol each has one hydroxyl group per molecule. An alcohol with two hydroxyl groups per molecule is called a **glycol.** Ethylene glycol is perhaps the best-known glycol since it is used as an antifreeze. An alcohol with three hydroxyl groups per molecule is called **glycerol** (or glycerin). Glycerol is a by-product in the making of soap. It is added to toothpastes, lotions, and some candies to retain moisture and softness. Ethanol, ethylene glycol, and glycerol are compared in figure 14.16.

Organic Chemistry

Figure 14.15

Gasoline is a mixture of hydrocarbons (C_8H_{18} for example), which contain no atoms of oxygen. Gasohol contains ethyl alcohol, C_2H_5OH, which does contain oxygen. The addition of alcohol to gasoline, therefore, adds oxygen to the fuel. Since carbon monoxide forms when there is an insufficient supply of oxygen, the addition of alcohol to gasoline helps cut down on carbon monoxide emissions.

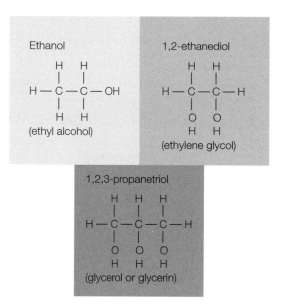

Figure 14.16

Common examples of alcohols with one, two, and three hydroxyl groups per molecule. The IUPAC name is given above each structural formula and the common name is given below.

Glycerol reacts with nitric acid in the presence of sulfuric acid to produce glyceryl trinitrate, commonly known as *nitroglycerine*. Nitroglycerine is a clear oil that is violently explosive, and when warmed, it is extremely unstable. In 1867, Alfred Nobel discovered that a mixture of nitroglycerine and siliceous earth was more stable than pure nitroglycerine but was nonetheless explosive. The mixture is packed in a tube and is called *dynamite*. Old dynamite tubes, however, leak pure nitroglycerine that is again sensitive to a slight shock.

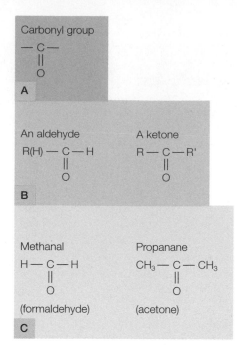

Figure 14.17

The carbonyl group (*a*) is present in both aldehydes and ketones as shown in (*b*). (*c*) The simplest example of each with the IUPAC name above and the common name below each formula.

Ethers, Aldehydes, and Ketones

An **ether** has a general formula of ROR', and the best-known ether is diethylether. In a molecule of diethylether, both the R and the R' are ethyl groups. Diethylether is a volatile, highly flammable liquid that was used as an anesthetic in the past. Today, it is used as an industrial and laboratory solvent.

Aldehydes and ketones both have a functional group of a carbon atom doubly bonded to an oxygen atom called a *carbonyl group*. The **aldehyde** has a hydrocarbon group, R (or a hydrogen in one case), and a hydrogen attached to the carbonyl group. A **ketone** has a carbonyl group with two hydrocarbon groups attached (figure 14.17).

The simplest aldehyde is *formaldehyde*. Formaldehyde is soluble in water, and a 40 percent concentration called *formalin* has been used as an embalming agent and to preserve biological specimens. Formaldehyde is also a raw material used to make plastics such as Bakelite. All the aldehydes have odors, and the odors of some aromatic hydrocarbons include the odors of almonds, cinnamon, and vanilla. The simplest ketone is *acetone*. Acetone has a fragrant odor and is used as a solvent in paint removers and nail polish removers.

Organic Acids and Esters

Mineral acids, such as hydrochloric and sulfuric acid, are made of inorganic materials. Acids that were derived from organisms are called **organic acids.** Because many of these organic acids can be formed from fats, they are sometimes called *fatty acids*. Chemically, they are known as the *carboxylic acids* because they contain the carboxyl functional group, $-COOH$, and have a general formula of RCOOH.

Figure 14.18

These red ants, as other ants, make the simplest of the organic acids, formic acid. The sting of bees, ants, and some plants contains formic acid, along with some other irritating materials. Formic acid is HCOOH.

The simplest carboxylic acid has been known since the Middle Ages, when it was isolated by the distillation of ants. The Latin word *formica* means "ant," so this acid was given the name *formic acid* (figure 14.18). Formic acid is

$$H - C - OH$$
$$\parallel$$
$$O$$

It is formic acid, along with other irritating materials, that causes the sting of bees, ants, and certain plants.

Acetic acid, the acid of vinegar, has been known since antiquity. Acetic acid forms from the oxidation of ethanol. An oxidized bottle of wine contains acetic acid in place of the alcohol, which gives the wine a vinegar taste. Before wine is served in a restaurant, the person ordering is customarily handed the bottle cork and a glass with a small amount of wine. You first break the cork in half to make sure it is dry, which tells you that the wine has been sealed from oxygen. The small sip is to taste for vinegar before the wine is served. If the wine has been oxidized, the reaction is

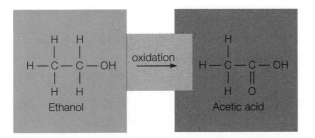

Organic acids are common in many foods. The juice of citrus fruit, for example, contains citric acid, which relieves a thirsty feeling by stimulating the flow of saliva. Lactic acid is found in sour milk, buttermilk, sauerkraut, and pickles. Lactic acid also forms in your muscles as a product of carbohydrate metabolism, causing a feeling of fatigue. Citric and lactic acids are small molecules compared to some of the carboxylic acids

Table 14.5

Flavors and esters

Ester Name	Formula	Flavor
Amyl Acetate	$CH_3 - \underset{\underset{O}{\parallel}}{C} - O - C_5H_{11}$	Banana
Octyl Acetate	$CH_3 - \underset{\underset{O}{\parallel}}{C} - O - C_8H_{17}$	Orange
Ethyl Butyrate	$C_3H_7 - \underset{\underset{O}{\parallel}}{C} - O - C_2H_5$	Pineapple
Amyl Butyrate	$C_3H_7 - \underset{\underset{O}{\parallel}}{C} - O - C_5H_{11}$	Apricot
Ethyl Formate	$H - \underset{\underset{O}{\parallel}}{C} - O - C_2H_5$	Rum

that are formed from fats. Palmitic acid, for example, is $C_{16}H_{32}O_2$ and comes from palm oil. The structure of palmitic acid is a chain of fourteen CH_2 groups with CH_3^- at one end and $-COOH$ at the other. Again, it is the functional carboxyl group, $-COOH$, that gives the molecule its acid properties. Organic acids are also raw materials used in the making of polymers of fabric, film, and paint.

Esters are common in both plants and animals, giving fruits and flowers their characteristic odor and taste. Esters are also used in perfumes and artificial flavorings. A few of the flavors that particular esters are responsible for are listed in table 14.5. These liquid esters can be obtained from natural sources or they can be chemically synthesized. Whatever the source, amyl acetate, for example, is the chemical responsible for what you identify as the flavor of banana. Natural flavors, however, are complex mixtures of these esters along with other organic compounds. Lower molecular weight esters are fragrant-smelling liquids, but higher molecular weight esters are odorless oils and fats. These are discussed in the next section along with carbohydrates and proteins.

Organic Compounds of Life

Aristotle and the later proponents of the vitalist theory were *partly* correct in their concept that living organisms are different from inorganic substances made of the same elements. Living organisms, for example, have the ability to (1) exchange matter and energy with their surroundings and (2) transform matter and energy into different forms as they (3) respond to changes in their surroundings. In addition, living organisms can use the transformed matter and energy to (4) grow and (5) reproduce. Living organisms are able to do these things through a great variety of organic reactions that are catalyzed by enzymes, however, and not through some mysterious "vital force." These enzyme-regulated organic reactions take place because living organisms are highly organized and have an incredible number of relationships between many different chemical processes.

The chemical processes regulated by living organisms begin with relatively small organic molecules and water. The organism uses energy and matter from the surroundings to build large macromolecules. A **macromolecule** is a very large molecule that is a combination of many smaller, similar molecules joined together in a chainlike structure. Macromolecules have molecular weights of thousands or millions of atomic mass units. There are three main types of macromolecules: (1) proteins, (2) carbohydrates, and (3) nucleic acids, in addition to fats. A living organism, even a single-cell organism such as a bacterium, contains six thousand or so different kinds of macromolecules. The basic unit of an organism is called a *cell*. Cells are made of macromolecules that are formed inside the cell. The cell decomposes organic molecules taken in as food and uses energy from the food molecules to build more macromolecules. The process of breaking down organic molecules and building up macromolecules is called *metabolism*. Through metabolism, the cell grows, then divides into two cells. Each cell is an exact duplicate of the other, even down to the number and kinds of macromolecules contained within. Each new cell continues the process of growth, then reproduces again, making more cells. This is the basic process of life. The complete process is complicated and very involved, easily filling a textbook in itself, so the details will not be presented here. The following discussion will be limited to three groups of organic molecules involved in the process: proteins, carbohydrates, and fats and oils.

Proteins

Proteins are macromolecular polymers made up of smaller molecules of amino acids. These very large macromolecules have molecular weights that vary from about six thousand to fifty million. Some proteins are simple straight-chain polymers of amino acids, but others contain metal ions such as Fe^{2+} or parts of organic molecules derived from vitamins. Proteins serve as major structural and functional materials in animals. *Structurally,* proteins are major components of muscles, connective tissue, and the skin, hair, and nails. *Functionally,* some proteins are enzymes, which catalyze metabolic reactions; hormones, which regulate body activities; hemoglobin, which carries oxygen to cells; and antibodies, which protect the body.

Proteins are formed from 20 **amino acids,** which are organic acid functional groups with the general formula of

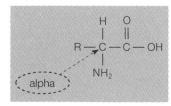

Note the carbon atom labeled "alpha" in the general formula. The amino functional group (NH_2) is attached to this carbon atom, which is next to the carboxylic group (COOH). This arrangement is called an *alpha-amino acid,* and the building blocks of proteins are all alpha-amino acids. The 20 amino acids differ in the nature of the R group, also called the *side chain*. It is the side chain that determines the properties of a protein. Figure 14.19 gives the structural formula for the 20 amino acids found in most proteins and the three-letter abbreviations of the name of each amino acid.

Amino acids are linked to form a protein by a peptide bond between the amino group of one amino acid and the carboxyl group of a second amino acid. A polypeptide is a polymer formed from linking many amino acid molecules. If the polypeptide has a role as a biological structure or function, it is called a *protein.* A protein chain can consist of different combinations of the 20 amino acids with hundreds or even thousands of amino acid molecules held together with peptide bonds (figure 14.20). The arrangement or sequence of these amino acid molecules determines the structure that gives the protein its unique set of biochemical properties. Insulin, for example, is a protein hormone that biochemically regulates the blood sugar level. Insulin contains 86 amino acids that begin (at the amino group) with phenylalanine, valine, asparagine, and then 83 other amino acid molecules in the chain. Hemoglobin is the protein that carries oxygen in the bloodstream, and its biochemical characteristics are determined by its chain of 146 amino acid molecules.

Carbohydrates

Carbohydrates are an important group of organic compounds that includes sugars, starches, and cellulose, and they are important in plants and animals for structure, protection, and food. Cellulose is the skeletal substance of plants and plant materials, and chitin is a similar material that forms the hard, protective covering of insects and shellfish such as crabs and lobsters. *Glucose,* $C_6H_{12}O_6$, is the most abundant carbohydrate and serves as a food and a basic building block for other carbohydrates.

Carbohydrates were named when early studies found that water vapor was given off and carbon remained when sugar was heated. The name *carbohydrate* literally means "watered carbon," and the empirical formulas for most carbohydrates indeed indicate carbon (C) and water (H_2O). Glucose, for example, could be considered to be six carbons with six waters, or $C_6(H_2O)_6$. However, carbohydrate molecules are more complex than just water attached to a carbon atom. They are polyhydroxyl aldehydes and ketones, two of which are illustrated in figure 14.21. The two carbohydrates in this illustration belong to a group of carbohydrates known as **monosaccharides,** or *simple sugars*. They are called simple sugars because they are mostly 6-carbon molecules such as glucose and fructose. Glucose (also called dextrose) is found in the sap of plants, and in the human bloodstream it is called *blood sugar*. Corn syrup, which is often used as a sweetener, is mostly glucose. Fructose, as its name implies, is the sugar that occurs in fruits and it is sometimes called *fruit sugar*. Both glucose and fructose have the same molecular formula, but glucose is an aldehyde sugar

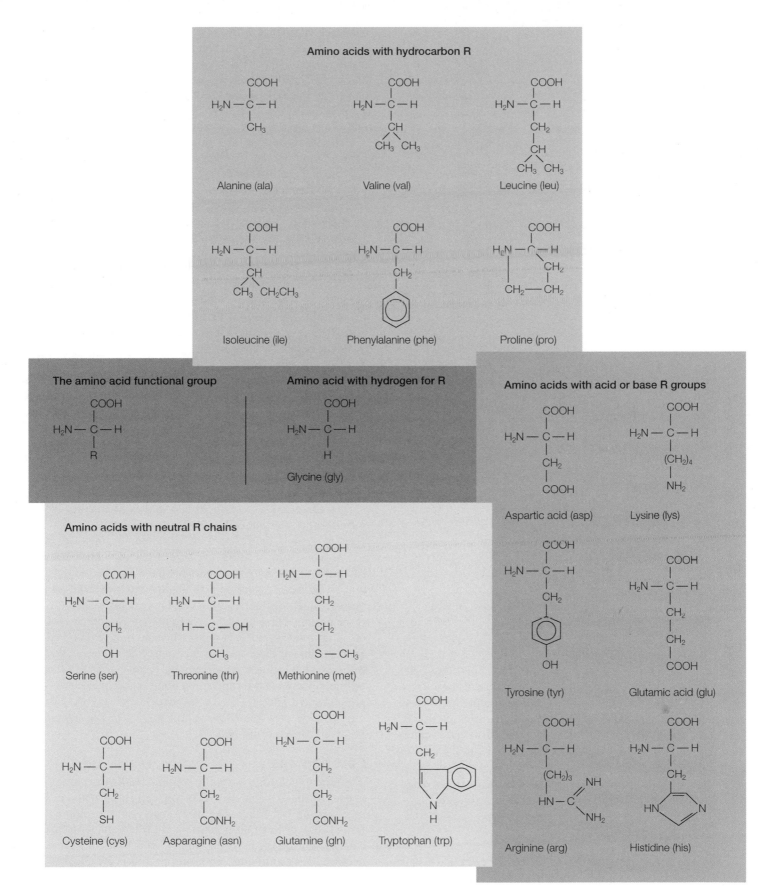

Figure 14.19

The twenty amino acids that make up proteins, with three-letter
abbreviations. The carboxyl group of one amino acid bonds with the
amino group of a second acid to yield a dipeptide and water. Proteins are
polypeptides.

Organic Chemistry

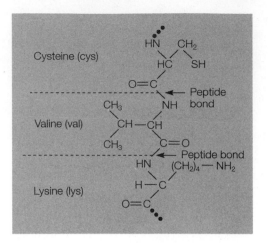

Figure 14.20

Part of a protein polypeptide made up of the amino acids cysteine (cys), valine (val), and lysine (lys). A protein can have from fifty to one thousand of these amino acid units, with each protein having its own unique sequence.

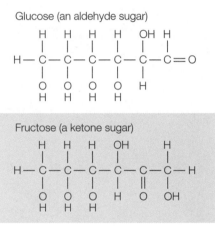

Figure 14.21

Glucose (blood sugar) is an aldehyde and fructose (fruit sugar) is a ketone. Both have a molecular formula of $C_6H_{12}O_6$.

and fructose is a ketone sugar (figure 14.21). A mixture of glucose and fructose is found in honey. This mixture also is formed when table sugar is reacted with water in the presence of an acid, a reaction that takes place in the preparation of canned fruit and candies. The mixture of glucose and fructose is called *invert sugar*. Invert sugar is about twice as sweet to the taste as the same amount of sucrose.

Two monosaccharides are joined together to form **disaccharides** with the loss of a water molecule, for example,

$$\underset{\text{glucose}}{C_6H_{12}O_6} + \underset{\text{fructose}}{C_6H_{12}O_6} \rightarrow \underset{\text{sucrose}}{C_{12}H_{22}O_{11}} + H_2O$$

The most common disaccharide is *sucrose,* or ordinary table sugar. Sucrose occurs in high concentrations in sugar cane and sugar beets. It is extracted by crushing the plant materials, then dissolving the sucrose from the materials with water. The water is evaporated and the crystallized sugar is decolorized with charcoal to produce white sugar. Other common disaccharides include *lactose* (milk sugar) and *maltose* (malt sugar). All three disaccharides have similar properties, but maltose tastes only about one-third as sweet as sucrose. Lactose tastes only about one-sixth as sweet as sucrose. No matter which sugar is consumed (sucrose, lactose, or maltose), it is converted into glucose and transported by the bloodstream for use by the body.

Polysaccharides are polymers consisting of monosaccharide units joined together in straight or branched chains. Polysaccharides are the energy-storage molecules of plants and animals (starch and glycogen) and the structural-building molecules of plants (cellulose). **Starch** is a group of complex carbohydrates that plants use as a stored food source. Potatoes, rice, corn, and wheat contain starch granules and serve as an important source of food for humans. The human body breaks down the starch molecules to glucose, which is transported by the bloodstream and utilized just like any other glucose. This digestive process begins with enzymes secreted with saliva in the mouth. You may have noticed a result of this enzyme catalyzed reaction as you eat bread. If you chew the bread for awhile it begins to taste sweet.

Plants store sugars in the form of starch polysaccharides, and animals store sugars in the form of the polysaccharide **glycogen.** Glycogen is a starchlike polysaccharide that is synthesized by the human body and stored in the muscles and liver. Glycogen, like starch, is a very high molecular weight polysaccharide but it is more highly branched. These highly branched polysaccharides serve as a direct reserve source of energy in the muscles. In the liver, they serve as a reserve source to maintain the blood sugar level.

Cellulose is a polysaccharide that is abundant in plants, forming the fibers in cell walls that preserves the structure of plant materials (figure 14.22). Cellulose molecules are straight chains, consisting of large numbers of glucose units. These glucose units are arranged very similar to the glucose units of starch but with differences in the bonding arrangement that holds the glucose units together (figure 14.23). This difference turns out to be an important one where humans are concerned, because enzymes that break down starches do not affect cellulose. Humans do not have the necessary enzymes to break down the cellulose chain (digest it), so humans receive no food value from cellulose. Cattle and termites that do utilize cellulose as a source of food have bacteria (with the necessary enzymes) in their digestive systems. Cellulose is still needed in the human diet, however, for fiber and bulk.

Fats and Oils

The human body can normally synthesize all of the amino acids needed to build proteins except for eight called the *essential amino acids*. An adequate diet must contain the eight essential amino acids or health problems result. Meat and dairy products usually provide the essential amino acids, but they can also be acquired by combining cereal grains (corn, wheat, rice, etc.) with a legume (beans, peanuts, etc.). Interestingly, many ethnic foods have such a combination, for example, corn and beans (Mexican), rice and soybeans (tofu) (Japanese), and rice and red beans (Cajun).

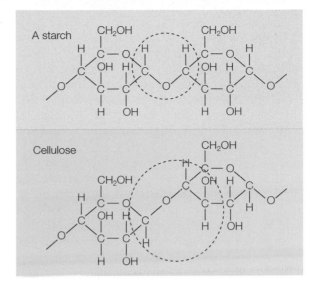

Figure 14.23

Starch and cellulose are both polymers of glucose, but humans cannot digest cellulose. The difference in the bonding arrangement might seem minor, but enzymes must fit a molecule very precisely. Thus, enzymes that break down starch do nothing to cellulose.

Figure 14.22

These plants and their flowers are made up of a mixture of carbohydrates that were manufactured from carbon dioxide and water, with the energy of sunlight. The simplest of the carbohydrates are the monosaccharides, simple sugars (fruit sugar) that the plant synthesizes. Food is stored as starches, which are polysaccharides made from the simpler monosaccharides. The plant structure is held upright by fibers of cellulose, another form of a polysaccharide.

Cereal grains and legumes also provide carbohydrates, the human body's preferred food for energy. When a sufficient amount of carbohydrates are consumed, the body begins to store some of its energy source in the form of glycogen in the muscles and liver. Beyond this storage for short-term needs, the body begins to store energy in a different chemical form for longer-term storage. This chemical form is called **fats** in animals and **oils** in plants. Fats and oils are esters formed from glycerol (1,2,3-trihydroxypropane) and three long-chain carboxylic acids (fatty acids). This ester is called a **triglyceride,** and its structural formula is shown in figure 14.24. Fats are solids and oils are liquids at room temperature, but they both have this same general structure.

Fats and oils usually have two or three different fatty acids, and several are listed in table 14.6. Animal fats can be either saturated or unsaturated but most are saturated. Oils are liquids at room temperature because they contain a higher number of unsaturated units. These unsaturated oils (called "poly" unsaturated in news and advertisements), such as safflower and corn oils, are used as liquid cooking oils because unsaturated oils are believed to lead to lower cholesterol levels in the bloodstream. Saturated fats, along with cholesterol, are believed to contribute to hardening of the arteries over time.

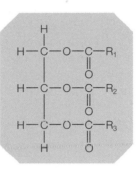

Figure 14.24

The triglyceride structure of fats and oils. Note the glycerol structure on the left and the ester structure on the right. Also notice that R_1, R_2, and R_3 are carboxylic acid (fatty acid) groups.

Table 14.6

Some fatty acids occurring in fats

Common Name	Condensed Structure	Source
Lauric Acid	$CH_3(CH_2)_{10}COOH$	Coconuts
Palmitic Acid	$CH_3(CH_2)_{14}COOH$	Palm oil
Stearic Acid	$CH_3(CH_2)_{16}COOH$	Animal fats
Oleic Acid	$CH_3(CH_2)_7CH = CH(CH_2)_7COOH$	Corn oil
Linoleic Acid	$CH_3(CH_2)_4CH = CHCH_2 = CH(CH_2)_7COOH$	Soybean oil
Linolenic Acid	$CH_3CH_2(CH = CHCH_2)_3(CH_2)_6COOH$	Fish oils

Organic Chemistry

Cooking oils from plants, such as corn and soybean oil, are hydrogenated to convert the double bonds of the unsaturated oil to the single bonds of a saturated one. As a result, the liquid oils are converted to solids at room temperature. For example, one brand of oleomargarine lists ingredients as "liquid soybean oil (nonhydrogenated) and partially hydrogenated cottonseed oil with water, salt, preservatives, and coloring." Complete hydrogenation would result in a hard solid, so the cottonseed oil is partially hydrogenated, then mixed with liquid soybean oil. Coloring is added because oleo is white, not the color of butter. Vegetable shortening is the very same product without added coloring. Reaction of a triglyceride with a strong base such as KOH or NaOH yields a fatty acid of salt and glycerol. A sodium or potassium fatty acid is commonly known as *soap*.

Excess food from carbohydrate, protein, or fat and oil sources is converted to fat for long-term energy storage in *adipose tissue*, which also serves to insulate and form a protective padding. In terms of energy storage, fats yield more than twice the energy per gram oxidized than carbohydrates or proteins.

Synthetic Polymers

Polymers are huge, chainlike molecules made of hundreds or thousands of smaller, repeating molecular units called *monomers*. Polymers occur naturally in plants and animals. Cellulose, for example, is a natural plant polymer made of glucose monomers. Wool and hair are natural animal polymers made of protein monomers. Synthetic polymers are now manufactured from a wide variety of substances, and you are familiar with these polymers as synthetic fibers such as nylon and the inexpensive light plastic used for wrappings and containers (figure 14.25).

The first synthetic polymer was a modification of the naturally existing cellulose polymer. Cellulose was chemically modified in 1862 to produce celluloid, the first *plastic*. The term "plastic" means that celluloid could be molded to any desired shape. Celluloid was produced by first reacting cotton with a mixture of nitric and sulfuric acids, which produced an ester of cellulose nitrate. This ester is an explosive compound known as "guncotton," or smokeless gunpowder. When made with ethanol and camphor, the product is less explosive and can be formed and molded into useful articles. This first plastic, celluloid, was used to make dentures, combs, glasses frames, and photographic film. Before the discovery of celluloid, many of these articles, including dentures, were made from wood. Today, only Ping-Pong balls are made from cellulose nitrate.

Cotton reacted with acetic acid and sulfuric acid produces a cellulose acetate ester. This polymer, through a series of chemical reactions, produces viscose rayon filaments when forced through small holes. The filaments are twisted together to form viscose rayon thread. When forced through a thin slit, a sheet is formed rather than filaments, and the transparent sheet is called *cellophane*. Both rayon and cellophane, as celluloid, are manufactured by modifying the natural polymer of cellulose.

The first truly synthetic polymer was produced in the early 1900s by reacting two chemicals with relatively small molecules rather than modification of a natural polymer. Phenol, an aromatic hydrocarbon, was reacted with formaldehyde, the simplest aldehyde, to produce the polymer named *Bakelite*. Bakelite is a *thermosetting* material that forms cross-links between the polymer chains. Once the links are formed during production, the plastic becomes permanently hardened and cannot be softened or made to flow. Some plastics are *thermoplastic* polymers and soften during heating and harden during cooling because they do not have cross-links.

Polyethylene is a familiar thermoplastic polymer used for vegetable bags, dry cleaning bags, grocery bags, and plastic squeeze bottles. Polyethylene is a polymer produced by a polymerization reaction of ethylene, which is derived from petroleum. Polyethylene was invented just before World War II and was used as an electrical insulating material during the war. Today, there are many variations of polyethylene that are produced by different reaction conditions or by substitution of one or more hydrogen atoms in the ethylene molecule. When soft polyethylene near the melting point is rolled in alternating perpendicular directions or expanded and compressed as it is cooled, the polyethylene molecules become ordered in such a way to improve the rigidity and tensile strength. This change in the microstructure produces *high-density polyethylene* with a superior rigidity and tensile strength compared to *low-density polyethylene*. High-density polyethylene is used as liners in screw-on jar tops, bottle caps, and as a material for toys.

The properties of polyethylene are changed by replacing one of the hydrogen atoms in a molecule of ethylene. If the hydrogen is replaced by a chlorine atom the compound is called vinyl chloride, and the polymer formed from vinyl chloride is

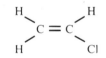

polyvinyl chloride (PVC). Polyvinyl chloride is used to make plastic water pipes, synthetic leather, and other vinyl products. It differs from the waxy plastic of polyethylene because of the chlorine atom that replaces hydrogen on each monomer.

Replacement of a hydrogen atom with a benzene ring makes a monomer called *styrene*. Styrene is

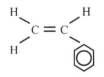

and polymerization of styrene produces *polystyrene*. Polystyrene is puffed full of air bubbles to produce the familiar Styrofoam coolers, cups, and insulating materials.

If all hydrogens of an ethylene molecule are replaced with atoms of fluorine, the product is polytetrafluorethylene, a tough

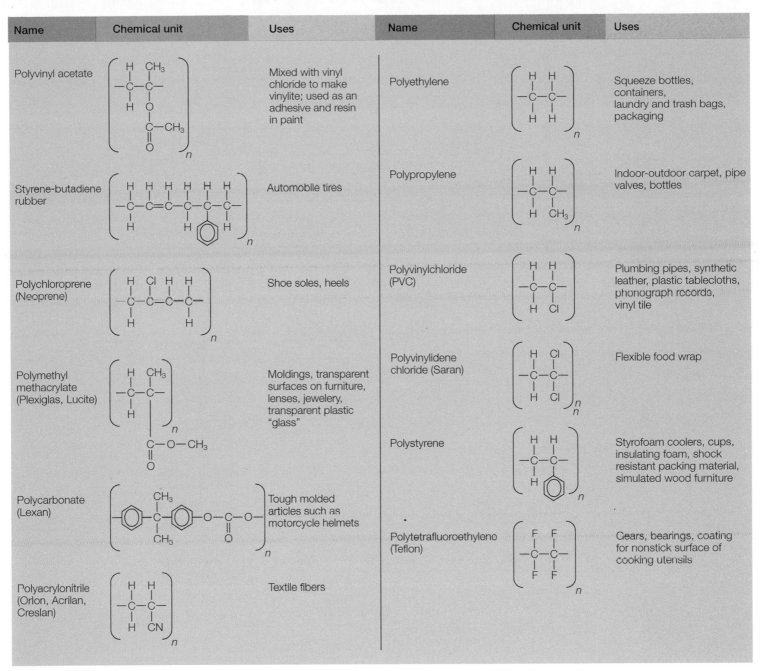

Figure 14.25

Synthetic polymers, the polymer unit, and some uses of each polymer.

plastic that resists high temperatures and acts more like a metal than a plastic. Since it has a low friction it is used for bearings, gears, and as a nonsticking coating on frying pans. You probably know of this plastic by its trade name of *Teflon.*

There are many different polymers in addition to PVC, Styrofoam, and Teflon, and the monomers of some of these are shown in figure 14.25. There are also polymers of isoprene, or synthetic rubber, in wide use. Fibers and fabrics may be polyamides (such as nylon), polyesters (such as Dacron), or polyacrylonitriles (Orlon, Acrilon, Creslon), which have a CN in place of a hydrogen atom on an ethylene molecule and are

called acrylic materials. All of these synthetic polymers have added much to practically every part of your life. It would be impossible to list all of their uses here, however, they present problems since (1) they are manufactured from raw materials obtained from coal and a dwindling petroleum supply (figure 14.26) and (2) they do not readily decompose when dumped into rivers, oceans, or other parts of the environment. However, research in the polymer sciences is beginning to reflect new understandings learned from research on biological tissues. This could lead to whole new molecular designs for synthetic polymers that will be more compatible with the ecosystem.

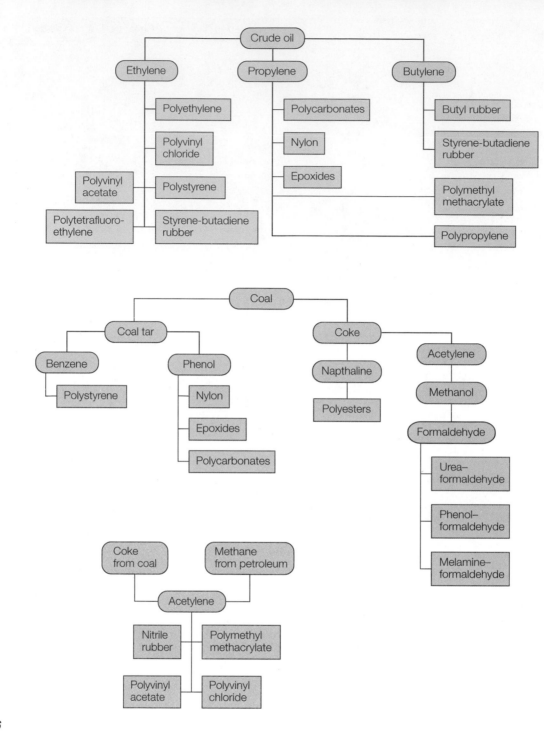

Figure 14.26

Petroleum and coal as sources of raw materials for manufacturing synthetic polymers.

How to Sort Plastic Bottles for Recycling

*P*lastic containers are made of different types of plastic resins, and some are suitable for recycling and some are not. How do you know which are suitable and how to sort them? Most all plastic containers have a code stamped on the bottom. The code is a number in the recycling arrow logo, sometimes appearing with some letters. Here is what the numbers and letters mean in terms of (1) the plastic, (2) how used, and (3) if it is usually recycled or not.

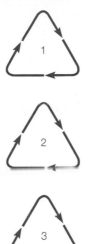

1. Polyethylene terephthalate (PET)
2. Large soft-drink bottles, salad dressing bottles
3. Frequently recycled

1. High-density polyethylene (HDPE)
2. Milk jugs, detergent and bleach bottles, others
3. Frequently recycled

1. Polyvinyl chloride (PVC or PV)
2. Shampoos, hair conditioners, others
3. Rarely recycled

1. Low density polyethylene (LDPE)
2. Plastic wrap, laundry and trash bags
3. Rarely recycled

1. Polypropylene (PP)
2. Food containers
3. Rarely recycled

1. Polystyrene (PS)
2. Styrofoam cups, burger boxes, plates
3. Occasionally recycled

1. Mixed resins
2. Catsup squeeze bottles, other squeeze bottles
3. Rarely recycled

Organic Chemistry

Organic Molecules and Heredity

*D*id you ever wonder how hereditary characteristics are passed on from parents to their children and how one protein becomes a muscle but another forms hair? Hereditary information and the synthesis of proteins are controlled and carried out by certain organic molecules. This Feature is but a brief introduction to the complicated and involved chemical processes that determine everything biological, from what kinds of proteins are formed as you grow to why you are you.

Nucleic acids are organic polymers that give the self-replicating ability to living cells (Box figure 14.1). There are two types of nucleic acid polymers, deoxyribonucleic acids (DNA) and ribonucleic acids (RNA). Both of these polymers consist of repeating units called *nucleotides*. The nucleotides of DNA are made up of (1) the organic base amines of thymine, adenine, guanine, and cytosine; (2) a phosphoric acid molecule; and (3) the simple sugar deoxyribose. RNA is different in structure in that thymine is replaced by a different organic base, uracil, and the deoxyribose is replaced by ribose.

In general, DNA is found in the nucleus of a cell. The linear sequences of nucleotides are the genetic codes of *chromosomes,* the cell structures that contain the DNA. A DNA molecule is very large, with molecular weights up to sixteen million, and is configured in a double helix. The phosphate and sugar groups are on the outside of this spiral arrangement and the amines are inside, bonded in a specific way. When a cell divides, the DNA strands are separated and each single strand takes other organic molecules to build two new, identical double helixes. The sequences of amines determine the hereditary information for the synthesis of proteins that work together to produce the particular organism being replicated.

Protein synthesis is not carried out by the DNA. The DNA only carries the genetic information. Information coded in the DNA is transmitted to structures called *ribosomes,* where the protein synthesis takes place, by RNA. Actually, there are three kinds of RNA: (1) messenger RNA, (2) transfer RNA, and (3) ribosomal RNA. The transfer RNA is a complementary strand of the DNA, which produces a smaller messenger RNA molecule. Messenger RNA carries the pattern of protein synthesis in a sequence of three

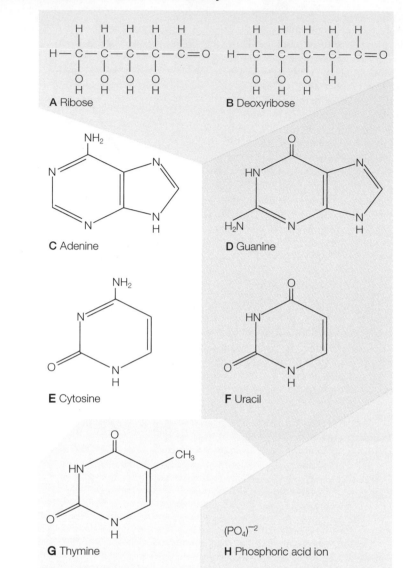

Box Figure 14.1

The building blocks of nucleotides are two simple sugars, (*a*) and (*b*), five amines, (*c*) through (*g*), and phosphoric acid molecules, (*h*).

amines. The messenger RNA molecules move about the cell, attaching to the ribosome, where they serve as a template for protein synthesis. The sequence of this template determines the kind of protein formed and thus the function and purpose of that protein. There are sixty-one different arrangements, or codes, that the RNA template can carry that determine if muscle proteins or the pro-

teins of hormones are produced, for example. It is a complicated and involved process and any small error, from the replication of DNA to the production of proteins, can lead to mutations or genetic diseases that result from faulty protein synthesis. Research in *genetic engineering,* the manipulation of DNA molecules, may someday provide answers to how to correct such genetic errors.

Summary

Organic chemistry is the study of compounds that have carbon as the principal element. Such compounds are called *organic compounds* and all the rest are *inorganic compounds*. There are millions of organic compounds because a carbon atom can link with other carbon atoms as well as atoms of other elements.

A *hydrocarbon* is an organic compound consisting of hydrogen and carbon atoms. The simplest hydrocarbon is one carbon atom and four hydrogen atoms, or CH_4. All hydrocarbons larger than CH_4 have one or more carbon atoms bonded to another carbon atom. The bond can be single, double, or triple, and this forms a basis for classifying hydrocarbons. A second basis is if the carbons are in a ring or not. The *alkanes* are hydrocarbons with single carbon-to-carbon bonds, the *alkenes* have a double carbon-to-carbon bond, and the *alkynes* have a triple carbon-to-carbon bond. The alkanes, alkenes, and alkynes can have straight- or branched-chain molecules. When the number of carbon atoms is greater than three, there are different arrangements that can occur for a particular number of carbon atoms. The different arrangements with the same molecular formula are called isomers. *Isomers* have different physical properties, so each isomer is given its own name. The name is determined by (1) identifying the longest continuous carbon chain as the base name, (2) locating the attachment of other atoms or hydrocarbon groups by counting from the direction that results in the smallest numbers, (3) identifying attached hydrocarbon groups by changing the "-ane" suffix of alkanes to "-yl," (4) identifying the number of these hydrocarbon groups with prefixes, and (5) identifying the location of the groups with the carbon atom number.

The alkanes have all the hydrogen atoms possible, so they are *saturated* hydrocarbons. The alkenes and the alkynes can add more hydrogens to the molecule, so they are *unsaturated* hydrocarbons. Unsaturated hydrocarbons are more chemically reactive than saturated molecules.

Hydrocarbons that occur in a ring or cycle structure are cyclohydrocarbons. A six-carbon cyclohydrocarbon with three double bonds has different properties than the other cyclohydrocarbons because the double bonds are not localized. This six-carbon molecule is *benzene*, the basic unit of the *aromatic hydrocarbons*.

Petroleum is a mixture of alkanes, cycloalkanes, and a few aromatic hydrocarbons that formed from the slow decomposition of buried marine plankton and algae. Petroleum from the ground, or *crude oil*, is distilled into petroleum products of *natural gas, LPG, petroleum ether, gasoline, kerosene, diesel fuel,* and *motor oils*. Each group contains a range of hydrocarbons and is processed according to use.

In addition to oxidation, hydrocarbons react by *substitution, addition,* and *polymerization* reactions. Reactions take place at sites of multiple bonds or lone pairs of electrons on the *functional groups*. The functional group determines the chemical properties of organic compounds. Functional group results in the *hydrocarbon derivatives* of *alcohols, ethers, aldehydes, ketones, organic acids, esters,* and *amines*.

Living organisms have an incredible number of highly organized chemical reactions that are catalyzed by *enzymes,* using food and energy to grow and reproduce. The process involves building large *macromolecules* from smaller molecules and units. The organic molecules involved in the process are proteins, carbohydrates, and fats and oils.

Proteins are macromolecular polymers of *amino acids* held together by *peptide bonds*. There are twenty amino acids that are used in various polymer combinations to build structural and functional proteins. *Structural proteins* are muscles, connective tissue, and the skin, hair, and nails of animals. *Functional proteins* are enzymes, hormones, and antibodies.

Carbohydrates are polyhydroxyl aldehydes and ketones that form three groups, the monosaccharides, disaccharides, and polysaccharides. The *monosaccharides* are simple sugars such as *glucose* and *fructose*. Glucose is *blood sugar,* a source of energy. The disaccharides are *sucrose* (table sugar), *lactose* (milk sugar), and *maltose* (malt sugar). The disaccharides are broken down (digested) to glucose for use by the body. The polysaccharides are polymers or glucose in straight or branched chains used as a near-term source of stored energy. Plants store the energy in the form of *starch,* and animals store it in the form of *glycogen. Cellulose* is a polymer similar to starch that humans cannot digest.

Fats and oils are esters formed from three fatty acids and glycerol into a *triglyceride. Fats* are usually solid triglycerides associated with animals, and *oils* are liquid triglycerides associated with plant life, but both represent a high-energy storage material.

Polymers are huge, chain-like molecules of hundreds or thousands of smaller, repeating molecular units called *monomers*. Polymers occur naturally in plants and animals, and many *synthetic polymers* are made today from variations of the ethylene-derived monomers. Among the more widely used synthetic polymers derived from ethylene are polyethylene, polyvinyl chloride, polystyrene, and Teflon. Problems with the synthetic polymers include that (1) they are manufactured from fossil fuels that are also used as the primary energy supply and (2) they do not readily decompose and tend to accumulate in the environment.

Key Terms

alcohol (p. 296)	hydrocarbon derivatives (p. 296)
aldehyde (p. 298)	inorganic chemistry (p. 287)
alkanes (p. 288)	isomers (p. 289)
alkenes (p. 292)	ketone (p. 298)
alkyne (p. 292)	macromolecule (p. 300)
amino acids (p. 300)	monosaccharides (p. 300)
aromatic hydrocarbons (p. 293)	oil field (p. 294)
carbohydrates (p. 300)	oils (p. 303)
cellulose (p. 302)	organic acids (p. 298)
crude oil (p. 294)	organic chemistry (p. 287)
disaccharides (p. 302)	petroleum (p. 293)
esters (p. 299)	polymer (p. 296)
ether (p. 298)	polysaccharides (p. 302)
fats (p. 303)	proof (p. 297)
functional group (p. 296)	proteins (p. 300)
gasohol (p. 297)	saturated (p. 292)
glycerol (p. 297)	starch (p. 302)
glycogen (p. 302)	triglyceride (p. 303)
glycol (p. 297)	unsaturated (p. 292)
hydrocarbon (p. 287)	

Applying the Concepts

1. An organic compound is a compound that
   a. contains carbon and was formed only by a living organism.
   b. is a natural compound that has not been synthesized.
   c. contains carbon, no matter if it was formed by a living thing or not.
   d. was formed by a plant.

2. There are millions of organic compounds but only thousands of inorganic compounds because
   a. organic compounds were formed by living things.
   b. there is more carbon on the earth's surface than any other element.

c. atoms of elements other than carbon never combine with themselves.

d. carbon atoms can combine with up to four other atoms, including other carbon atoms.

3. You know for sure that the compound named decane has
   a. more than 10 isomers.
   b. 10 carbon atoms in each molecule.
   c. only single bonds.
   d. all of the above.

4. An alkane with 4 carbon atoms would have how many hydrogen atoms in each molecule?
   a. 4
   b. 8
   c. 10
   d. 16

5. Isomers are compounds with the same
   a. molecular formula with different structures.
   b. molecular formula with different atomic masses.
   c. atoms, but different molecular formulas.
   d. structures, but different formulas.

6. Isomers have
   a. the same chemical and physical properties.
   b. the same chemical, but different physical properties.
   c. the same physical, but different chemical properties.
   d. different physical and chemical properties.

7. The organic compound 2,2,4-trimethylpentane is an isomer of
   a. propane.
   b. pentane.
   c. heptane.
   d. octane.

8. The hydrocarbons with a double covalent carbon-carbon bond are called
   a. alkanes.
   b. alkenes.
   c. alkynes.
   d. none of the above.

9. According to their definitions, which of the following would not occur as unsaturated hydrocarbons?
   a. alkanes
   b. alkenes
   c. alkynes
   d. None of the above are correct.

10. Petroleum is believed to have formed mostly from the anaerobic decomposition of buried
    a. dinosaurs.
    b. fish.
    c. pine trees.
    d. plankton and algae.

11. The label on a container states that the product contains "petroleum distillates." Which of the following hydrocarbons is probably present?
    a. CH_4
    b. C_5H_{12}
    c. $C_{16}H_{34}$
    d. $C_{40}H_{82}$

12. Tetraethyl lead ("ethyl") was added to gasoline to increase the octane rating by
    a. increasing the power by increasing the heat of combustion.
    b. increasing the burning rate of the gasoline.
    c. absorbing the pings and knocks.
    d. decreasing the burning rate.

13. The reaction of $C_2H_2 + Br_2 \rightarrow C_2H_2Br_2$ is a
    a. substitution reaction.
    b. addition reaction.
    c. addition polymerization reaction.
    d. substitution polymerization reaction.

14. Ethylene molecules can add to each other in a reaction to form a long chain called a
    a. monomer.
    b. dimer.
    c. trimer.
    d. polymer.

15. Chemical reactions usually take place on an organic compound at the site of (a)
    a. double bond.
    b. lone pair of electrons.
    c. functional group.
    d. any of the above.

16. The R in ROH represents
    a. a functional group.
    b. a hydrocarbon group with a name ending in "-yl."
    c. an atom of an inorganic element.
    d. a polyatomic ion that does not contain carbon.

17. The OH in ROH represents
    a. a functional group.
    b. a hydrocarbon group with a name ending in "-yl."
    c. the hydroxide ion, which ionizes to form a base.
    d. the site of chemical activity in a strong base.

18. What is the proof of a "wine cooler" that is 5 percent alcohol by volume?
    a. 2.5 proof
    b. 5 proof
    c. 10 proof
    d. 50 proof

19. An alcohol with two hydroxyl groups per molecules is called
    a. ethanol.
    b. glycerol.
    c. glycerin.
    d. glycol.

20. A bottle of wine that has "gone bad" now contains
    a. CH_3OH.
    b. CH_3OCH_3.
    c. CH_3COOH.
    d. CH_3COOCH_3.

21. A protein is a polymer formed from the linking of many
    a. glucose units.
    b. DNA molecules.
    c. amino acid molecules.
    d. monosaccharides.

22. Which of the following is *not* converted to blood sugar by the human body?
    a. lactose
    b. dextrose
    c. cellulose
    d. glycogen

23. Fats from animals and oils from plants have the general structure of a (an)
    a. aldehyde.
    b. ester.
    c. amine.
    d. ketone.

24. Liquid oils from plants can be converted to solids by adding what to the molecule?
    a. metal ions
    b. carbon
    c. polyatomic ions
    d. hydrogen

25. The basic difference between a monomer of polyethylene and a monomer of polyvinyl chloride is
    a. the replacement of a hydrogen by a chlorine.
    b. the addition of four fluorines.
    c. the elimination of double bonds.
    d. the removal of all hydrogens.

26. Many synthetic polymers become a problem in the environment because they
    a. decompose to nutrients, which accelerates plant growth.
    b. do not readily decompose and tend to accumulate.
    c. do not contain vitamins as natural materials do.
    d. become a source of food for fish, but ruin the flavor of fish meat.

Answers

1. c   2. d   3. d   4. c   5. a   6. b   7. d   8. b   9. a
10. d   11. b   12. d   13. b   14. d   15. d   16. b   17. b
18. c   19. d   20. c   21. c   22. c   23. b   24. d   25. a
26. b

Questions for Thought

1. What is an organic compound?

2. There are millions of organic compounds but only thousands of inorganic compounds. Explain why this is the case.

3. What is cracking and reforming? For what purposes are either or both used by the petroleum industry?

4. Is it possible to have an isomer of ethane? Explain.

5. Suggest a reason ethylene is an important raw material used in the production of plastics but ethane is not.

6. What is the size of the "barrel of oil" that is described in news reports?

7. What is (a) natural gas, (b) LPG, and (c) petroleum ether?

8. What does the octane number of gasoline describe? On what is the number based?

9. Why is unleaded gasoline more expensive than gasoline with lead additives?

10. What is a functional group? What is it about the nature of a functional group that makes it the site of chemical reactions?

11. Draw a structural formula for alcohol. Describe how alcohols are named.

12. A soft drink is advertised to "contain no sugar." The label lists ingredients of carbonated water, dextrose, corn syrup, fructose, and flavorings. Evaluate the advertising and the list of ingredients.

13. What are fats and oils? What are saturated and unsaturated fats and oils?

14. What is a polymer? Give an example of a naturally occurring plant polymer. Give an example of a synthetic polymer.

15. Explain why a small portion of wine is customarily poured before a bottle of wine is served. Sometimes the cork is handed to the person doing the ordering with the small portion of wine. What is the person supposed to do with the cork and why?

Exercises

1. Draw the structural formulas for (a) *n*-pentane, and (b) an isomer of pentane with the maximum possible branching. (c) Give the IUPAC name of this isomer.

2. Write structural formulas for all the hexane isomers you can identify. Write the IUPAC name for each isomer.

3. Write structural formulas for
   (a) 3,3,4-trimethyloctane.
   (b) 2-methyl-1-pentene.
   (c) 5,5-dimethyl-3-heptyne.

4. Write the IUPAC name for each of the following

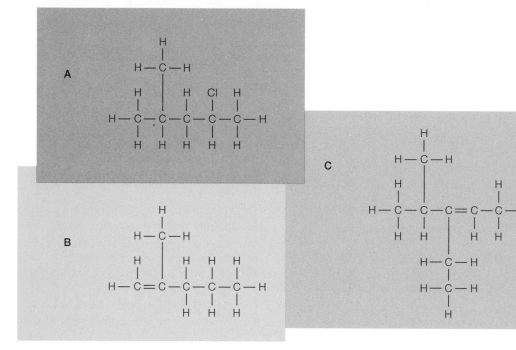

5. Which would have the higher octane rating, 2,2,3-trimethylbutane or 2,2-dimethylpentane? Explain with an illustration.

6. Use the information in table 14.4 to classify the following as an alcohol, ether, organic acid, ester, or amide.

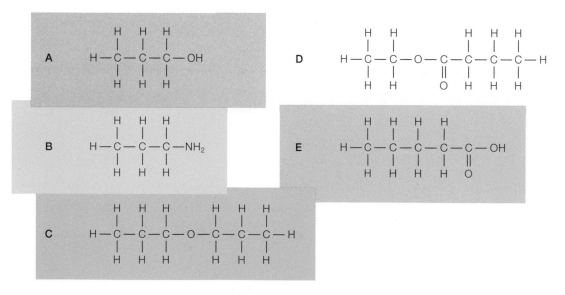

1. Write structural formulas for (a) *n*-octane, (b) an isomer of octane with the maximum possible branching. (c) Give the IUPAC name of this isomer.

2. Write the structural formulas for all the heptane isomers you can identify. Write the IUPAC name for each isomer.

3. Write structural formulas for
   (a) 2,3-dimethylpentane.
   (b) 1-butene.
   (c) 2-methyl-3-ethyl-3-hexene.

4. Write the IUPAC name for each of the following

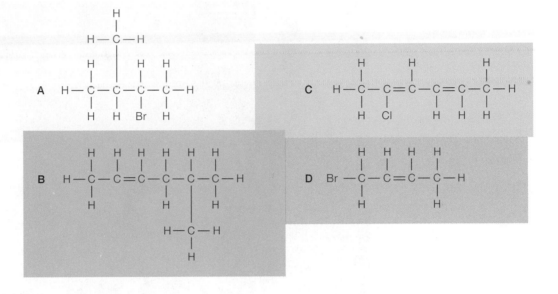

5. Which would have the higher octane rating, 2-methyl-butane or dimethylpropane? Explain with an illustration.

6. Classify the following as an alcohol, ether, organic acid, ester, or amide.

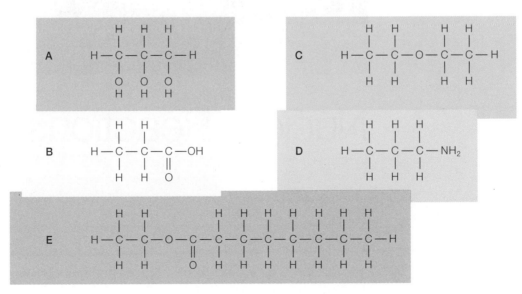

Chapter
15

Outline

Nuclear Reactions

With the top half of the steel vessel and control rods removed, fuel rod bundles can be replaced in the water-flooded nuclear reactor.

*T*HE ancient alchemist dreamed of changing one element into another, such as lead into gold. The alchemist was never successful, however, because such changes were attempted with chemical reactions. Chemical reactions are reactions that involve only the electrons of atoms. Electrons are shared or transferred in chemical reactions and the internal nucleus of the atom is unchanged. Elements thus retain their identity during the sharing or transferring of electrons. This chapter is concerned with a different kind of reaction, one that involves the *nucleus* of the atom. In nuclear reactions, the nucleus of the atom is often altered, changing the identity of the elements involved. The ancient alchemist's dream of changing one element into another was actually a dream of achieving a nuclear change, that is, a nuclear reaction.

Understanding nuclear reactions is important because although fossil fuels are the major source of energy today, there are growing concerns about (1) air pollution from fossil fuel combustion, (2) increasing levels of CO_2 from fossil fuel combustion, which may be warming the earth (the greenhouse effect), and (3) the dwindling fossil fuel supply itself, which cannot last forever. Energy experts see nuclear energy as a means of meeting rising energy demands in an environmentally acceptable way. However, the topic of nuclear energy is controversial, and discussions about the topic often result in strong emotional responses. Decisions about the use of nuclear energy require some understandings about nuclear reactions and some facts about radioactivity and radioactive materials (figure 15.1). These understandings and facts are the topics of this chapter.

Natural Radioactivity

Radioactivity was discovered in 1896 by Henri Becquerel, a French scientist who was very interested in the recent discovery of X rays. Becquerel was experimenting with fluorescent minerals, minerals that give off visible light after being exposed to sunlight. He wondered if fluorescent minerals emitted X rays in addition to visible light. From previous work with X rays, Becquerel knew that they would penetrate a wrapped, light-tight photographic plate, exposing it as visible light exposes an unprotected plate. Thus, Becquerel decided to place a fluorescent uranium mineral on a protected photographic plate while the mineral was exposed to sunlight. Sure enough, he found a silhouette of the mineral on the plate when it was developed. Believing the uranium mineral emitted X rays, he continued his studies until the weather turned cloudy. Storing a wrapped, protected photographic plate and the uranium mineral together during the cloudy weather, Becquerel returned to the materials later and developed the photographic plate to again find an image of the mineral (figure 15.2). He concluded that the mineral was emitting an "invisible radiation" that was not induced by sunlight. Becquerel named the emission of invisible radiation *radioactivity*. Materials that have the property of radioactivity are called *radioactive* materials.

Becquerel's discovery led to the beginnings of the modern atomic theory and to the discovery of new elements. Ernest Rutherford studied the nature of radioactivity and found that there are three kinds, which are today known by the first three letters of the Greek alphabet—alpha (α), beta (β), and gamma (γ). These Greek letters were used at first before the nature of the radiation was known. Today, an **alpha particle** (sometimes

Figure 15.1

Decisions about nuclear energy require some understanding about nuclear reactions and the nature of radioactivity. This is one of the three units of the Palo Verde Nuclear Generating Station in Arizona. With all three units running, enough power is generated to meet the electrical needs of nearly four million people.

called an alpha ray) is known to be the nucleus of a helium atom, that is, two protons and two neutrons. A **beta particle** (or beta ray) is a high-energy electron. A **gamma ray** is electromagnetic radiation, as is light, but of very short wavelength (figure 15.3).

It was Rutherford's work with alpha particles that resulted in the discovery of the nucleus and the proton (see chapter 9). At Becquerel's suggestion, Madame Marie Curie searched for other radioactive materials and in the process discovered two new elements, polonium and radium. More radioactive elements were discovered since that time, and, in fact, all the isotopes of all the elements with an atomic number greater than 83 (bismuth) are radioactive. Today, **radioactivity** is defined as the *spontaneous emission of particles or energy from an atomic nucleus* as it disintegrates. As a result of the disintegration, the nucleus of an atom often undergoes a change of identity, becoming a simpler nucleus. The spontaneous disintegration of a given nucleus is a purely natural process and cannot be controlled or influenced. The natural spontaneous disintegration or decomposition of a nucleus is also called **radioactive decay**. Although it is impossible to know *when* a given nucleus will undergo radioactive decay, as you will see later, it is possible to deal with the *rate* of decay for a given radioactive material with precision.

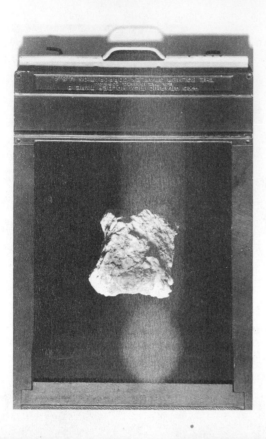

A

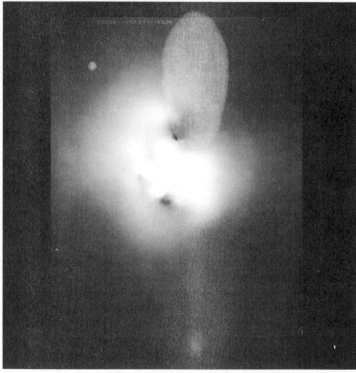

B

Figure 15.2

Radioactivity was discovered by Henri Becquerel when he exposed a light-tight photographic plate to a radioactive mineral, then developed the plate. (a) A photographic film is exposed to an uranite ore sample. (b) The film, developed normally after a four-day exposure to uranite. Becquerel found an image like this one and deduced that the mineral gave off ''invisible radiation'' he called radioactivity.

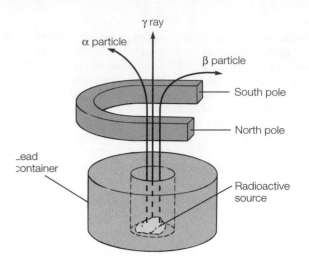

Figure 15.3

Radiation passing through a magnetic field shows that massive, positively charged alpha particles are deflected one way and less massive beta particles with their negative charge are greatly deflected in the opposite direction. Gamma rays, like light, are not deflected.

Nuclear Equations

There are two main subatomic particles in the nucleus, the proton and the neutron. The proton and neutron are called **nucleons.** Recall that the number of protons, the *atomic number,* determines what element an atom is, and that all atoms of a given element have the same number of protons. The number of neutrons varies in *isotopes,* which are atoms with the same atomic number but different numbers of neutrons (figure 15.4). The number of protons and neutrons together determines the *mass number,* so different isotopes of the same element are identified with their mass numbers. Thus, the two most common, naturally occurring isotopes of uranium are referred to as uranium-238 and uranium-235, and the 238 and 235 are the mass numbers of these isotopes. Isotopes are also represented by the symbol:

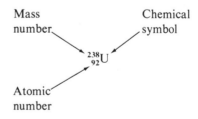

Subatomic particles involved in nuclear reactions are represented by symbols with the following form:

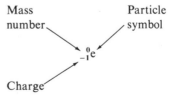

Symbols for these particles are illustrated in table 15.1.

Symbols are used in an equation for a nuclear reaction that is written much like a chemical reaction with reactants and products. When a uranium-238 nucleus emits an alpha particle

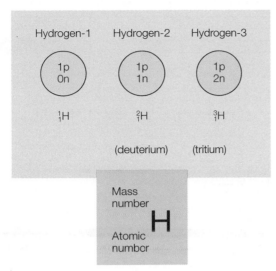

Hydrogen-1   Hydrogen-2   Hydrogen-3

1p   0n   1p   1n   1p   2n

$^{1}_{1}H$ $^{2}_{1}H$ $^{3}_{1}H$

(deuterium)   (tritium)

Mass number

H

Atomic number

Figure 15.4
The three isotopes of hydrogen have the same number of protons but different numbers of neutrons. Hydrogen-1 is the most common isotope. Hydrogen-2, with an additional neutron, is named deuterium, and hydrogen-3 is called tritium. Neutrons and protons are called *nucleons* because they are in the nucleus.

($^{4}_{2}He$), for example, it loses two protons and two neutrons. The nuclear reaction is written in equation form as

$$^{238}_{92}U \rightarrow {}^{234}_{90}Th + {}^{4}_{2}He$$

The *products* of this nuclear reaction from the decay of a uranium-238 nucleus are (1) the alpha particle ($^{4}_{2}He$) given off and (2) the nucleus, which remains after the alpha particle leaves the original nucleus. What remains is easily determined since all nuclear equations must show conservation of charge and conservation of the total number of nucleons. Therefore: (1) the number of protons (positive charge) remains the same, and the sum of the subscripts (atomic number, or numbers of protons) in the reactants must equal the sum of the subscripts in the products; and (2) the total number of nucleons remains the same, and the sum of the superscripts (atomic mass, or number of protons plus neutrons) in the reactants must equal the sum of the superscripts in the products. The new nucleus remaining after the emission of an alpha particle, therefore, has an atomic number of 90 (92 − 2 = 90). According to the table of atomic numbers on the inside back cover of this text, this new nucleus is thorium (Th). The mass of the thorium isotope is 238 minus 4, or 234. The emission of an alpha particle thus decreases the number of protons by 2 and the mass number by 4. From the subscripts, you can see that the total charge is conserved (92 = 90 + 2). From the superscripts, you can see that the total number of nucleons is also conserved (238 = 234 + 4). The mass numbers (superscripts) and the atomic numbers (subscripts) are *balanced* in a correctly written nuclear equation. Such nuclear equations are considered to be independent of any chemical form or chemical reaction. Nuclear reactions are independent and separate from chemical reactions, whether or not the atom is in the pure element or in a compound. The particles that are involved in nuclear reactions each has its own symbol with a superscript indicating mass number and a subscript in-

Table 15.1

Names, symbols, and properties of particles in nuclear equations

Name	Symbol	Mass Number	Charge
Proton	$^{1}_{1}H$ (or $^{1}_{1}p$)	1	1+
Electron	$^{0}_{-1}e$ (or $^{0}_{-1}\beta$)	0	1−
Neutron	$^{1}_{0}n$	1	0
Gamma Photon	$^{0}_{0}\gamma$	0	0

dicating the charge. These symbols, names, and numbers are given in table 15.1.

Example 15.1

A plutonium-242 nucleus undergoes radioactive decay, emitting an alpha particle. Write the nuclear equation for this nuclear reaction.

Solution

Step 1: The table of atomic weights on the inside back cover gives the atomic number of plutonium as 94. Plutonium-242 therefore has a symbol of $^{242}_{94}Pu$. The symbol for an alpha particle is ($^{4}_{2}He$), so the nuclear equation so far is

$$^{242}_{94}Pu \rightarrow {}^{4}_{2}He + ?$$

Step 2: From the subscripts, you can see that 94 = 2 + 92, so the new nucleus has an atomic number of 92. The table of atomic weights identifies element 92 as uranium with a symbol of U.

Step 3: From the superscripts, you can see that the mass number of the uranium isotope formed is 242 − 4 = 238, so the product nucleus is $^{238}_{92}U$ and the complete nuclear equation is

$$^{242}_{94}Pu \rightarrow {}^{4}_{2}He + {}^{238}_{92}U$$

Step 4: Checking the subscripts (94 = 2 + 92) and the superscripts (242 = 4 + 238), you can see that the nuclear equation is balanced.

Example 15.2

What is the product nucleus formed when radium emits an alpha particle? (Answer: Radon-222, a chemically inert, radioactive gas)

The Nature of the Nucleus

The modern atomic theory does not picture the nucleus as a group of stationary protons and neutrons clumped together by some "nuclear glue." The protons and neutrons are understood to be held together by a **nuclear force,** a strong fundamental force of attraction that is functional only at very short distances, on the order of 10^{-15} m or less. At distances greater than about 10^{-15} m the nuclear force is negligible, and the weaker **electromagnetic force,** the force of attraction between like charges, is the operational force. Thus, like charged protons experience a repulsive force when they are farther apart than about 10^{-15} m. When closer together than 10^{-15} m, the short-range, stronger nuclear force predominates and the protons experience a strong attractive force. This explains why the like charged protons of the nucleus are not repelled by their like electric charges.

Observations of radioactive decay reactants and products and experiments with nuclear stability have led to a **shell model of the nucleus.** This model considers the protons and neutrons moving in energy levels, or shells, in the nucleus analogous to the shell structure of electrons in the outermost part of the atom. As in the electron shells, there are certain configurations of nuclear shells that have a greater stability than others. Considering electrons, filled and half-filled shells are more stable than other arrangements, and maximum stability occurs with the noble gases and their 2, 10, 18, 36, 54, and 86 electrons. Considering the nucleus, atoms with 2, 8, 20, 28, 50, 82, or 126 protons or neutrons have a maximum nuclear stability. The stable numbers are not the same for electrons and nucleons because of differences in nuclear and electromagnetic forces.

Isotopes of uranium, radium, and plutonium, as well as other isotopes, emit an alpha particle during radioactive decay to a simpler nucleus. The alpha particle is a helium nucleus, $^4_2$He. The alpha particle contains two protons as well as two neutrons, which is one of the nucleon numbers of stability, so you would expect the helium nucleus (or alpha particle) to have a stable nucleus and it does. Stable means it does not undergo radioactive decay. Pairs of protons and pairs of neutrons have increased stability just as pairs of electrons in a molecule. As a result, nuclei with an *even number* of both protons and neutrons are, in general, more stable than nuclei with odd numbers of protons and neutrons. There are a little more than 150 isotopes with an even number of protons and an even number of neutrons, but there are 5 stable isotopes with odd numbers of each. Just as in the case of electrons, there are other factors that come into play as the nucleus becomes larger and larger with increased numbers of nucleons.

The results of some of these factors are shown in figure 15.5, which is a graph of the number of neutrons versus the number of protons in nuclei. As the number of protons increases, the neutron-to-proton ratio of the *stable nuclei* also increases in a **band of stability.** Within the band the neutron-to-proton ratio increases from about 1:1 at the bottom left to about 1½:1 at the top right. The increased ratio of neutrons is needed to produce a stable nucleus as the number of protons increases. Neutrons provide additional attractive *nuclear* (not electrical) forces, which counters the increased electrical repulsion from a larger number of positively charged protons. Thus, more neutrons are required in larger nuclei to produce a stable nucleus. However, there is a limit to the additional attractive forces that can be provided by more and more neutrons, and all isotopes of all elements with more than 83 protons are unstable and thus undergo radioactive decay.

The generalizations about nuclear stability provide a means of predicting if a particular nucleus is radioactive. The generalizations are as follows:

1. All isotopes with an atomic number greater than 83 have an unstable nucleus.
2. Isotopes that contain 2, 8, 20, 28, 50, 82, or 126 protons or neutrons in their nucleus are generally more stable than other numbers of protons or neutrons.

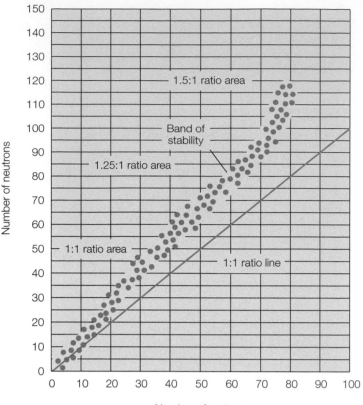

Figure 15.5

The dots indicate stable nuclei, which group in a band of stability according to their neutron-to-proton ratio. As the size of nuclei increases, so does the neutron-to-proton ratio that represents stability. Nuclei outside this band of stability are radioactive.

3. Pairs of protons and pairs of neutrons have increased stability, so isotopes that have nuclei with even numbers of both protons and neutrons are generally more stable than nuclei with odd numbers of both protons and neutrons.
4. Isotopes with an atomic number less than 83 are stable when the ratio of neutrons to protons in the nucleus is about 1:1 in isotopes with up to 20 protons, but the ratio increases in larger nuclei in a band of stability (figure 15.5). Isotopes with a ratio to the left or right of this band are unstable and thus will undergo radioactive decay.

Example 15.3

Would you predict the following isotopes to be radioactive or stable?

(a) $^{60}_{27}$Co

(b) $^{222}_{86}$Rn

(c) $^3_1$H

(d) $^{40}_{20}$Ca

Solution

(a) Cobalt-60 has 27 protons and 33 neutrons, both odd numbers, so you might expect $^{60}_{27}$Co to be radioactive.

(b) Radon has an atomic number of 86, and all isotopes of all elements beyond atomic number 83 are radioactive. Radon-222 is therefore radioactive.

(c) Hydrogen-3 has an odd number of protons and an even number of neutrons, but its 2:1 neutron-to-proton ratio places it outside the band of stability. Hydrogen-3 is radioactive.

(d) Calcium-40 has an even number of protons and an even number of neutrons, containing 20 of each. The number 20 is a particularly stable number of protons or neutrons, and calcium-40 has 20 of each. In addition, the neutron-to-proton ratio is 1:1, placing it within the band of stability. All indications are that calcium-40 is stable, not radioactive.

Example 15.4

Which of the following would you predict to be radioactive?

(a) $^{127}_{53}\text{I}$

(b) $^{131}_{53}\text{I}$

(c) $^{206}_{82}\text{Pb}$

(d) $^{214}_{82}\text{Pb}$

(Answer: (b) and (d))

Types of Radioactive Decay

Through the process of radioactive decay, an unstable nucleus becomes a more stable one with less energy. The three more familiar types of radiation emitted—alpha, beta, and gamma—were introduced earlier. There are five common types of radioactive decay, and three of these involve alpha, beta, and gamma radiation.

1. *Alpha emission.* Alpha (α) emission is the expulsion of an alpha particle (^4_2He) from an unstable, disintegrating nucleus. The alpha particle, a helium nucleus, travels from 2 to 12 cm through the air, depending on the energy of emission from the source. An alpha particle is easily stopped by a sheet of paper close to the nucleus. As an example of alpha emission, consider the decay of a radon-222 nucleus,

$$^{222}_{86}\text{Rn} \rightarrow {}^{218}_{84}\text{Po} + {}^4_2\text{He}$$

The spent alpha particle eventually acquires two electrons and becomes an ordinary helium atom.

2. *Beta emission.* Beta (β^-) emission is the expulsion of a different particle, a beta particle, from an unstable disintegrating nucleus. A beta particle is simply an electron ($_{-1}^0\text{e}$) ejected from the nucleus at a high speed. The emission of a beta particle *increases the number of protons* in a nucleus. It is as if a neutron changed to a proton by emitting an electron, or

$$^1_0\text{n} \rightarrow {}^1_1\text{p} + {}^{0}_{-1}\text{e}$$

Carbon-14 is a carbon isotope that decays by beta emission:

$$^{14}_6\text{C} \rightarrow {}^{14}_7\text{N} + {}^{0}_{-1}\text{e}$$

Note that the number of protons increased from six to seven, but the mass number remained the same. The mass number is unchanged because the mass of the expelled electron (beta particle) is negligible.

Beta particles are more penetrating than alpha particles and may travel several hundred centimeters through the air. They can be stopped by a thin layer of metal close to the emitting nucleus, such as a 1 cm thick piece of aluminum. A spent beta particle may eventually join an ion to become part of an atom, or it may remain a free electron.

Table 15.2

Radioactive decay

Unstable Condition	Type of Decay	Emitted	Product Nucleus
More than 83 protons	Alpha emission	^4_2He	Lost 2 protons and 2 neutrons
Neutron-to-proton ratio too large	Beta emission	$_{-1}^{0}\text{e}$	Gained 1 proton, no mass change
Excited nucleus	Gamma emission	$^0_0\gamma$	No change
Neutron-to-proton ratio too small	Other emission	^0_1e	Lost 1 proton, no mass change

3. *Gamma emission.* Gamma (γ) emission is a high-energy burst of electromagnetic radiation from an excited nucleus. It is a burst of light (photon) of a wavelength much too short to be detected by the eye. Other types of radioactive decay, such as alpha or beta emission, sometimes leave the nucleus with an excess of energy, a condition called an *excited state.* As in the case of excited electrons, the nucleus returns to a lower energy state by emitting electromagnetic radiation. From a nucleus, this radiation is in the high-energy portion of the electromagnetic spectrum. Gamma is the most penetrating of the three common types of nuclear radiation. Like X rays, gamma rays can pass completely through a person, but all gamma radiation can be stopped by a 5 cm thick piece of lead close to the source. As with other types of electromagnetic radiation, gamma radiation is absorbed by and gives its energy to materials. Since the product nucleus changed from an excited state to a lower energy state, there is no change in the number of nucleons. For example, radon-222 is an isotope that emits gamma radiation:

$$^{222}_{86}\text{Rn*} \rightarrow {}^{222}_{86}\text{Rn} + {}^0_0\gamma$$
(* denotes excited state)

Alpha, beta, and gamma emission radioactive decay are summarized in table 15.2, which also lists the unstable nuclear conditions that lead to the particular type of emission. Just as electrons seek a state of greater stability, a nucleus undergoes radioactive decay to achieve a balance between nuclear attractions, electromagnetic repulsions, and a low quantum of nuclear shell energy. The key to understanding the types of reactions that occur is found in the band of stable nuclei illustrated in figure 15.5. The isotopes within this band have achieved the state of stability, and other isotopes above, below, or beyond the band are unstable and thus radioactive.

Nuclei that have a neutron-to-proton ratio beyond the upper right part of the band are unstable because of an imbalance between the proton-proton electromagnetic repulsions and all the combined proton and neutron nuclear attractions. Recall that the neutron-to-proton ratio increases from about 1:1 to about 1½:1 in the larger nuclei. The additional neutron provided additional nuclear attractions to hold the nucleus together, but atomic number 83 appears to be the upper limit to this additional stabilizing contribution. Thus, all nuclei with an atomic number greater than 83 are outside the upper right limit of the

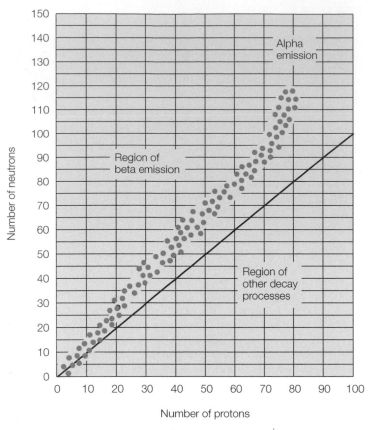

Figure 15.6

Unstable nuclei undergo different types of radioactive decay to obtain a more stable nucleus. What type of decay depends, in general, on the neutron-to-proton ratio as shown.

band of stability. Emission of an alpha particle reduces the number of protons by two and the number of neutrons by two, moving the nucleus more toward the band of stability. Thus, you can expect a nucleus that lies beyond the upper right part of the band of stability to be an alpha emitter (figure 15.6).

A nucleus that has a neutron-to-proton ratio that is too large will be on the left side of the band of stability. Emission of a beta particle decreases the number of neutrons and increases the number of protons, so a beta emission will lower the neutron-to-proton ratio. Thus, you can expect a nucleus with a large neutron-to-proton ratio, that is, one to the left of the band of stability, to be a beta emitter.

A nucleus that has a neutron-to-proton ratio that is too small will be on the right side of the band of stability. These nuclei can increase the number of neutrons and reduce the number of protons in the nucleus by other types of radioactive decay. As is usually the situation when dealing with broad generalizations and trends, there are exceptions to the summarized relationships between neutron-to-proton ratios and radioactive decay.

Example 15.5

Refer to figure 15.6 and predict the type of radioactive decay for each of the following unstable nuclei:

(a) $^{131}_{53}I$

(b) $^{241}_{94}Pu$

Solution

(a) Iodine-131 has a nucleus with 53 protons and 131 minus 53, or 78 neutrons, so it has a neutron-to-proton ratio of 1.47:1. This places iodine-131 on the left side of the band of stability, with a high neutron-to-proton ratio that can be reduced by beta emission. The nuclear equation is

$$^{131}_{53}I \rightarrow ^{131}_{54}Xe + ^{0}_{-1}e$$

(b) Plutonium-241 has 94 protons and 241 minus 94, or 147 neutrons, in the nucleus. This nucleus is to the upper right, beyond the band of stability. It can move back toward stability by emitting an alpha particle, losing 2 protons and 2 neutrons from the nucleus. The nuclear equation is

$$^{241}_{94}Pu \rightarrow ^{238}_{92}U + ^{4}_{2}He$$

Radioactive Decay Series

A radioactive decay reaction produces a simpler, more stable nucleus than the reactant nucleus. As discussed in the previous section, large nuclei with an atomic number greater than 83 decay by alpha emission, giving up two protons and two neutrons with each alpha particle. A nucleus with an atomic number greater than 86, however, will emit an alpha particle and *still* have an atomic number greater than 83, which means the product nucleus will also be radioactive. This nucleus will also undergo radioactive decay, and the process will continue through a series of decay reactions until a stable nucleus is achieved. Such a series of decay reactions that (1) begins with one radioactive nucleus, which (2) decays to a second nucleus, which (3) then decays to a third nucleus, and so on until (4) a stable nucleus is reached is called a **radioactive decay series.** There are three naturally occurring radioactive decay series. One begins with thorium-232 and ends with lead-208, another begins with uranium-235 and ends with lead-207, and the third series begins with uranium-238 and ends with lead-206. Figure 15.7 shows the uranium-238 radioactive decay series.

As figure 15.7 illustrates, the uranium-238 begins with uranium-238 decaying to thorium-234 by alpha emission. Thorium has a new position on the graph because it now has a new atomic number and a new mass number. Thorium-234 is unstable and decays to protactinium-234 by beta emission, which is also unstable and decays by beta emission to uranium-234. The process continues with five sequential alpha emissions, then two beta-beta-alpha decay steps before the series terminates with the stable lead-206 nucleus.

Uranium-238 is radioactive and decays to thorium-234 by emitting an alpha particle. Yet not all uranium-238 has decayed to lead-206, and, in fact, a sample of uranium-238 will continue to give off alpha particles for millions of years. Uranium-238, as uranium-235 and thorium-232, undergoes radioactive decay very slowly, and any given nucleus may disintegrate today or it may disintegrate millions of years later. It is not possible to predict when a nucleus will decay because it is a random process. It is possible, however, to deal with nuclear disintegration statistically since the rate of decay is not changed by any external conditions of temperature, pressure, or any chemical state. When dealing with a large number of nuclei, the ratio of the rate of

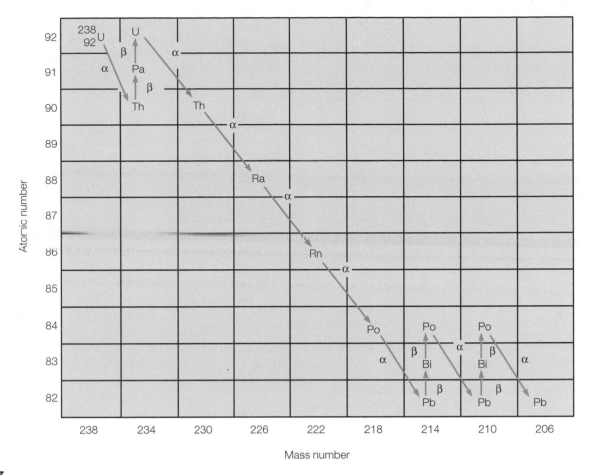

Figure 15.7

The radioactive decay series for uranium-238. This is one of three
naturally occurring series.

nuclear disintegration per unit of time to the total number of
radioactive nuclei will be a constant, or

$$\text{radioactive decay constant} = \frac{\text{decay rate}}{\text{number of nuclei}}$$

or, in symbols,

$$k = \frac{\text{rate}}{n} \qquad \textbf{equation 15.1}$$

The **radioactive decay constant,** k, is a specific constant for a
particular isotope, and each isotope has its own decay constant.
For example, a 238 g sample of uranium-238 (1 mole) that has
2.93×10^6 disintegrations per second would have a decay constant of

$$k = \frac{\text{rate}}{n} = \frac{2.93 \times 10^6 \text{ nuclei/sec}}{6.02 \times 10^{23} \text{ nuclei}}$$

$$= 4.87 \times 10^{-18}/\text{sec}$$

The rate of radioactive decay is usually described in terms
of its *half-life*. The **half-life** is the time required for one-half of
the unstable nuclei to decay. Since each isotope has a characteristic decay constant, then each isotope has its own characteristic half-life. Half-lives of some highly unstable isotopes are

Table 15.3

Half-lives of some radioactive isotopes

Isotope	Half-Life	Mode of Decay
$_{1}^{3}\text{H}$ (tritium)	12.26 years	Beta
$_{6}^{14}\text{C}$	5,930 years	Beta
$_{38}^{90}\text{Sr}$	28 years	Beta
$_{53}^{131}\text{I}$	8 days	Beta
$_{54}^{133}\text{Xe}$	5.27 days	Beta
$_{92}^{238}\text{U}$	4.51×10^9 years	Alpha
$_{94}^{242}\text{Pu}$	3.79×10^5 years	Alpha
$_{94}^{240}\text{Pu}$	6,760 years	Alpha
$_{94}^{239}\text{Pu}$	24,360 years	Alpha
$_{19}^{40}\text{K}$	1.3×10^9 years	Alpha

measured in fractions of seconds, and other isotopes have half-
lives measured in seconds, minutes, hours, days, months, years,
or billions of years. Table 15.3 lists half-lives of some of the isotopes, and the process is illustrated in figure 15.8.

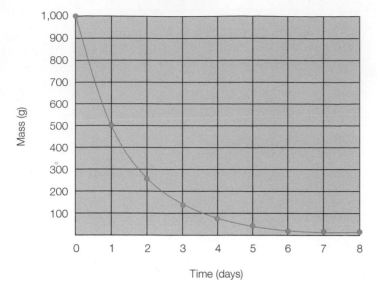

Figure 15.8

Radioactive decay of a hypothetical isotope with a half-life of one day. Each sample decays by one-half to some other element in each half-life of one day. Actual half-lives may be in seconds, minutes, or any time unit up to billions of years.

As an example of the half-life measure, consider a hypothetical isotope that has a half-life of one day. The half-life is independent of the amount of the isotope being considered, but suppose you start with a 1.0 kg sample of this element with a half-life of one day. One day later, you will have half of the original sample, or 500 g. The other half did not disappear, but it is now the decay products, that is, some new element. During the next day half of the remaining nuclei will disintegrate and only 250 g of the initial sample is still the original element. One-half of the remaining sample will disintegrate each day until the original sample no longer exists.

The half-life of a radioactive nucleus is related to its radioactive decay constant by

$$\text{half-life} = \frac{\text{a mathematical constant}}{\text{decay constant}}$$

or

$$t_{1/2} = \frac{0.693}{k} \qquad \textbf{equation 15.2}$$

For example, the radioactive decay constant for uranium-238 was determined earlier to be $4.87 \times 10^{-18}/\text{sec}$. The half-life of uranium-238 is therefore

$$t_{1/2} = \frac{0.693}{4.87 \times 10^{-18}/\text{sec}} = 1.42 \times 10^{17} \text{ sec}$$

This is the half-life of uranium-238 in seconds. There are $60 \times 60 \times 24 \times 365$, or 3.15×10^{7} sec in a year, so

$$\frac{1.42 \times 10^{17} \text{ sec}}{3.15 \times 10^{7} \text{ sec/yr}} = 4.5 \times 10^{9} \text{ yr}$$

The half-life of uranium-238 is thus 4.5 billion years. Figure 15.9 gives the half-life for each step in the uranium-238 decay series.

As you can see from equations 15.1 and 15.2, the half-life of a radioactive isotope is directly proportional to its rate of disintegration. Thus, isotopes with a shorter half-life are more active and are disintegrating at a faster rate. On the other hand, longer half-lives mean less activity and lower rates of radiation.

Measurement of Radiation

The measurement of radiation is important in determining the half-life of radioactive isotopes, as you learned in the previous section. Radiation measurement is also important in considering biological effects, which will be discussed in the next section. As is the case with electricity, it is not possible to make direct measurements on things as small as electrons and other parts of atoms. Indirect measurement methods are possible, however, by considering the effects of the radiation.

Measurement Methods

As Becquerel discovered, radiation affects photographic film, exposing it as visible light does. Since the amount of film exposure is proportional to the amount of radiation, photographic film can be used as an indirect measure of radiation. Today, people who work around radioactive materials or X rays carry light-tight film badges. The film is replaced periodically and developed. The optical density of the developed film provides a record of the worker's exposure to radiation.

There are also devices that indirectly measure radiation by measuring an effect of the radiation. An **ionization counter** is one type of device and measures ions produced by radiation. A second type of device is called a **scintillation counter.** "Scintillate" is a word meaning "sparks or flashes," and a scintillation counter measures the flashes of light produced when radiation strikes a phosphor.

The most common example of an ionization counter is known as a **Geiger counter** (figure 15.11). The working components of a Geiger counter are illustrated in figure 15.11. Radiation is received in a metal tube filled with an inert gas, such as argon, through a thin plastic window that is transparent to alpha, beta, and gamma radiation. An insulated wire inside the tube is connected to the positive terminal of a direct current source. The metal cylinder around the insulated wire is connected to the negative terminal. There is not a current between the center wire and the metal cylinder because the gas acts as an insulator. When radiation passes through the window, however, it ionizes some of the gas atoms, releasing free electrons. These electrons are accelerated by the field between the wire and cylinder, and the accelerated electrons ionize more gas molecules, which results in an *avalanche* of free electrons. The avalanche creates a pulse of current that is amplified, then measured. More radiation means more avalanches, so the pulses are an indirect means of measuring radiation. When connected to a speaker or earphone, each avalanche produces a "pop" or "click."

Some materials are *phosphors,* substances that emit a flash of light when excited by radiation. Zinc sulfide, for example, is used in television screens and luminous watches, and it was used by Rutherford to detect alpha particles. A luminous watch dial has a mixture of zinc sulfide and a small amount of radium sul-

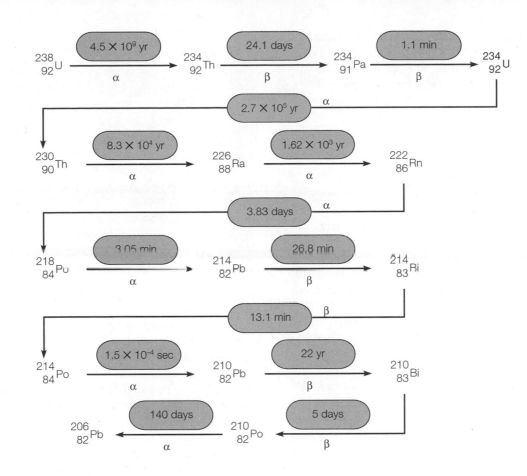

Figure 15.9

The half-life of each step in the uranium-238 radioactive decay series.

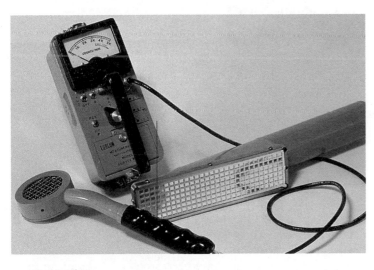

Figure 15.10

This is a beta-gamma probe, which can measure beta and gamma radiation in millirems per unit of time.

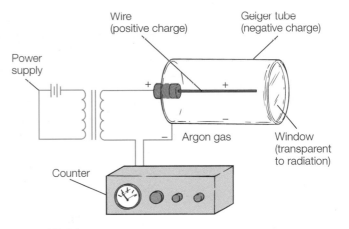

Figure 15.11

The working parts of a Geiger counter.

fate. A zinc sulfide atom gives off a tiny flash of light when struck by radiation from a disintegrating radium nucleus. A scintillation counter measures the flashes of light through the photoelectric effect, producing free electrons that are accelerated to produce a pulse of current. Again, the pulses of current are used as an indirect means to measure radiation.

Radiation Units

You have learned that *radioactivity* is a property of isotopes with unstable, disintegrating nuclei and *radiation* is emitted particles (alpha or beta) or energy traveling in the form of photons (gamma). Radiation can be measured (1) at the source of radioactivity or (2) at a place of reception, where the radiation is absorbed.

The *activity* of a radioactive source is a measure of the number of nuclear disintegrations per unit of time. The unit of

Nuclear Reactions

activity at the source is called a **curie** (Ci), which is defined as 3.70×10^{10} nuclear disintegrations per second. The radioactivity can be measured by a radiation counter, or it can be calculated from the radioactive decay rate. For example, a 238 g sample of uranium-238 has a decay rate of 2.93×10^6 disintegrations per second, so the activity in curies is

$$\frac{2.93 \times 10^6}{3.70 \times 10^{10}} = 7.92 \times 10^{-5} \text{ Ci}$$

A unit frequently mentioned is a *picocurie,* which is a millionth of a millionth of a curie.

As radiation from a source moves out and strikes a material it gives the material energy. The amount of energy released by radiation striking living tissue is usually very small, but it can cause biological damage nonetheless. Chemical bonds are broken and free polyatomic ions are produced by radiation, and the broken bonds and free polyatomic ions are the damaging results.

One measure of radiation received by a material is called the **rad.** The term *rad* is from *r*adiation *a*bsorbed *d*ose, and one rad releases 1×10^{-5} J/g (1×10^{-3} J/kg). Another measure of radiation received considers the biological effect from a rad. This unit is called a **rem,** which takes into account the possible biological damage produced by different types of radiation. The term *rem* is from *r*oentgen *e*quivalent *m*an (a roentgen is another measure of radiation). The equivalent measure is needed because alpha radiation, for example, has a greater ionizing power than beta or gamma radiation, so fewer rads of alpha are required to produce a rem. Beta and gamma radiation are the most penetrating, however, and alpha radiation barely penetrates the skin. Alpha radiation can be very damaging if the source gets inside the body, which is the reason so many people are concerned about exposure to radon gas. Radon is chemically inert and cannot be filtered or absorbed by a gas mask or any other means. Most common isotopes of radon are alpha emitters.

Overall, there are many factors and variables that affect the possible damage from radiation, including the distance from the source and what shielding materials are between a person and a source. A *millirem* is 1/1,000 of a rem and is the unit of choice when low levels of radiation are discussed.

Radiation Exposure

Natural radioactivity is a part of your environment, and you receive between 100 and 500 millirems each year from natural sources. This radiation from natural sources is called **background radiation.** Background radiation comes from outer space in the form of cosmic rays and from unstable isotopes in the ground, building materials, and foods. Many activities and situations will increase your yearly exposure to radiation. For example, the atmosphere absorbs some of the cosmic rays from space, so the less atmosphere above you, the more radiation you will receive. You are exposed to one additional millirem per year for each 100 feet you live above sea level. You receive approximately 0.3 millirem for each hour spent on a jet flight. Airline crews receive an additional 300 to 400 millirems per year because they spend so much time high in the atmosphere. Addi-

Table 15.4

Approximate single dose, whole body effects of radiation exposure

Level	Comment
0.130 rem	Average annual exposure to natural background radiation
0.500 rem	Upper limit of annual exposure to general public
25.0 rem	Threshold for observable effects such as blood count changes
100.0 rem	Fatigue and other symptoms of radiation sickness
200.0 rem	Definite radiation sickness, bone marrow damage, possibility of developing leukemia
500.0 rem	Lethal dose for 50 percent of individuals
1,000.0 rem	Lethal dose for all

tional radiation exposure comes from medical X rays, television sets, and luminous objects such as watch and clock dials. In general, the background radiation exposure for the average person is about 130 millirems per year.

What are the consequences of radiation exposure? Radiation can be a hazard to living organisms because it produces ionization along its path of travel. This ionization can (1) disrupt chemical bonds in essential macromolecules such as DNA and (2) produce molecular fragments, which are free polyatomic ions that can interfere with enzyme action and other essential cell functions. Tissues with highly active cells are more vulnerable to radiation damage than others, such as blood-forming tissue. Thus, one of the symptoms of an excessive radiation exposure is an altered blood count. Table 15.4 compares the estimated results of various levels of acute radiation exposure.

Radiation is not a mysterious, unique health hazard. It is a hazard that should be understood and neither ignored nor exaggerated. Excessive radiation exposure should be avoided, just as you avoid excessive exposure to other hazards such as certain chemicals, electricity, or even sunlight. Everyone agrees that *excessive* radiation exposure should be avoided, but there is some controversy about long-term, low-level exposure and its possible role in cancer. Some claim that tolerable low-level exposure does not exist because that is not possible. Others point to many studies comparing high and low background radioactivity with cancer mortality data. For example, no cancer mortality differences could be found between people receiving 500 or more millirems a year and those receiving less than 100 millirems a year. The controversy continues, however, because of lack of knowledge about long-term exposure. Two models of long-term, low-level radiation exposure have been proposed: (1) a linear model and (2) a threshold model. The *linear model* proposes that any radiation exposure above zero is damaging and can produce cancer and genetic damage. The *threshold model* proposes that the human body can repair damage and get rid of damaging free polyatomic ions up to a certain exposure level called the threshold (figure 15.12). The controversy over long-term, low-level radiation exposure will probably continue until there is clear evidence about which model is correct. Whichever is correct will not lessen the need for rational risks versus cost-benefit analyses of all energy alternatives.

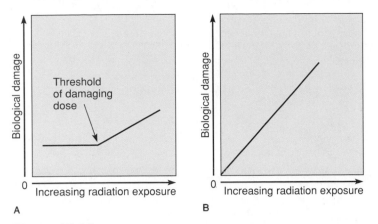

Figure 15.12
Graphic representation of the (a) threshold model and (b) linear model of low-level radiation exposure. The threshold model proposes that the human body can repair damage up to a threshold. The linear model proposes that any radiation exposure is damaging.

Nuclear Energy

As discussed, some nuclei are unstable because they are too large or because they have an unstable neutron-to-proton ratio. These unstable nuclei undergo radioactive decay, forming products of greater stability. An example of this radioactive decay is the alpha emission reaction of uranium-238 to thorium-234,

$$^{238}_{92}U \rightarrow {}^{234}_{90}Th + {}^{4}_{2}He$$
$$238.0003 \text{ u} \rightarrow 233.9942 \text{ u} + 4.00150 \text{ u}$$

The numbers below the nuclear equation are the *nuclear* masses (u) of the reactant and products. As you can see, there seems to be a loss of mass in the reaction,

$$233.9942 + 4.00150 - 238.0003 = -0.0046 \text{ u}$$

This change in mass is related to the energy change according to the relationship that was formulated by Albert Einstein in 1905. The relationship is

$$E = mc^2 \qquad \qquad \textbf{equation 15.3}$$

where E is a quantity of energy, m is a quantity of mass, and c is a constant of the speed of light in a vacuum, 3.00×10^8 m/sec. According to this relationship, matter and energy are the same thing, and energy can be changed to matter and vice versa. Since the mass of a mole in grams is numerically equal to the atomic mass unit (u) of a nucleus, the mass change for a mole of decaying uranium-238 is -0.0046 g, or -4.6×10^{-6} kg. Using this mass loss (Δm) in equation 15.3, you can calculate the energy change (ΔE),

$$\Delta E = \Delta mc^2$$
$$= (-4.6 \times 10^{-6} \text{ kg})\left(3.00 \times 10^8 \frac{\text{m}}{\text{sec}}\right)^2$$
$$= (-4.6 \times 10^{-6} \text{ kg})\left(9.00 \times 10^{16} \frac{\text{m}^2}{\text{sec}^2}\right)$$
$$= (-4.6 \times 9.00) \times 10^{(-6 + 16)} \frac{\text{kg} \cdot \text{m}^2}{\text{sec}^2}$$
$$= -4.14 \times 10^{11} \text{ J}$$

Thus, the products of a mole of uranium-238 decaying to more stable products (1) have a lower energy of 4.14×10^{11} J and (2) lost a mass of 4.6×10^{-6} kg. As you can see, a very small amount of matter was converted into a large amount of energy in the process, forming products of lower energy.

The relationship between mass and energy explains why the mass of a nucleus is always *less* than the sum of the masses of the individual particles of which it is made. For example, the masses of the particles making up a helium-4 nucleus are

$$2 \text{ protons } = 2(1.00728 \text{ u}) = 2.01456 \text{ u}$$
$$2 \text{ neutrons} = 2(1.00867 \text{ u}) = \underline{2.01734 \text{ u}}$$
$$4.03190 \text{ u}$$

But the mass of a helium-4 nucleus is 4.00150 u, a difference of 0.03040. The difference between (1) the mass of the individual nucleons making up a nucleus and (2) the actual mass of the nucleus is called the **mass defect** of the nucleus. The explanation for the mass defect is again found in $E = mc^2$. When nucleons join to make a nucleus, energy is released as the more stable nucleus is formed. A mole of helium-4 nuclei would release a very large amount of energy,

$$\Delta E = \Delta mc^2$$
$$= (-3.04 \times 10^{-5} \text{ kg})\left(3.00 \times 10^8 \frac{\text{m}}{\text{sec}}\right)^2$$
$$= (-3.04 \times 10^{-5} \text{ kg})\left(9.00 \times 10^{16} \frac{\text{m}^2}{\text{sec}^2}\right)$$
$$= (-3.04 \times 9.00) \times 10^{(-5 + 16)} \frac{\text{kg} \cdot \text{m}^2}{\text{sec}^2}$$
$$= -2.74 \times 10^{12} \text{ J}$$

By comparison, hydrogen atoms coming together to form one mole of H_2 molecules release about 4.3×10^5 J of chemical energy, or about 10,000,000 times less energy per mole.

The energy equivalent released when a nucleus is formed is the same as the **binding energy**, the energy required to break the nucleus into individual protons and neutrons. The binding energy of the nucleus of any isotope can be calculated from the mass defect of the nucleus.

The ratio of binding energy to nucleon number is a reflection of the stability of a nucleus (figure 15.13). The greatest binding energy per nucleon occurs near mass number 56, with about 1.5×10^{-12} J per nucleon, then decreases for both more massive and less massive nuclei. This means that more massive nuclei can gain stability by splitting into smaller nuclei with the release of energy. It also means that less massive nuclei can gain stability by joining together with the release of energy. The slope also shows that more energy is released in the coming-together process than in the splitting process.

The nuclear reaction of splitting a massive nucleus into more stable, less massive nuclei with the release of energy is **nuclear fission** (figure 15.14). Nuclear fission occurs rapidly in an atomic bomb explosion and occurs relatively slowly in a nuclear reactor. The nuclear reaction of less massive nuclei, coming together to form more stable, and more massive, nuclei with the release of energy is **nuclear fusion.** Nuclear fusion occurs rap-

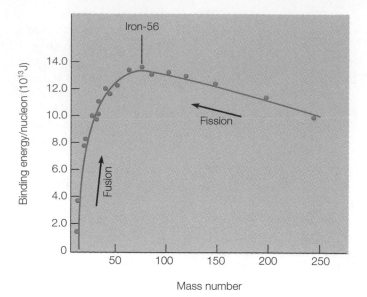

Figure 15.13

The maximum binding energy per nucleon occurs around mass number 56, then decreases in both directions. As one result, fission of massive nuclei and fusion of less massive nuclei both release energy.

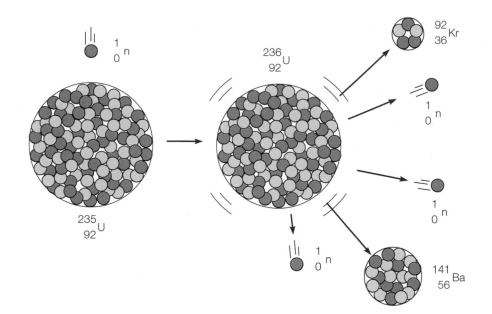

Figure 15.14

The fission reaction occurring when a neutron is absorbed by a uranium-235 nucleus. The deformed nucleus splits any number of ways into lighter nuclei, releasing neutrons in the process.

idly in a hydrogen bomb explosion and occurs continually in the sun, releasing the energy essential for the continuation of life on the earth. Nuclear fission and nuclear fusion are the topics of the next sections.

Nuclear Fission

Nuclear fission was first accomplished in the late 1930s when researchers were attempting to produce isotopes by bombarding massive nuclei with neutrons. In 1938 two German scientists, Otto Hahn and Fritz Strassman, identified the element barium in a uranium sample that had been bombarded with neutrons. Where the barium came from was a puzzle at the time, but soon afterward Lise Meitner, an associate who had moved to Sweden, deduced that uranium nuclei had split, producing barium. The reaction might have been

$$\ _{0}^{1}n + \ _{92}^{235}U \rightarrow \ _{56}^{141}Ba + \ _{36}^{92}Kr + 3 \ _{0}^{1}n$$

The phrase "might have been" is used because a massive nucleus can split in many different ways, producing different products. About thirty-five different, less massive nuclei have been

Table 15.5

Fragments and products from fission of uranium-235

Isotope	Major Mode of Decay	Half-Life	Isotope	Major Mode of Decay	Half-Life
Tritium	Beta	12.26 years	Cerium-144	Beta, gamma	285 days
Carbon-14	Beta	5,930 years	Promethium-147	Beta	2.6 years
Argon-41	Beta, gamma	1.83 hours	Samarium-151	Beta	90 years
Iron-55	Electron capture	2.7 years	Europium-154	Beta, gamma	16 years
Cobalt-58	Beta, gamma	71 days	Lead-210	Beta	22 years
Cobalt-60	Beta, gamma	5.26 years	Radon-222	Alpha	3.8 days
Nickel-63	Beta	92 years	Radium-226	Alpha, gamma	1,620 years
Krypton-85	Beta, gamma	10.76 years	Thorium-229	Alpha	7,300 years
Strontium-89	Beta	5.4 days	Thorium-230	Alpha	26,000 years
Strontium-90	Beta	28 years	Uranium-234	Alpha	2.48×10^5 years
Yttrium-91	Beta	59 days	Uranium-235	Alpha, gamma	7.13×10^8 years
Zirconium-93	Beta	9.5×10^5 years	Uranium-238	Alpha	4.51×10^9 years
Zirconium-95	Beta, gamma	65 days	Neptunium-237	Alpha	2.14×10^6 years
Niobium-95	Beta, gamma	35 days	Plutonium-238	Alpha	89 years
Technetium-99	Beta	2.1×10^5 years	Plutonium-239	Alpha	24,360 years
Ruthenium-106	Beta	1 year	Plutonium-240	Alpha	6,760 years
Iodine-129	Beta	1.6×10^7 years	Plutonium-241	Beta	13 years
Iodine-131	Beta, gamma	8 days	Plutonium-242	Alpha	3.79×10^5 years
Xenon-133	Beta, gamma	5.27 days	Americium-241	Alpha	458 years
Cesium-134	Beta, gamma	2.1 years	Americium-243	Alpha	7,650 years
Cesium-135	Beta	2×10^6 years	Curium-242	Alpha	163 days
Cesium-137	Beta	30 years	Curium-244	Alpha	18 years
Cerium-141	Beta	32.5 days			

identified among the fission products of uranium-235. Some of these products are fission fragments and some are produced as unstable fragments that undergo radioactive decay. These fission fragments are listed in table 15.5, together with their major modes of radioactive decay and half-lives. Some of the isotopes are the focus of concern about nuclear wastes, the topic of the Feature at the end of this chapter.

Fissioning of a uranium-235 nucleus produces two or three neutrons along with other products. These neutrons can each move to other uranium-235 nuclei where they are absorbed, causing fission with the release of more neutrons, which move to other uranium-235 nuclei to continue the process. A reaction where the products are able to produce more reactions in a self-sustaining series is called a **chain reaction.** A chain reaction is self-sustaining until all the uranium-235 nuclei have fissioned or until the neutrons fail to strike a uranium-235 nucleus (figure 15.15).

You might wonder why all the uranium in the universe does not fission in a chain reaction. Natural uranium is mostly uranium-238, an isotope that does not fission easily. Only about 0.7 percent of natural uranium is the highly fissionable uranium-235. This low ratio of readily fissionable uranium-235

nuclei makes it unlikely that a stray neutron would be able to achieve a chain reaction.

In order to achieve a chain reaction, there must be (1) a sufficient mass with (2) a sufficient concentration of fissionable nuclei. When the mass and concentration are sufficient to sustain a chain reaction the amount is called a **critical mass.** Likewise, a mass too small to sustain a chain reaction is called a *subcritical mass.* A mass of sufficiently pure uranium-235 (or plutonium-239) that is large enough to produce a rapidly accelerating chain reaction is called a *supercritical mass.* An atomic bomb is simply a device that uses a small, conventional explosive to push subcritical masses of fissionable material into a supercritical mass. Fissioning occurs almost instantaneously in the supercritical mass, and tremendous energy is released in a violent explosion.

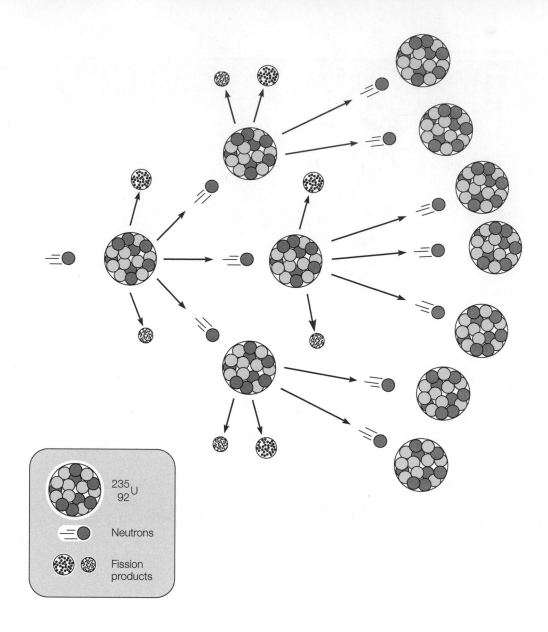

235 92 U

Neutrons

Fission products

Figure 15.15

A schematic representation of a chain reaction. Each fissioned nucleus releases neutrons, which move out to fission other nuclei. The number of neutrons can increase quickly with each series.

Nuclear Power Plants

The nuclear part of a nuclear power plant is the **nuclear reactor,** a steel vessel in which a controlled chain reaction of fissionable material releases energy (figure 15.16). In the most popular design, called a pressurized light-water reactor, the fissionable material is enriched 3 percent uranium-235 and 97 percent uranium-238 that has been fabricated in the form of small ceramic pellets. The pellets are encased in a long zirconium alloy tube called a **fuel rod** (figure 15.17). The fuel rods are locked into a *fuel rod assembly* by locking collars, arranged in a way to permit pressurized water to flow around each fuel rod and to allow the insertion of *control rods* between the fuel rods. **Control rods** are constructed of materials, such as cadmium, that absorb neutrons. The lowering or raising of control rods within the fuel rod assemblies slows or increases the chain reaction by

varying the amount of neutrons absorbed. When lowered completely into the assembly, enough neutrons are absorbed to stop the chain reaction.

It is physically impossible for the low-concentration fuel pellets to form a supercritical mass. A nuclear reactor in a power plant can only release energy at a comparatively slow rate, and it is impossible for a nuclear power plant to produce a nuclear explosion. In a pressurized water reactor the energy released is carried away from the reactor by pressurized water in a closed pipe called the **primary loop** (figure 15.18). The water is pressurized at about 150 atmospheres (about 2,200 lb/in^2) to keep the water from boiling, which may be 350°C (about 660°F).

In the pressurized light-water (ordinary water) reactor the circulating pressurized water acts as a coolant, carrying heat away from the reactor. The water also acts as a **moderator,** a

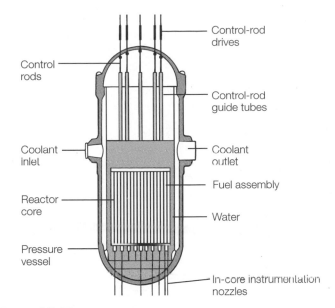

Figure 15.16

A schematic representation of the basic parts of a nuclear reactor. The largest commercial nuclear power plant reactors are nine- to eleven-inch thick steel vessels with a stainless steel liner, standing about forty feet high with a diameter of sixteen feet. Such a reactor has four pumps, which move 440,000 gallons of water per minute through the primary loop.

A

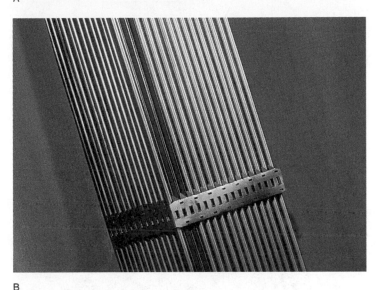

B

Figure 15.17

(a) These are uranium oxide fuel pellets that are stacked inside fuel rods, which are then locked together in a fuel rod assembly. (b) A fuel rod assembly. See also figure 15.19, where you can see a fuel rod assembly being loaded into a reactor.

substance that slows neutrons so they are more readily absorbed by uranium-235 nuclei. Other reactor designs use heavy water (deuterium dioxide) or graphite as a moderator.

Water from the closed primary loop is circulated through a heat exchanger called a **steam generator** (figure 15.19). The pressurized high-temperature water from the reactor moves through hundreds of small tubes inside the generator as *feedwater* from the **secondary loop** flows over the tubes. The water in the primary loop heats feedwater in the steam generator, then returns to the nuclear reactor to become heated again. The feedwater is heated to steam at about 235°C (455°F) with a pressure of about 68 atmospheres (1,000 lb/in²). This steam is piped to the turbines, which turn an electric generator (figure 15.20).

After leaving the turbines, the spent steam is condensed back to liquid water in a second heat exchanger receiving water from the cooling towers. Again, the cooling water does not mix with the closed secondary loop water. The cooling tower water enters the condensing heat exchanger at about 32°C (90°F) and leaves at about 50°C (about 120°F) before returning to a cooling tower, where it is cooled by evaporation. The feedwater is preheated, then recirculated back to the steam generator to start the cycle over again. The steam is condensed back to liquid water because of the difficulty of pumping and reheating steam.

After a period of time the production of fission products in the fuel rods begins to interfere with effective neutron transmission, so the reactor is shut down annually for refueling. During refueling about one-third of the fuel that had the longest exposure in the reactor is removed as "spent" fuel. New fuel rod assemblies are inserted to make up for the part removed. However, only about 4 percent of the "spent" fuel is unusable waste, about 94 percent is uranium-238, 0.8 percent is uranium-235, and about 0.9 percent is plutonium (figure 15.21). Thus, "spent" fuel rods contain an appreciable amount of usable uranium and plutonium. For now, spent reactor fuel rods are mostly stored in cooling pools at the nuclear plant sites. In the future, a decision will be made either to reprocess the spent fuel, recovering the uranium and plutonium through chemical reprocessing, or put the fuel in terminal storage. Concerns about reprocessing are based on the fact that plutonium-239 and uranium-235 are fissionable and could possibly be used by terrorist groups to construct nuclear explosive devices. Six other

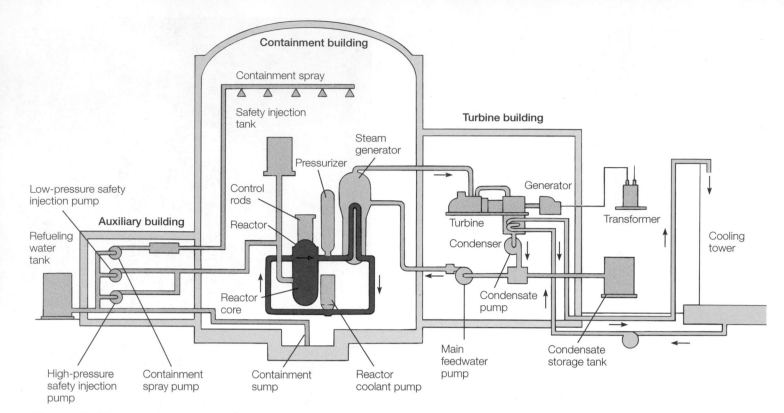

Figure 15.18

A schematic general system diagram of a pressurized water nuclear power plant, not to scale. The containment building is designed to withstand an internal temperature of 300° F at a pressure of 60 lbs/in² and still maintain its leak-tight integrity.

Figure 15.19

Spent fuel rod assemblies are removed and new ones are added to a reactor head during refueling. This shows an initial fuel load to a reactor, which has the upper part removed and set aside for the loading.

Figure 15.20

The turbine deck of a nuclear generating station. There is one large generator in line with four steam turbines in this nonnuclear part of the plant. The large silver tanks are separators that remove water from the steam after it has left the high-pressure turbine and before it is recycled back into the low-pressure turbines.

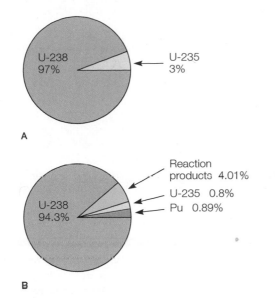

A

B

Figure 15.21

The composition of the nuclear fuel in a fuel rod (a) before and (b) after use over a three-year period in a nuclear reactor.

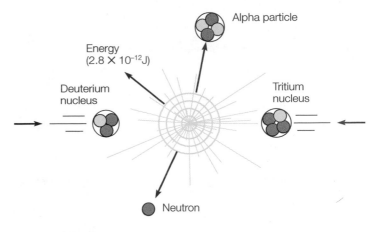

Figure 15.22

A fusion reaction between a tritium nucleus and a deuterium nucleus requires a certain temperature, density, and time of containment to take place.

countries do have reprocessing plants, however, and the spent fuel rods represent an energy source that will accumulate by the year 2000 to an amount equivalent to more than 25 billion barrels of petroleum. Some energy experts say that it would be inappropriate to dispose of such an energy source.

The technology to dispose of fuel rods exists if the decision is made to do so. The longer half-life waste products are mostly alpha emitters. These metals could be converted to oxides, mixed with powdered glass (or a ceramic), melted, and then poured into stainless steel containers. The solidified canisters would then be buried in a stable geologic depository. The glass technology is used in France for disposal of high-level wastes. Buried at two-thousand- to three-thousand-foot depths in solid granite, the only significant means of the radioactive wastes reaching the surface would be through groundwater dissolving the stainless steel, glass, and waste products and then transporting them back to the surface. Many experts believe that if such groundwater dissolving were to take place it would require thousands of years. The radioactive isotopes would thus undergo natural radioactive decay by the time they could reach the surface. Nonetheless, research is continuing on nuclear waste and its disposal. In the meantime, the question of whether it is best to reprocess fuel rods or place them in permanent storage remains unanswered.

What is the volume of nuclear waste under question? If all the spent fuel rods from all the commercial nuclear plants accumulated up to the year 2000 were reprocessed, then mixed with glass, the total amount of glassified waste would make a pile on one football field an estimated 4 m (about 13 ft) high.

Nuclear Fusion

As the graph of nuclear binding energy versus mass numbers shows (see figure 15.13), nuclear energy is released when (1) massive nuclei such as uranium-235 undergo fission and (2) when less massive nuclei come together to form more massive nuclei through nuclear fusion. Nuclear fusion is responsible for the energy released by the sun and other stars. The sun is composed of about 73 percent helium, 26 percent hydrogen, and about 1 percent other elements. Through fusion, the sun converts about 600 million tons of hydrogen to 596 million tons of helium every second. The other 4 million tons of matter are converted into energy. Even at this rate, the sun has enough hydrogen to continue the process for an estimated 5 billion years. There are several fusion reactions that take place between hydrogen and helium isotopes including the following:

$$^1_1H + {}^1_1H \rightarrow {}^2_1H + {}^0_1e$$

$$^2_1H + {}^2_1H \rightarrow {}^3_2He + {}^1_0n$$

$$^3_2He + {}^3_2He \rightarrow {}^4_2He + 2\ {}^1_1H$$

The fusion process would seem to be a desirable energy source on earth because (1) two isotopes of hydrogen, deuterium (^2_1H) and tritium (^3_1H), undergo fusion at a relatively low temperature; (2) the supply of deuterium is practically unlimited, with each gallon of seawater containing about a teaspoonful of deuterium dioxide; and (3) enormous amounts of energy are released with no radioactive by-products.

The oceans contain enough deuterium to generate electricity for the entire world for millions of years, and tritium can be constantly produced by a fusion device. Researchers know what needs to be done to tap this tremendous energy source. The problem is *how* to do it in an economical, continuous energy-producing fusion reactor. The problem, one of the most difficult engineering tasks ever attempted, is meeting three basic fusion reaction requirements of (1) temperature, (2) density, and (3) time (figure 15.22):

1. *Temperature.* Nuclei contain protons and are positively charged, so they experience the electromagnetic repulsion of like charges. This force of repulsion can be overcome, moving the nuclei close enough to fuse together, by giving the nuclei sufficient kinetic energy. The fusion reaction of deuterium and tritium, which has the lowest temperature requirements of any fusion reaction known at the present time, requires temperatures on the order of 100 million °C.

Nuclear Waste

There are two general categories of nuclear wastes: (1) low-level wastes and (2) high-level wastes. The *low-level wastes* are produced by the normal operation of a nuclear reactor. Radioactive isotopes sometimes escape from fuel rods in the reactor and in the spent fuel storage pools. These isotopes are removed from the water by ion-exchange resins and from the air by filters. The used resins and filters will contain the radioactive isotopes and will become low-level wastes. In addition, any contaminated protective clothing, tools, and discarded equipment also become low-level wastes.

Low-level liquid wastes are evaporated, mixed with cement, then poured into fifty-five-gallon steel drums. Solid wastes are compressed and placed in similar drums. The drums are currently disposed of by burial in government-licensed facilities. In general, low-level waste has an activity of less than 1.0 curie per cubic foot. Contact with the low-level waste could expose a person to up to 20 millirems per hour of contact.

High-level wastes from nuclear power plants are spent nuclear fuel rods. At the present time most of the commercial nuclear power plants have these rods in temporary storage at the plant site. These rods are "hot" in the radioactive sense, producing about

Box Figure 15.1

This a standard warning sign for a possible radioactive hazard. Such warning signs would have to be maintained around a nuclear waste depository for thousands of years.

100,000 curies per cubic foot. They are also hot in the thermal sense, continuing to generate heat for months after removal from the reactor. The rods are cooled by heat exchangers connected to storage pools; they could

otherwise achieve an internal temperature as high as 800°C for several decades. In the future, these spent fuel rods will be reprocessed or disposed of through terminal storage.

Agencies of the United States federal government have also accumulated millions of gallons of high-level wastes from the manufacture of nuclear weapons and nuclear research programs. These liquid wastes are stored in million-gallon stainless steel containers that are surrounded by concrete. The containers are located in the states of Washington, Idaho, and South Carolina. The future of this large amount of high-level wastes may be evaporation to a solid form or mixture with a glass or ceramic matrix, which is melted and poured into stainless steel containers. These containers would be buried in solid granite rock in a stable geologic depository. Such high-level wastes must be contained for thousands of years as they undergo natural radioactive decay. Burial at a depth of two thousand to three thousand feet in solid granite would provide protection from exposure by explosives, meteorite impact, or erosion. One major concern about this plan is that a hundred generations later, people might lose track of what is buried in the nuclear garbage dump.

---

2. *Density.* There must be a sufficiently dense concentration of heavy hydrogen nuclei, on the order of 10^{14}/cm^3, so many reactions occur in a short time.
3. *Time.* The nuclei must be confined at the appropriate density up to a second or longer at pressures of a least 10 atmospheres to permit a sufficient number of reactions to take place.

The temperature, density, and time requirements of a fusion reaction are interrelated. A short time of confinement, for example, requires an increased density, and a longer confinement time requires less density. The primary problems of fusion research are the high-temperature requirements and confinement. No material in the world can stand up to a temperature of 100 million °C, and any material container would be instantly vaporized. Thus, research has centered on meeting the fusion reaction requirements without a material container. Two approaches are being tested, *magnetic confinement* and *inertial confinement.*

Magnetic confinement utilizes a **plasma,** a very hot gas consisting of atoms that have been stripped of their electrons because of the high kinetic energies. The resulting positively and negatively charged particles respond to electrical and magnetic forces, enabling researchers to develop a "magnetic bottle,"

that is, magnetic fields that confine the plasma and avoid the problems of material containers that would vaporize. A magnetically confined plasma is very unstable, however, and researchers have compared the problem to trying to carry a block of jello on a pair of rubber bands. Different magnetic field geometries and magnetic "mirrors" are the topics of research in attempts to stabilize the hot, wobbly plasma. Electric currents, injection of fast ions, and radio frequency (microwave) heating methods are also being studied.

Inertial confinement is an attempt to heat and compress small frozen pellets of deuterium and tritium with energetic laser beams or particle beams, producing fusion. The focus of this research is new and powerful lasers, light ion and heavy ion beams. If successful, magnetic or inertial confinement will provide a long-term solution for future energy requirements.

The Source of Nuclear Energy

When elements undergo the natural radioactive decay process, energy is released and the decay products have less energy than the original reactant nucleus. When massive nuclei undergo fission, much energy is rapidly released along with fission products that continue to release energy through radioactive decay. What is the source of all this nuclear energy? The answer to this question is found in current theories about how the universe started and in theories about the life cycle of the stars. Theories about the life cycle of stars are discussed in chapters 16 and 17. For now, consider just a brief introduction to the life cycle of a star in order to understand the ultimate source of nuclear energy.

The current universe is believed to have started with a "big bang" of energy, which created a plasma of protons and neutrons. This primordial plasma cooled rapidly and, after several minutes, began to form hydrogen nuclei. Throughout the newly formed universe massive numbers of hydrogen atoms—on the order of 10^{57} nuclei—were gradually pulled together by gravity into masses that would become the stars. As the hydrogen atoms fell toward the center of each mass of gas they accelerated, just like any other falling object. As they accelerated the contracting mass began to heat up because the average kinetic energy of the atoms increased from acceleration. Eventually, after say ten million years or so of collapsing and heating, the mass of hydrogen condensed to a sphere with a diameter of 1.5 million miles or so, or about twice the size of the sun today. At the same time the temperature increased to millions of degrees, reaching the critical points of density, temperature, and containment for a fusion reaction to begin. Thus, a star was born as hydrogen nuclei fused into helium nuclei, releasing enough energy that the star began to shine.

Hydrogen nuclei in the newborn star had a higher energy per nucleon than helium nuclei, and helium nuclei had more energy per nucleon than other nuclei up to around iron. The fusion process continued for billions of years, releasing energy as heavier and heavier nuclei were formed. Eventually, the star materials were fused into nuclei around iron, the elements with the lowest amount of energy per nucleon, and the star used up its energy source. Larger, more massive dying stars explode into a supernova (discussed in chapter 16). Such an explosion releases a flood of neutrons, which bombard medium-weight nuclei and build them up to more massive nuclei, all the way from iron up to uranium. Thus, the more massive elements were born from an exploding supernova, then spread into space as dust. In a process to be discussed later, this dust became the materials of which planets were made, including the earth. The point for the present discussion, however, is that the energy of naturally radioactive elements, and the energy released during fission, can be traced back to the force of gravitational attraction, which provided the initial energy for the whole process.

Summary

Radioactivity is the spontaneous emission of particles or energy from an unstable atomic nucleus. The modern atomic theory pictures the nucleus as protons and neutrons held together by a short-range *nuclear force* that has moving *nucleons* (protons and neutrons) in *energy shells* analogous to the shell structure of electrons. A graph of the number of neutrons to the number of protons in a nucleus reveals that stable nuclei have a certain neutron-to-proton ratio in a *band of stability*. Nuclei that are above or below the band of stability, and nuclei that are beyond atomic number 83, are radioactive and undergo *radioactive decay*.

Three common examples of radioactive decay involve the emission of an *alpha particle,* a *beta particle,* and a *gamma ray.* An alpha particle is a helium nucleus, consisting of two protons and two neutrons. A beta particle is a high-speed electron that is ejected from the nucleus. A gamma ray is a short wavelength electromagnetic radiation from an excited nucleus. In general, nuclei with an atomic number of 83 or larger become more stable by alpha emission. Nuclei with a neutron-to-proton ratio that is too large become more stable by beta emission. Gamma ray emission occurs from a nucleus that was left in a high-energy state by the emission of an alpha or beta particle.

Each radioactive isotope has its own specific *radioactive decay constant* (k), a ratio of the rate of nuclear disintegration to the total number of nuclei (n), or k = rate/n. The rate is usually described in terms of *half-life,* the time required for one-half the unstable nuclei to decay. Half-life is related to the decay constant by half-life $= 0.693/k$, where 0.693 is a mathematical constant for exponential decay (the natural log of 2).

Radiation is measured by (1) its effects on photographic film, (2) the number of ions it produces, or (3) the flashes of light produced on a phosphor. It is measured at a source in units of a *curie,* defined as 3.70×10^{10} nuclear disintegrations per second. It is measured where received in units of a *rad,* defined as 1×10^{-5} J. A *rem* is a measure of radiation that takes into account the biological effectiveness of different types of radiation damage. In general, the natural environment exposes everyone to 100 to 500 millirems per year, an exposure called *background radiation.* Life-style and location influence the background radiation received, but the average is 130 millirems per year.

Energy and mass are related by Einstein's famous equation of $E = mc^2$, which means that *matter can be converted to energy and energy to matter.* The mass of a nucleus is always less than the sum of the masses of the individual particles of which it is made. This *mass defect* of a nucleus is equivalent to the energy released when the nucleus was formed according to $E = mc^2$. It is also the *binding energy,* the energy required to break the nucleus apart into nucleons.

When the binding energy is plotted against the mass number, the greatest binding energy per nucleon is seen to occur for an atomic number near iron. More massive nuclei therefore release energy by fission, or splitting to more stable nuclei. Less massive nuclei release energy by fusion, the joining of less massive nuclei to produce a more stable, more massive nucleus. Nuclear fission provides the energy for atomic explosions and nuclear power plants. Nuclear fusion is the energy source of the sun and other stars and also holds promise as a future energy source. The source of the energy of a nucleus can be traced back to the gravitational attraction that formed a star.

Summary of Equations

15.1

$$\text{radioactive decay constant} = \frac{\text{decay rate}}{\text{number of nuclei}}$$

$$k = \frac{\text{rate}}{\text{n}}$$

15.2

$$\text{half-life} = \frac{\text{a mathematical constant}}{\text{decay constant}}$$

$$t_{1/2} = \frac{0.693}{k}$$

15.3

$$\text{energy} = \text{mass} \times \text{the speed of light squared}$$

$$E = mc^2$$

Key Terms

alpha particle (p. 315)
background radiation (p. 324)
band of stability (p. 318)
beta particle (p. 315)
binding energy (p. 325)
chain reaction (p. 327)
control rods (p. 328)
critical mass (p. 327)
curie (p. 324)
electromagnetic force (p. 317)
fuel rod (p. 328)
gamma ray (p. 315)
Geiger counter (p. 322)
half-life (p. 321)
ionization counter (p. 322)
mass defect (p. 325)
moderator (p. 328)
nuclear fission (p. 325)

nuclear force (p. 317)
nuclear fusion (p. 325)
nuclear reactor (p. 328)
nucleons (p. 316)
plasma (p. 332)
primary loop (p. 328)
rad (p. 324)
radioactive decay (p. 315)
radioactive decay constant
 (p. 321)
radioactive decay series (p. 320)
radioactivity (p. 315)
rem (p. 324)
scintillation counter (p. 322)
secondary loop (p. 329)
shell model of the nucleus
 (p. 318)
steam generator (p. 329)

Applying the Concepts

1. A high-speed electron ejected from a nucleus during radioactive decay is called a (an)
   a. alpha particle.
   b. beta particle.
   c. gamma ray.
   d. None of the above are correct.

2. The ejection of an alpha particle from a nucleus results in
   a. an increase in the atomic number by one.
   b. an increase in the atomic mass by four.
   c. a decrease in the atomic number by two.
   d. none of the above.

3. The emission of a gamma ray from a nucleus results in
   a. an increase in the atomic number by one.
   b. an increase in the atomic mass by four.
   c. a decrease in the atomic number by two.
   d. none of the above.

4. An atom of radon-222 loses an alpha particle to become a more stable atom of
   a. radium.
   b. bismuth.
   c. polonium.
   d. radon.

5. The nuclear force is
   a. attractive when nucleons are closer than 10^{-15} m.
   b. repulsive when nucleons are closer than 10^{-15} m.
   c. attractive when nucleons are farther than 10^{-15} m.
   d. repulsive when nucleons are farther than 10^{-15} m.

6. Which of the following is more likely to be radioactive?
   a. nuclei with an even number of protons and neutrons
   b. nuclei with an odd number of protons and neutrons
   c. nuclei with the same number of protons and neutrons
   d. Number of protons and neutrons have nothing to do with radioactivity.

7. Which of the following isotopes is more likely to be radioactive?
   a. magnesium-24
   b. calcium-40
   c. astatine-210
   d. ruthenium-101

8. Hydrogen-3 is a radioactive isotope of hydrogen. Which type of radiation would you expect an atom of this isotope to emit?
   a. an alpha particle
   b. a beta particle
   c. either of the above
   d. neither of the above

9. A sheet of paper will stop a (an)
   a. alpha particle.
   b. beta particle.
   c. gamma ray.
   d. None of the above.

10. The most penetrating of the three common types of nuclear radiation is the
    a. alpha particle.
    b. beta particle.
    c. gamma ray.
    d. All have equal penetrating ability.

11. An atom of an isotope with an atomic number greater than 83 will probably emit a (an)
    a. alpha particle.
    b. beta particle.
    c. gamma ray.
    d. None of the above.

12. An atom of an isotope with a large neutron-to-proton ratio will probably emit a (an)
    a. alpha particle.
    b. beta particle.
    c. gamma ray.
    d. None of the above.

13. All of the naturally occurring radioactive decay series end when the radioactive elements have decayed to
    a. lead.
    b. bismuth.
    c. uranium.
    d. hydrogen.

14. The rate of radioactive decay can be increased by increasing the
    a. temperature.
    b. pressure.
    c. size of the sample.
    d. None of the above are correct.

15. The radioactive decay constant is a specific constant only for (a)
    a. particular isotope.
    b. certain temperature.
    c. certain sample size.
    d. all of the above.

16. Isotope A has a half-life of seconds and isotope B has a half-life of millions of years. Which isotope is more radioactive?
    a. It depends on the sample size.
    b. isotope A
    c. isotope B
    d. Unknown, from the information given.

17. A Geiger counter indirectly measures radiation by measuring
    a. ions produced.
    b. flashes of light.
    c. speaker static.
    d. Curies.

18. A measure of radioactivity at the *source* is (the)
    a. Curie.
    b. rad.
    c. rem.
    d. any of the above.

19. A measure of radiation received that considers the biological effect resulting from the radiation is (the)
    a. Curie.
    b. rad.
    c. rem.
    d. any of the above.

20. The mass of a nucleus is always _____ the sum of the masses of the individual particles of which it is made.
    a. equal to
    b. less than
    c. more than
    d. Unable to say without more information.

21. When protons and neutrons join together to make a nucleus, energy is
    a. released.
    b. absorbed.
    c. neither released nor absorbed.
    d. unpredictably absorbed or released.

22. Used fuel rods from a nuclear reactor contain about
    a. 96% usable uranium and plutonium.
    b. 33% usable uranium and plutonium.
    c. 4% usable uranium and plutonium.
    d. 0% usable uranium and plutonium.

23. The source of energy from the sun is
    a. chemical (burning).
    b. fission.
    c. fusion.
    d. radioactive decay.

24. The energy released by radioactive decay and the energy released by nuclear reactions can be traced back to the energy that isotopes acquired from
    a. fusion.
    b. the sun.
    c. gravitational attraction.
    d. the big bang.

Answers

1. b   2. c   3. d   4. c   5. a   6. b   7. c   8. b   9. a
10. c   11. a   12. b   13. a   14. d   15. a   16. b   17. a
18. a   19. c   20. b   21. a   22. a   23. c   24. c

Questions for Thought

1. How is a radioactive material different from a material that is not radioactive?

2. What is radioactive decay? Describe how the radioactive decay rate can be changed if this is possible.

3. Describe three kinds of radiation emitted by radioactive materials. Describe what eventually happens to each kind of radiation after it is emitted.

4. How are positively charged protons able to stay together in a nucleus since like charges repel?

5. What is half-life? Give an example of the half-life of an isotope, describing the amount remaining and the time elapsed after five half-life periods.

6. Would you expect an isotope with a long half-life to be more, the same, or less radioactive than an isotope with a short half-life? Explain.

7. What is (a) a curie? (b) a rad? (c) a rem?

8. What is meant by background radiation? What is the normal radiation dose for the average person from background radiation?

9. Why is there controversy about the effects of long-term, low levels of radiation exposure?

10. What is a mass defect? How is it related to the binding energy of a nucleus? How can both be calculated?

11. Compare and contrast nuclear fission and nuclear fusion.

12. What is a chain reaction?

13. Is it possible for a nuclear power plant to produce a nuclear explosion? Explain.

14. How much of a used nuclear fuel rod is really waste? Describe several different futures for used nuclear fuel rods that might take place.

Nuclear Reactions

Exercises

Note: You will need the Table of Atomic weights inside the back cover of this text.

1. Give the number of protons and the number of neutrons in the nucleus of each of the following isotopes:
   (a) cobalt-60
   (b) potassium-40
   (c) neon-24
   (d) lead-204

2. Write the nuclear symbols for each of the nuclei in exercise 1.

3. Predict if the nuclei in exercise 1 are radioactive or stable, giving your reasoning behind each prediction.

4. Write a nuclear equation for the decay of the following nuclei as they give off a beta particle:
   (a) $^{56}_{26}Fe$
   (b) $^{7}_{4}Be$
   (c) $^{64}_{29}Cu$
   (d) $^{24}_{11}Na$
   (e) $^{214}_{82}Pb$
   (f) $^{32}_{15}P$

5. Write a nuclear equation for the decay of the following nuclei as they undergo alpha emission:
   (a) $^{235}_{92}U$
   (b) $^{226}_{88}Ra$
   (c) $^{239}_{94}Pu$
   (d) $^{214}_{83}Bi$
   (e) $^{230}_{90}Th$
   (f) $^{210}_{84}Po$

6. The half-life of iodine-131 is 8 days. How much of a 1.0 oz sample of iodine-131 will remain after 32 days?

7. If the half-life of strontium-90 is 27.6 years, what is the decay constant for strontium-90?

8. Using the decay constant for strontium-90 obtained in exercise 7, find the number of nuclear disintegrations over a period of time for a molar mass of strontium-90.

9. What is the activity in curies of the molar mass of strontium-90 described in exercise 8?

10. How much energy must be supplied to break a single iron-56 nucleus into separate protons and neutrons? (The mass of an iron-56 nucleus is 55.9206 u, one proton is 1.00728 u, and one neutron is 1.00867 u.)

Note: You will need the Table of Atomic weights inside the back cover of this text.

1. Give the number of protons and the number of neutrons in the nucleus of each of the following isotopes:
   (a) aluminum-25
   (b) technetium-95
   (c) tin-120
   (d) mercury-200

2. Write the nuclear symbols for each of the nuclei in exercise 1.

3. Predict if the nuclei in exercise 1 are radioactive or stable, giving your reasoning behind each prediction.

4. Write a nuclear equation for the beta emission decay of each of the following:
   (a) $^{14}_{6}C$
   (b) $^{60}_{27}Co$
   (c) $^{24}_{11}Na$
   (d) $^{241}_{94}Pu$
   (e) $^{131}_{53}I$
   (f) $^{210}_{82}Pb$

5. Write a nuclear equation for each of the following alpha emission decay reactions:
   (a) $^{241}_{95}Am$
   (b) $^{232}_{90}Th$
   (c) $^{223}_{88}Ra$
   (d) $^{234}_{92}U$
   (e) $^{242}_{96}Cm$
   (f) $^{237}_{93}Np$

6. If the half-life of cesium-137 is 30 years, how much time will be required to reduce a 1.0 kg sample to 1.0 g?

7. The half-life of tritium ($^{3}_{1}H$) is 12.26 years. What is the radioactive decay constant for tritium?

8. What is the number of disintegrations per unit of time for a molar mass of tritium? The decay constant is obtained from exercise 7.

9. Calculate the activity in curies of the molar mass of tritium described in exercise 8.

10. How much energy is needed to separate the nucleons in a single lithium-7 nucleus? (The mass of a lithium-7 nucleus is 7.01435 u, one proton is 1.00728 u, and one neutron is 1.00867 u.)

Appendix A
Mathematical Review

Working with Equations

Many of the problems of science involve an equation, a short-hand way of describing patterns and relationships that are observed in nature. Equations are also used to identify properties and to define certain concepts, but all uses have well-established meanings, symbols that are used by convention, and allowed mathematical operations. This appendix will assist you in better understanding equations and the reasoning that goes with the manipulation of equations in problem-solving activities.

Background

In addition to a knowledge of rules for carrying out mathematical operations, an understanding of certain quantitative ideas and concepts can be very helpful when working with equations. Among these helpful concepts are (1) the meaning of inverse and reciprocal, (2) the concept of a ratio, and (3) fractions.

The term *inverse* means the opposite, or reverse, of something. For example, addition is the opposite, or inverse, of subtraction, and division is the inverse of multiplication. A *reciprocal* is defined as an inverse multiplication relationship between two numbers. For example, if the symbol n represents any number (except zero), then the reciprocal of n is $1/n$. The reciprocal of a number $(1/n)$ multiplied by that number (n) always gives a product of 1. Thus, the number multiplied by 5 to give 1 is $1/5$ ($5 \times 1/5 = 5/5 = 1$). So $1/5$ is the reciprocal of 5 and 5 is the reciprocal of $1/5$. Each number is the *inverse* of the other.

The fraction $1/5$ means 1 divided by five, and if you carry out the division it gives the decimal 0.2. Calculators that have a $1/x$ key will do the operation automatically. If you enter 5, then press the $1/x$ key, the answer of 0.2 is given. If you press the $1/x$ key again, the answer of 5 is given. Each of these numbers is a reciprocal of the other.

A *ratio* is a comparison between two numbers. If the symbols m and n are used to represent any two numbers, then the ratio of the number m to the number n is the fraction m/n. This expression means to divide m by n. For example, if m is 10 and n is 5, the ratio of 10 to 5 is $10/5$, or 2:1.

Working with *fractions* is sometimes necessary in problem-solving exercises, and an understanding of these operations is needed to carry out unit calculations. It is helpful in many of these operations to remember that a number (or a unit) divided by itself is equal to 1, for example,

$$\frac{5}{5} = 1 \qquad \frac{\text{inch}}{\text{inch}} = 1 \qquad \frac{5 \text{ inches}}{5 \text{ inches}} = 1$$

When one fraction is divided by another fraction, the operation commonly applied is to "invert the denominator and multiply." For example, $2/5$ divided by $1/2$ is

$$\frac{\frac{2}{5}}{\frac{1}{2}} = \frac{2}{5} \times \frac{2}{1} = \frac{4}{5}$$

What you are really doing when you invert the denominator of the larger fraction and multiply is making the denominator $(1/2)$ equal to 1. Both the numerator $(2/5)$ and the denominator $(1/2)$ are multiplied by $2/1$, which does not change the value of the overall expression. The complete operation is

$$\frac{\frac{2}{5}}{\frac{1}{2}} \times \frac{\frac{2}{1}}{\frac{2}{1}} = \frac{\frac{2}{5} \times \frac{2}{1}}{\frac{1}{2} \times \frac{2}{1}} = \frac{\frac{4}{5}}{\frac{2}{2}} = \frac{\frac{4}{5}}{1} = \frac{4}{5}$$

Symbols and Operations

The use of symbols seems to cause confusion for some students because it seems different from their ordinary experiences with arithmetic. The rules are the same for symbols as they are for numbers, but you cannot do the operations with the symbols until you know what values they represent. The operation signs, such as $+$, $\div$, $\times$, and $-$ are used with symbols to indicate the operation that you *would* do if you knew the values. Some of the mathematical operations are indicated several ways. For example, a $\times$ b, a$\cdot$b, and ab all indicate the same thing, that a is to be multiplied by b. Likewise, a $\div$ b, a/b, and a $\times$ 1/b all indicate that a is to be divided by b. Since it is not possible to carry out the operations on symbols alone, they are called *indicated operations*.

Operations in Equations

An equation is a shorthand way of expressing a simple sentence with symbols. The equation has three parts: (1) a left side, (2) an equal sign ($=$), which indicates the equivalence of the two sides, and (3) a right side. The left side has the same value and units as the right side, but the two sides may have a very different appearance. The two sides may also have the symbols that indicate mathematical operations ($+$, $-$, $\times$, and so forth) and may be in certain forms that indicate operations (a/b, ab, and so forth). In any case, the equation is a complete expression that states the left side has the same value and units as the right side.

Equations may contain different symbols, each representing some unknown quantity. In science, the term "solve the equation" means to perform certain operations with one symbol (which represents some variable) by itself on one side of the equation. This single symbol is usually, but not necessarily, on the left side and is not present on the other side. For example, the equation $F = ma$ has the symbol F on the left side. In science, you would say that this equation is solved for F. It could also be solved for m or for a, which will be considered shortly. The equation $F = ma$ is solved for F, and the *indicated operation* is to multiply m by a because they are in the form ma, which means the same thing as $m \times a$. This is the only indicated operation in this equation.

A solved equation is a set of instructions that has an order of indicated operations. For example, the equation for the relationship between a Fahrenheit and Celsius temperature, solved for °C, is $°C = 5/9(F - 32°)$. A list of indicated operations in this equation is as follows:

1. Subtract 32° from the given Fahrenheit temperature.
2. Multiply the result of (1) by 5.
3. Divide the result of (2) by 9.

Why are the operations indicated in this order? Because the bracket means 5/9 of the *quantity* $(F - 32°)$. In its expanded form, you can see that $5/9(F - 32°)$ actually means $5/9(F) - 5/9(32°)$. Thus, you cannot multiply by 5 or divide by 9 until you have found the quantity of $(F - 32°)$. Once you have figured out the order of operations, finding the answer to a problem becomes almost routine as you complete the needed operations on both the numbers and the units.

Solving Equations

Sometimes it is necessary to rearrange an equation to move a different symbol to one side by itself. This is known as solving an equation for an unknown quantity. But you cannot simply move a symbol to one side of an equation. Since an equation is a statement of equivalence, the right side has the same value as the left side. If you move a symbol, you must perform the operation in a way that the two sides remain equivalent. This is accomplished by "canceling out" symbols until you have the unknown on one side by itself. One key to understanding the canceling operation is to remember that a fraction with the same number (or unit) over itself is equal to 1. For example, consider the equation $F = ma$, which is solved for F. Suppose you are considering a problem in which F and m are given, and the unknown is a. You need to solve the equation for a so it is on one side by itself. To eliminate the m, you do the *inverse* of the indicated operation on m, dividing both sides by m. Thus,

$$F = ma$$

$$\frac{F}{m} = \frac{ma}{m}$$

$$\frac{F}{m} = a$$

Since m/m is equal to 1, the a remains by itself on the right side. For convenience, the whole equation may be flipped to move the unknown to the left side,

$$a = \frac{F}{m}$$

Thus, a quantity that indicated a multiplication (ma) was removed from one side by an inverse operation of dividing by m.

Consider the following inverse operations to "cancel" a quantity from one side of an equation, moving it to the other side:

If the Indicated Operation of the Symbol You Wish to Remove Is:	Perform This Inverse Operation on Both Sides of the Equation:
multiplication	division
division	multiplication
addition	subtraction
subtraction	addition
squared	square root
square root	square

Example

The equation for finding the kinetic energy of a moving body is $KE = 1/2mv^2$. You need to solve this equation for the velocity, v.

Solution

The order of indicated operations in the equation is as follows:

1. Square v.
2. Multiply v^2 by m.
3. Divide the result of (2) by 2.

To solve for v, this order is *reversed* as the "canceling operations" are used:

Step 1: Multiply both sides by 2

$$KE = \frac{1}{2}mv^2$$

$$2KE = \frac{2}{2}mv^2$$

$$2KE = mv^2$$

Step 2: Divide both sides by m

$$\frac{2KE}{m} = \frac{mv^2}{m}$$

$$\frac{2KE}{m} = v^2$$

Step 3: Take the square root of both sides

$$\sqrt{\frac{2KE}{m}} = \sqrt{v^2}$$

$$\sqrt{\frac{2KE}{m}} = v$$

or

$$v = \sqrt{\frac{2KE}{m}}$$

The equation has been solved for v, and you are now ready to substitute quantities and perform the needed operations (see example 1.3 in chapter 1 for information on this topic).

Significant Figures

The numerical value of any measurement will always contain some uncertainty. Suppose, for example, that you are measuring one side of a square piece of paper as shown in figure A.1. You could say that the paper is *about* 2.5 cm wide and you would be correct. This measurement, however, would be unsatisfactory for many purposes. It does not approach the true value of the length and contains too much uncertainty. It seems clear that the paper width is larger than 2.4 cm but shorter than 2.5 cm. But how much larger than 2.4 cm? You cannot be certain if the paper is 2.44, 2.45, or 2.46 cm wide. As your best estimate, you might say that the paper is 2.45 cm wide. Everyone would agree that you can be certain about the first two numbers (2.4) and they should be recorded. The last number (0.05) has been estimated and is not certain. The two certain numbers, together with one uncertain number, represent the greatest accuracy possible with the ruler being used. The paper is said to be 2.45 cm wide.

A *significant figure* is a number that is believed to be correct with some uncertainty only in the last digit. The value of the width of the paper, 2.45 cm, represents three significant figures. As you can see, the number of significant figures can be determined by the degree of accuracy of the measuring instrument being used. But suppose you need to calculate the area of the paper. You would multiply 2.45 cm × 2.45 cm and the product for the area would be 15.24635 cm². This is a greater accuracy than you were able to obtain with your measuring instrument. The result of a calculation can be no more accurate than the values being treated. Because the measurement had only three significant figures (two certain, one uncertain), then the answer can have only three significant figures. The area is correctly expressed as 15.2 cm².

There are a few simple rules that will help you determine how many significant figures are contained in a reported measurement:

1. All digits reported as a direct result of a measurement are significant.
2. Zero is significant when it occurs between nonzero digits. For example, 607 has three significant figures, and the zero is one of the significant figures.

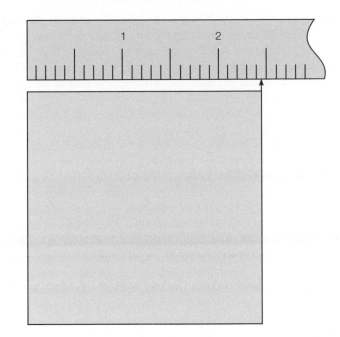

3. In figures reported as *larger than the digit one,* the digit zero is not significant when it follows a nonzero digit to indicate place. For example, in a report that "23,000 people attended the rock concert," the digits 2 and 3 are significant but the zeros are not significant. In this situation the 23 is the measured part of the figure and the three zeros tell you an estimate of how many attended the concert, that is, 23 thousand. If the figure is a measurement rather than an estimate, then it is written *with a decimal point after the last zero* to indicate that the zeros *are* significant. Thus 23,000 has *two* significant figures (2 and 3), but 23,000. has *five* significant figures. The figure 23,000 means "about 23 thousand," but 23,000. means 23,000. and not 22,999 or 23,001.

4. In figures reported as *smaller than the digit one,* zeros after a decimal point that come before nonzero digits *are not* significant and serve only as place holders. For example, 0.0023 has two significant figures, 2 and 3. Zeros alone after a decimal point or zeros after a nonzero digit indicate a measurement, however, so these zeros *are* significant. The figure 0.00230, for example, has three significant figures since the 230 means 230 and not 229 or 231. Likewise, the figure 3.000 cm has four significant figures because the presence of the three zeros means that the measurement was actually 3.000 and not 2.999 or 3.001.

Multiplication and Division

When multiplying or dividing measurement figures, the answer may have no more significant figures than the *least* number of significant figures in the figures being multiplied or divided. This simply means that an answer can be no more accurate than the least accurate measurement entering into the calculation. You cannot improve the accuracy of a measurement by doing a calculation. For example, in multiplying 54.2 mi/hr × 4.0 hr to find out the total distance traveled, the first figure (54.2) has three significant figures but the second (4.0) has only two significant figures. The answer can contain only two significant figures since this is the weakest number of those involved in the calculation. The correct answer is therefore 220 mi, not 216.8 mi. This may seem strange since multiplying the two numbers together gives the answer of 216.8 mi. This answer, however, means a greater accuracy than is possible and the accuracy cannot be improved over the weakest number involved in the calculation. Since the weakest number (4.0) has only two significant figures the answer must also have only two significant figures, which would be 220 mi.

The result of a calculation is *rounded off* to have the same least number of significant figures as the least number of a measurement involved in the calculation. When rounding numbers the last significant figure is increased by one if the number after it is five or larger. If the number after the last significant figure is four or less, the nonsignificant figures are simply dropped. Thus, if two significant figures are called for in an answer as in the preceding example, 216.8 is rounded up to 220 because the last number after the two significant figures is 6 (a number larger than 5). If the calculation result had been 214.8, the rounded number would be 210 miles.

Note that *measurement figures* are the only figures involved in the number of significant figures in the answer. Numbers used as a part of a counting process are not included in the determination of significant figures in the answer. In dividing by 2 to find an average, for example, the 2 is not a measurement so it is ignored when considering the number of significant figures.

Addition and Subtraction

Addition and subtraction operations involving measurements, as multiplication and division, cannot result in an answer that implies greater accuracy than the measurements had before the calculation. Recall that the last digit to the right in a measurement is uncertain, that is, it is the result of an estimate. The answer to an addition or subtraction calculation can have this uncertain number *no farther from the decimal place than it was in the weakest number involved in the calculation.* Thus, when 8.4 is added to 4.926, the weakest number is 8.4 and the uncertain number is .4, one place to the right of the decimal. The sum of 13.326 is therefore rounded to 13.3, reflecting the placement of this weakest doubtful figure.

Example

In an example concerning percentage error, an experimental result of 511 Hz was found for a tuning fork with an accepted frequency value of 522 Hz. The error calculation is

$$\frac{(522 \text{ Hz} - 511 \text{ Hz})}{522 \text{ Hz}} \times 100\% = 2.1\%$$

Since 522 − 511 is 11, the least number of significant figures of measurements involved in this calculation is *two*. Note that the "100" does not enter into the determination since it is not a measurement number. The calculated result (from a calculator) is 2.1072797, which is rounded off to have only two significant figures, so the answer is recorded as 2.1%.

Conversion of Units

The measurement of most properties results in both a numerical value and a unit. The statement that a glass contains 50 cm³ of a liquid conveys two important concepts—the numerical value of 50 and the referent unit of cubic centimeters. Both the numerical value and the unit are necessary to communicate correctly the volume of the liquid.

When working with calculations involving measurement units, *both* the numerical value and the units are treated mathematically. As in other mathematical operations, there are general rules to follow.

1. Only properties with *like units* may be added or subtracted. It should be obvious that adding quantities such as 5 dollars and 10 dimes is meaningless. You must first convert to like units before adding or subtracting.
2. Like or unlike units may be multiplied or divided and treated in the same manner as numbers. You have used this rule when dealing with area (length × length = length², for example, cm × cm = cm²) and when dealing with volume (length × length × length = length³, for example, cm × cm × cm = cm³).

You can use these two rules to create a *conversion ratio* that will help you change one unit to another. Suppose you need to convert 2.3 kg to grams. First, write the relationship between kilograms and grams:

$$1,000 \text{ g} = 1.000 \text{ kg}$$

Next, divide both sides by what you wish to convert *from* (kilograms in this example):

$$\frac{1,000 \text{ g}}{1.000 \text{ kg}} = \frac{1.000 \text{ kg}}{1.000 \text{ kg}}$$

One kilogram divided by 1 kg equals 1, just as 10 divided by 10 equals 1. Therefore, the right side of the relationship becomes 1:

$$\frac{1,000 \text{ g}}{1.000 \text{ kg}} = 1$$

The 1 is usually understood, that is, not stated and the operation is called *canceling*. Canceling leaves you with the fraction 1,000. g/1.000 kg, which is a conversion ratio that can be used to convert from kilograms to grams. You simply multiply the conversion ratio by the numerical value and unit you wish to convert:

$$= 2.3 \text{ kg} \times \frac{1,000 \text{ g}}{1.000 \text{ kg}}$$

$$= \frac{2.3 \times 1,000}{1.000} \frac{\text{kg} \times \text{g}}{\text{kg}}$$

$$= \boxed{2,300 \text{ g}}$$

The kilogram units cancel. Showing the whole operation with units only, you can see how you end up with the correct unit of grams:

$$\text{kg} \times \frac{\text{g}}{\text{kg}} = \frac{\text{kg} \cdot \text{g}}{\text{kg}} = \text{g}$$

Since you did obtain the correct unit, you know that you used the correct conversion ratio. If you had blundered and used an inverted conversion ratio, you would obtain

$$2.3 \text{ kg} \times \frac{1.000 \text{ kg}}{1,000 \text{ g}} = .0023 \frac{\text{kg}^2}{\text{g}}$$

which yields the meaningless, incorrect units of kg^2/g. Carrying out the mathematical operations on the numbers and the units will always tell you whether or not you used the correct conversion ratio.

Example

A distance is reported as 100.0 km and you want to know how far this is in miles.

Solution

First, you need to obtain a *conversion factor* from a textbook or reference book, which usually lists the conversion factors by properties in a table. Such a table will show two conversion factors for kilometers and miles: (1) 1.00 km = 0.621 mi and (2) 1.00 mi = 1.609 km. You select the factor that is in the same form as your problem, for example, your problem is 100.0 km = ? mi. The conversion factor in this form is 1.00 km = 0.621 mi.

Second, you convert this conversion factor into a *conversion ratio* by dividing the factor by what you wish to convert *from:*

conversion factor:	1.00 km = 0.621 mi
divide factor by what you want to convert from:	$\dfrac{1.00 \text{ km}}{1.00 \text{ km}} = \dfrac{0.621 \text{ mi}}{1.00 \text{ km}}$
resulting conversion ratio:	$\dfrac{0.621 \text{ mi}}{\text{km}}$

Note that if you had used the 1.00 mi = 1.609 km factor, the resulting units would be meaningless. The conversion ratio is now multiplied by the numerical value *and unit* you wish to convert:

$$100.0 \text{ km} \times \frac{0.621 \text{ mi}}{\text{km}}$$

$$(100.0)(0.621) \frac{\text{km} \cdot \text{mi}}{\text{km}}$$

$$62.10 \text{ mi}$$

Example

A service station sells gasoline by the liter and you fill your tank with 72.0 liters. How many gallons is this? (Answer: 19.0 gal)

Scientific Notation

Most of the properties of things that you might measure in your everyday world can be expressed with a small range of numerical values together with some standard unit of measure. The range of numerical values for most everyday things can be dealt with by using units (1s), tens (10s), hundreds (100s), or perhaps thousands (1,000s). But the actual universe contains some objects of incredibly large size that require some very big numbers to describe. The sun, for example, has a mass of about 1,970,000,000,000,000,000,000,000,000,000 kg. On the other hand, very small numbers are needed to measure the size and parts of an atom. The radius of a hydrogen atom, for example, is about 0.00000000005 m. Such extremely large and small numbers are cumbersome and awkward since there are so many zeros to keep track of, even if you are successful in carefully counting all the zeros. A method does exist to deal with extremely large or small numbers in a more condensed form. The method is called *scientific notation,* but it is also sometimes called *powers of ten* or *exponential notation,* since it is based on exponents of 10. Whatever it is called, the method is a compact way of dealing with numbers that not only helps you keep track of zeros but provides a simplified way to make calculations as well.

In algebra you save a lot of time (as well as paper) by writing (a $\times$ a $\times$ a $\times$ a $\times$ a) as a^5. The small number written to the right and above a letter or number is a superscript called an *exponent.* The exponent means that the letter or number is to be multiplied by itself that many times, for example, a^5 means "a" multiplied by itself five times, or a $\times$ a $\times$ a $\times$ a $\times$ a. As you can see, it is much easier to write the exponential form of this operation than it is to write it out in the long form. Scientific notation uses an exponent to indicate the power of the base 10. The exponent tells how many times the base, 10, is multiplied by itself. For example,

$$10,000 = 10^4$$
$$1,000 = 10^3$$
$$100 = 10^2$$
$$10 = 10^1$$
$$1 = 10^0$$
$$0.1 = 10^{-1}$$
$$0.01 = 10^{-2}$$
$$0.001 = 10^{-3}$$
$$0.0001 = 10^{-4}$$

This table could be extended indefinitely, but this somewhat shorter version will give you an idea of how the method works. The symbol 10^4 is read as "ten to the fourth power" and means $10 \times 10 \times 10 \times 10$. Ten times itself four times is 10,000, so 10^4 is the scientific notation for 10,000. It is also equal to the number of zeros between the 1 and the decimal point, that is, to write the longer form of 10^4 you simply write 1, then move the decimal point four places to the *right;* ten to the fourth power is 10,000.

The power of ten table also shows that numbers smaller than one have negative exponents. A negative exponent means a reciprocal:

$$10^{-1} = \frac{1}{10} = 0.1$$

$$10^{-2} = \frac{1}{100} = 0.01$$

$$10^{-3} = \frac{1}{1000} = 0.001$$

To write the longer form of 10^{-4}, you simply write 1 then move the decimal point four places to the *left;* ten to the negative fourth power is 0.0001.

Scientific notation usually, but not always, is expressed as the product of two numbers: (1) a number between 1 and 10 that is called the *coefficient* and (2) a power of ten that is called the *exponent.* For example, the mass of the sun that was given in long form earlier is expressed in scientific notation as

$$1.97 \times 10^{30} \text{ kg}$$

and the radius of a hydrogen atom is

$$5.0 \times 10^{-11} \text{ m}$$

In these expressions, the coefficients are 1.97 and 5.0 and the power of ten notations are the exponents. Note that in both of these examples, the exponent tells you where to place the decimal point if you wish to write the number all the way out in the long form. Sometimes scientific notation is written without a coefficient, showing only the exponent. In these cases the coefficient of 1.0 is understood, that is, not stated. If you try to enter a scientific notation in your calculator, however, you will need to enter the understood 1.0 or the calculator will not be able to function correctly. Note also that 1.97×10^{30} kg and the expressions 0.197×10^{31} kg and 19.7×10^{29} kg are all correct expressions of the mass of the sun. By convention, however, you will use the form that has one digit to the left of the decimal.

Example

What is 26,000,000 in scientific notation?

Solution

Count how many times you must shift the decimal point until one digit remains to the left of the decimal point. For numbers larger than the digit 1, the number of shifts tells you how much the exponent is increased, so the answer is

$$2.6 \times 10^{7}$$

which means the coefficient 2.6 is multiplied by 10 seven times.

Example

What is 0.000732 in scientific notation? (Answer: 7.32×10^{-4})

It was stated earlier that scientific notation provides a compact way of dealing with very large or very small numbers, but it provides a simplified way to make calculations as well. There are a few mathematical rules that will describe how the use of scientific notation simplifies these calculations.

To *multiply* two scientific notation numbers, the coefficients are multiplied as usual and the exponents are *added* al-

gebraically. For example, to multiply (2×10^2) by (3×10^3), first separate the coefficients from the exponents,

$$(2 \times 3) \times (10^2 \times 10^3),$$

then multiply the coefficients and add the exponents,

$$6 \times 10^{(2+3)} = 6 \times 10^5$$

Adding the exponents is possible because $10^2 \times 10^3$ means the same thing as $(10 \times 10) \times (10 \times 10 \times 10)$, which equals $(100) \times (1,000)$, or 100,000, which is expressed as 10^5 in scientific notation. Note that two negative exponents add algebraically, for example $10^{-2} \times 10^{-3} = 10^{[(-2)+(-3)]} = 10^{-5}$. A negative and a positive exponent also add algebraically, as in $10^5 \times 10^{-3} = 10^{[(+5)+(-3)]} = 10^2$.

If the result of a calculation involving two scientific notation numbers does not have the conventional one digit to the left of the decimal, move the decimal point so it does, changing the exponent according to which way and how much the decimal point is moved. Note that the exponent increases by one number for each decimal point moved to the left. Likewise, the exponent decreases by one number for each decimal point moved to the right. For example, $938. \times 10^3$ becomes 9.38×10^5 when the decimal point is moved two places to the left.

To *divide* two scientific notation numbers, the coefficients are divided as usual and the exponents are *subtracted.* For example, to divide (6×10^6) by (3×10^2), first separate the coefficients from the exponents,

$$(6 \div 3) \times (10^6 \div 10^2)$$

then divide the coefficients and subtract the exponents,

$$2 \times 10^{(6-2)} = 2 \times 10^4$$

Note that when you subtract a negative exponent, for example, $10^{[(3)-(-2)]}$, you change the sign and add, $10^{(3+2)} = 10^5$.

Example

Solve the following problem concerning scientific notation:

$$\frac{(2 \times 10^4) \times (8 \times 10^{-6})}{8 \times 10^4}$$

Solution

First, separate the coefficients from the exponents,

$$\frac{2 \times 8}{8} \times \frac{10^4 \times 10^{-6}}{10^4}$$

then multiply and divide the coefficients and add and subtract the exponents as the problem requires,

$$2 \times 10^{[(4)+(-6)]-(4)]}$$

solving the remaining additions and subtractions of the coefficients gives

$$2 \times 10^{-6}$$

Appendix B
Solubilities Chart

	Acetate	Bromide	Carbonate	Chloride	Fluoride	Hydroxide	Iodide	Nitrate	Oxide	Phosphate	Sulfate	Sulfide
Aluminum	S	S	—	S	s	i	S	S	i	i	S	d
Ammonium	S	S	S	S	S	S	S	S	—	S	S	S
Barium	S	S	i	S	s	S	S	S	S	i	i	d
Calcium	S	S	i	S	i	s	S	S	s	i	s	d
Copper (I)	—	s	i	s	i	—	i	—	i	—	d	i
Copper (II)	S	S	i	S	S	i	S	S	i	i	S	i
Iron (II)	S	S	i	S	s	i	S	S	i	i	S	i
Iron (III)	S	S	i	S	s	i	S	S	i	i	S	d
Lead	S	s	i	s	i	i	s	S	i	i	i	i
Magnesium	S	S	i	S	i	i	S	S	i	i	S	d
Mercury (I)	s	i	i	i	d	d	i	S	i	i	i	i
Mercury (II)	S	s	i	S	d	i	i	S	i	i	i	i
Potassium	S	S	S	S	S	S	S	S	S	S	S	i
Silver	s	i	i	i	S	—	i	S	i	i	i	i
Sodium	S	S	S	S	S	S	S	S	d	S	S	S
Strontium	S	S	s	S	i	s	S	S	—	i	i	i
Zinc	S	S	i	S	S	i	S	S	i	i	S	i

S—soluble
i—insoluble
s—slightly soluble
d—decomposes

Appendix C

Relative Humidity (%)

Dry-Bulb Temperature (°C)	Difference between Wet-Bulb and Dry-Bulb Temperatures (°C)																			
	1	2	3	4	5	6	7	8	9	10	11	12	13	14	15	16	17	18	19	20
0	81	64	46	29	13															
1	83	66	49	33	17															
2	84	68	52	37	22	7														
3	84	70	55	40	26	12														
4	86	71	57	43	29	16														
5	86	72	58	45	33	20	7													
6	86	73	60	48	35	24	11													
7	87	74	62	50	38	26	15													
8	87	75	63	51	40	29	19	8												
9	88	76	64	53	42	32	22	12												
10	88	77	66	55	44	34	24	15	6											
11	89	78	67	56	46	36	27	18	9											
12	89	78	68	58	48	39	29	21	12											
13	89	79	69	59	50	41	32	23	15	7										
14	90	79	70	60	51	42	34	26	18	10										
15	90	80	71	61	53	44	36	27	20	13	6									
16	90	81	71	63	54	46	38	30	23	15	8									
17	90	81	72	64	55	47	40	32	25	18	11									
18	91	82	73	65	57	49	41	34	27	20	14	7								
19	91	82	74	65	58	50	43	36	29	22	16	10								
20	91	83	74	66	59	51	44	37	31	24	18	12	6							
21	91	83	75	67	60	53	46	39	32	26	20	14	9							
22	92	83	76	68	61	54	47	40	34	28	22	17	11	6						
23	92	84	76	69	62	55	48	42	36	30	24	19	13	8						
24	92	84	77	69	62	56	49	43	37	31	26	20	15	10	5					
25	92	84	77	70	63	57	50	44	39	33	28	22	17	12	8					
26	92	85	78	71	64	58	51	46	40	34	29	24	19	14	10	5				
27	92	85	78	71	65	58	52	47	41	36	31	26	21	16	12	7				
28	93	85	78	72	65	59	53	48	42	37	32	27	22	18	13	9	5			
29	93	86	79	72	66	60	54	49	43	38	33	28	24	19	15	11	7			
30	93	86	79	73	67	61	55	50	44	39	35	30	25	21	17	13	9	5		
31	93	86	80	73	67	61	56	51	45	40	36	31	27	22	18	14	11	7		
32	93	86	80	74	68	62	57	51	46	41	37	32	28	24	20	16	12	9	5	
33	93	87	80	74	68	63	57	52	47	42	38	33	29	25	21	17	14	10	7	
34	93	87	81	75	69	63	58	53	48	43	39	35	30	28	23	19	15	12	8	5
35	94	87	81	75	69	64	59	54	49	44	40	36	32	28	24	20	17	13	10	7

Appendix D
Solutions for Group A Chapter Exercises

Chapter 1

1. Answers will vary but should have the relationship of 100 cm in 1 m, for example, 178 cm = 1.78 m.

2. Since mass density is given by the relationship $\rho = m/V$, then

$$\rho = \frac{m}{V} = \frac{272 \text{ g}}{20.0 \text{ cm}^3}$$

$$= \frac{272}{20.0} \frac{\text{g}}{\text{cm}^3}$$

$$= \boxed{13.6 \frac{\text{g}}{\text{cm}^3}}$$

3. The volume of a sample of lead is given and the problem asks for the mass. From the relationship of $\rho = m/V$, solving for the mass (m) tells you that the mass density (ρ) times the volume (V), or $m = \rho V$. The mass density of lead, 11.4 g/cm³, can be obtained from table 1.3, so

$$\rho = \frac{m}{V}$$

$$V\rho = \frac{m\cancel{V}}{\cancel{V}}$$

$$m = \rho V$$

$$m = \left(11.4 \frac{\text{g}}{\text{cm}^3}\right)(10.0 \text{ cm}^3)$$

$$11.4 \times 10.0 \frac{\text{g}}{\text{cm}^3} \times \text{cm}^3$$

$$114 \frac{\text{g} \cdot \cancel{\text{cm}^3}}{\cancel{\text{cm}^3}}$$

$$= \boxed{114 \text{ g}}$$

4. Solving the relationship $\rho = m/V$ for volume gives $V = m/\rho$, and

$$\rho = \frac{m}{V}$$

$$V\rho = \frac{m\cancel{V}}{\cancel{V}}$$

$$\frac{V\cancel{\rho}}{\cancel{\rho}} = \frac{m}{\rho}$$

$$V = \frac{m}{\rho}$$

$$V = \frac{600 \text{ g}}{3.00 \frac{\text{g}}{\text{cm}^3}}$$

$$= \frac{600}{3.00} \frac{\text{g}}{1} \times \frac{\text{cm}^3}{\text{g}}$$

$$= 200 \frac{\cancel{\text{g}} \cdot \text{cm}^3}{\cancel{\text{g}}}$$

$$= \boxed{200 \text{ cm}^3}$$

5. A 50 cm³ sample with a mass of 34 grams has a density of

$$\rho = \frac{m}{V} = \frac{34.0 \text{ g}}{50.0 \text{ cm}^3}$$

$$= \frac{34.0}{50.0} \frac{\text{g}}{\text{cm}^3}$$

$$= \boxed{0.680 \frac{\text{g}}{\text{cm}^3}}$$

According to table 1.3, 0.680 g/cm³ is the mass density of gasoline, so the substance must be gasoline.

6. The problem asks for a mass and gives a volume, so you need a relationship between mass and volume. Table 1.3 gives the mass density of water as 1.00 g/cm³, which is a density that is easily remembered. The volume is given in liters (L), which should first be converted to cm³ because this is the unit in which density is expressed. The relationship of $\rho = m/V$ solved for mass is ρV, so the solution is

$$\rho = \frac{m}{V} \therefore m = \rho V$$

$$m = \left(1.00 \frac{\text{g}}{\text{cm}^3}\right)(40{,}000 \text{ cm}^3)$$

$$= 1.00 \times 40{,}000 \frac{\text{g}}{\text{cm}^3} \times \text{cm}^3$$

$$= 40{,}000 \frac{\text{g} \cdot \cancel{\text{cm}^3}}{\cancel{\text{cm}^3}}$$

$$= 40{,}000 \text{ g}$$

$$= \boxed{40 \text{ kg}}$$

7. From table 1.3, the mass density of aluminum is given as 2.70 g/cm³. Converting 2.1 kg to the same units as the density gives 2,100 g. Solving $\rho = m/V$ for the volume gives

$$V = \frac{m}{\rho} = \frac{2{,}100 \text{ g}}{2.70 \dfrac{\text{g}}{\text{cm}^3}}$$

$$= \frac{2{,}100}{2.70} \frac{\text{g}}{1} \times \frac{\text{cm}^3}{\text{g}}$$

$$= 777.78 \frac{\cancel{\text{g}} \cdot \text{cm}^3}{\cancel{\text{g}}}$$

$$= \boxed{780 \text{ cm}^3}$$

8. The length of one side of the box is 0.1 m. Reasoning: Since the density of water is 1.00 g/cm³, then the volume of 1,000 g of water is 1,000 cm³. A cubic box with a volume of 1,000 cm³ is 10 cm (since 10 × 10 × 10 = 1,000). Converting 10 cm to m units, the cube is 0.1 m on each edge.

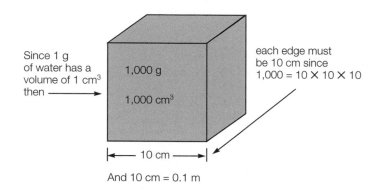

Since 1 g of water has a volume of 1 cm³ then →   1,000 g 1,000 cm³

each edge must be 10 cm since 1,000 = 10 × 10 × 10

|← 10 cm →|

And 10 cm = 0.1 m

9. The relationship between mass, volume, and density is $\rho = m/V$. The problem gives a volume, but not a mass. The mass, however, can be assumed to remain constant during the compression of the bread so the mass can be obtained from the original volume and density, or

$$\rho = \frac{m}{V} \therefore m = \rho V$$

$$m = \left(0.2 \frac{\text{g}}{\text{cm}^3}\right)(3{,}000 \text{ cm}^3)$$

$$= 0.2 \times 3{,}000 \frac{\text{g}}{\text{cm}^3} \times \text{cm}^3$$

$$= 600 \frac{\text{g} \cdot \cancel{\text{cm}^3}}{\cancel{\text{cm}^3}}$$

$$= 600 \text{ g}$$

A mass of 600 g and the new volume of 1,500 cm³ means that the new density of the crushed bread is

$$\rho = \frac{m}{V}$$

$$= \frac{600 \text{ g}}{1{,}500 \text{ cm}^3}$$

$$= \frac{600}{1{,}500} \frac{\text{g}}{\text{cm}^3}$$

$$= \boxed{0.4 \frac{\text{g}}{\text{cm}^3}}$$

10. According to table 1.3, lead has a density of 11.4 g/cm³. Therefore, a 1.00 cm³ sample of lead would have a mass of

$$\rho = \frac{m}{V} \therefore m = \rho V$$

$$m = \left(11.4 \frac{\text{g}}{\text{cm}^3}\right)(1.00 \text{ cm}^3)$$

$$= 11.4 \times 1.00 \frac{\text{g}}{\text{cm}^3} \times \text{cm}^3$$

$$= 11.4 \frac{\text{g} \cdot \cancel{\text{cm}^3}}{\cancel{\text{cm}^3}}$$

$$= 11.4 \text{ g}$$

Also according to table 1.3, copper has a density of 8.96 g/cm³. To balance a mass of 11.4 g of lead, a volume of this much copper would be required:

$$\rho = \frac{m}{V} \therefore V = \frac{m}{\rho}$$

$$V = \frac{11.4 \text{ g}}{8.96 \dfrac{\text{g}}{\text{cm}^3}}$$

$$= \frac{11.4}{8.96} \frac{\text{g}}{1} \times \frac{\text{cm}^3}{\text{g}}$$

$$= 1.27 \frac{\cancel{\text{g}} \cdot \text{cm}^3}{\cancel{\text{g}}}$$

$$= \boxed{1.27 \text{ cm}^3}$$

Chapter 2

1. The distance (d) and the time (t) quantities are given in the problem, and

$$\bar{v} = \frac{d}{t}$$

$$= \frac{285 \text{ mi}}{5.0 \text{ hr}}$$

$$= \frac{285}{5.0} \frac{\text{mi}}{\text{hr}}$$

$$= \boxed{57 \text{ mi/hr}}$$

The units cannot be simplified further. Note two significant figures in the answer, which is the least number of significant figures involved in the division operation.

2. **(a)** The sprinter's average speed is

$$\bar{v} = \frac{d}{t}$$

$$= \frac{200.0 \text{ m}}{21.4 \text{ sec}}$$

$$= \frac{200.0}{21.4} \frac{\text{m}}{\text{sec}}$$

$$= \boxed{9.35 \text{ m/sec}}$$

(Note the *calculator answer* of 9.3457944 m/sec is incorrect. The least number of significant figures involved in the division is three, so *your answer* should have three significant figures. This speed is approximately equivalent to 30 ft/sec or 20 mi/hr.)

(b) To find the time involved in maintaining a speed for a certain distance the relationship between average speed ($\bar{v}$), distance (d), and time (t) can be solved for t:

$$\bar{v} = \frac{d}{t}$$

$$\bar{v}t = d$$

$$t = \frac{d}{\bar{v}}$$

$$t = \frac{420,000 \text{ m}}{9.35 \frac{\text{m}}{\text{sec}}}$$

$$= \frac{420,000}{9.35} \times \frac{\text{m} \cdot \text{sec}}{\text{m}}$$

$$= 44,900 \text{ sec}$$

The seconds can be converted to hours by dividing by 3,600 sec/hr (60 sec in a minute times 60 minutes in an hour):

$$\frac{44,900 \text{ sec}}{3,600 \frac{\text{sec}}{\text{hr}}} = \frac{44,900 \text{ sec}}{3,600} \times \frac{\text{hr}}{\text{sec}}$$

$$= 12.5 \frac{\text{sec} \cdot \text{hr}}{\text{sec}}$$

$$= \boxed{12.5 \text{ hr}}$$

3. Average speed is represented by the symbol $\bar{v}$, and the relationship between the quantities is

(a) $\bar{v} = \frac{d}{t} = \frac{400.0 \text{ mi}}{8.00 \text{ hr}} = \boxed{50.0 \text{ mi/hr}}$

(b) $\bar{v} = \frac{d}{t} = \frac{400.0 \text{ mi}}{7.00 \text{ hr}} = \boxed{57.1 \text{ mi/hr}}$

4. Again, the relationship between average velocity ($\bar{v}$), distance (d), and time (t) can be solved for time:

$$\bar{v} = \frac{d}{t}$$

$$\bar{v}t = d$$

$$t = \frac{d}{\bar{v}}$$

$$t = \frac{5280 \text{ ft}}{2360 \frac{\text{ft}}{\text{sec}}}$$

$$= \frac{5280}{2360} \frac{\text{ft}}{1} \times \frac{\text{sec}}{\text{ft}}$$

$$= 2.24 \frac{\text{ft} \cdot \text{sec}}{\text{ft}}$$

$$= \boxed{2.24 \text{ sec}}$$

5. The relationship between average velocity ($\bar{v}$), distance (d), and time (t) can be solved for distance:

$$\bar{v} = \frac{d}{t} \therefore d = \bar{v}t$$

$$d = \left(40.0 \frac{\text{m}}{\text{sec}}\right)(0.4625 \text{ sec})$$

$$= 40.0 \times 0.4625 \frac{\text{m} \cdot \text{sec}}{\text{sec}}$$

$$= \boxed{18.5 \text{ m}}$$

(A distance of 18.5 m is equivalent to about 20 yards (60 feet) in the English system of measurement.)

6. "How many minutes . . . ," is a question about time and the distance is given. Since the distance is given in km and the speed in m/sec, a unit conversion is needed. The easiest thing to do is to convert km to m. There are 1,000 m in a km, and

$$(1.50 \times 10^8 \text{ km}) \times (1 \times 10^3 \text{ m/km}) = 1.50 \times 10^{11} \text{ m}$$

The relationship between average velocity ($\bar{v}$), distance (d), and time (t) can be solved for time:

$$\bar{v} = \frac{d}{t} \therefore t = \frac{d}{\bar{v}}$$

$$t = \frac{1.50 \times 10^{11} \text{ m}}{3.00 \times 10^8 \frac{\text{m}}{\text{sec}}}$$

$$= \frac{1.50}{3.00} \times 10^{11-8} \frac{\text{m}}{1} \times \frac{\text{sec}}{\text{m}}$$

$$= 0.500 \times 10^3 \frac{\text{m} \cdot \text{sec}}{\text{m}}$$

$$= 5.00 \times 10^2 \text{ sec}$$

$$\frac{500 \text{ sec}}{60 \frac{\text{sec}}{\text{min}}} = \frac{500 \text{ sec}}{60} \frac{\text{min}}{1} \times \frac{\text{min}}{\text{sec}}$$

$$= 8.33 \frac{\text{sec} \cdot \text{min}}{\text{sec}}$$

$$= \boxed{8.33 \text{ min}}$$

(Information on how to use scientific notation (also called powers of ten or exponential notation) is located in the Mathematical Review of Appendix A.)

7. The initial velocity (V_i) is given as 100.0 m/sec, the final velocity (V_f) is given as 51.0 m/sec, and the time is given as 5.00 sec. Acceleration, including a deceleration or negative acceleration, is found from a change of velocity during a given time. Thus,

$$\bar{a} = \frac{v_f - v_i}{t}$$

$$= \frac{\left(51.0\frac{m}{sec}\right) - \left(100.0\frac{m}{sec}\right)}{5.00 \text{ sec}}$$

$$= \frac{-49.0\frac{m}{sec}}{5.00 \text{ sec}}$$

$$= -9.80\frac{m}{sec} \times \frac{1}{sec}$$

$$= \boxed{-9.80\frac{m}{sec^2}}$$

(The negative sign means a negative acceleration, or deceleration.)

8. The initial velocity (V_i) is given as 0 ft/sec ("from rest"), the final velocity (V_f) is given as 235 ft/sec, and the time is given as 5.0 sec. Acceleration is a change of velocity during a given time period, so

$$\bar{a} = \frac{v_f - v_i}{t}$$

$$= \frac{\left(235\frac{ft}{sec}\right) - \left(0\frac{ft}{sec}\right)}{5.0 \text{ sec}}$$

$$= \frac{235\frac{ft}{sec}}{5.0 \text{ sec}}$$

$$= 47\frac{ft}{sec} \times \frac{1}{sec}$$

$$= \boxed{47\frac{ft}{sec^2}}$$

(An acceleration of 47 ft/sec² is equivalent to increasing the speed about 32 mi/hr every second.)

9. In this problem the initial speed is something other than zero. The change of velocity is the difference between the final velocity and the initial velocity, or

$$\bar{a} = \frac{v_f - v_i}{t}$$

$$= \frac{\left(220\frac{ft}{sec}\right) - \left(145\frac{ft}{sec}\right)}{11 \text{ sec}}$$

$$= \frac{75\frac{ft}{sec}}{11 \text{ sec}}$$

$$= 6.8\frac{ft}{sec} \times \frac{1}{sec}$$

$$= \boxed{6.8\frac{ft}{sec^2}}$$

(A velocity of 145 ft/sec is equivalent to about 100 mph and 220 ft/sec is equivalent to about 150 mph. An acceleration of 6.8 ft/sec² is a change of speed of about 5 mph every second.)

10. The question is asking for a final velocity (v_f) of an accelerating car. The acceleration ($\bar{a} = 9.0$ ft/sec²) and the time (t = 8.0 sec) are given. The initial velocity (v_i) is zero since the car accelerates from "rest." Therefore, the question involves a relationship between acceleration ($\bar{a}$), final velocity (v_f), and time (t), and

$$\bar{a} = \frac{v_f - v_i}{t}$$

$$\bar{a}t = v_f - v_i$$

$$v_f = \bar{a}t + v_i$$

$$v_f = \left(9.0\frac{ft}{sec^2}\right)(8.0 \text{ sec}) + 0\frac{ft}{sec}$$

$$= 9.0 \times 8.0 \frac{ft \cdot sec}{sec \cdot sec}$$

$$= \boxed{72 \text{ ft/sec}}$$

(A velocity of 72 ft/sec is about 50 mph. An acceleration of 9.0 ft/sec² means that you are increasing your speed about 6 mph every second.)

11. For this type of problem, note that the velocity, distance, and time relationship calls for $\bar{v}$, or *average* velocity. The 88.0 ft/sec is the *initial* velocity (v_i) that the car had before stopping ($v_f = 0$). The average velocity must be obtained before solving for the time required:

(a)

$$\bar{v} = \frac{d}{t}$$

$$\bar{v}t = d$$

$$t = \frac{d}{\bar{v}}$$

$$\bar{v} = \frac{v_f + v_i}{2} = \frac{0 + 88.0 \text{ ft/sec}}{2} = 44.0 \text{ ft/sec}$$

$$t = \frac{100.0 \text{ ft}}{44.0 \text{ ft/sec}}$$

$$= \frac{100.0}{44.0} \text{ ft} \times \frac{sec}{ft}$$

$$= 2.27 \frac{ft \cdot sec}{ft}$$

$$= \boxed{2.27 \text{ sec}}$$

(The 2 is not considered in significant figures since it is not the result of a measurement. A velocity of 88.0 ft/sec is exactly 60 miles per hour.)

(b) $\bar{a} = \dfrac{v_f - v_i}{t} = \dfrac{0 - 88.0 \text{ ft/sec}}{2.27 \text{ sec}}$

$= \dfrac{-88.0}{2.27} \dfrac{\text{ft}}{\text{sec}} \times \dfrac{1}{\text{sec}}$

$= -38.8 \dfrac{\text{ft}}{\text{sec} \cdot \text{sec}}$

$= \boxed{-38.8 \dfrac{\text{ft}}{\text{sec}^2}}$

(The negative sign simply means that the car is slowing, or decelerating from some given speed. The deceleration is equivalent to about 26 mph every second.)

One g is 32.0 ft/sec², so the deceleration was

$\dfrac{38.8 \text{ ft/sec}^2}{32.0 \text{ ft/sec}^2} = \dfrac{38.8 \text{ ft/sec}^2}{32.0 \text{ ft/sec}^2} = \boxed{1.21 \text{ g's}}$

12. A ball thrown straight up decelerates to a velocity of zero, then accelerates back to the surface, just as a dropped ball would do from the height reached. Thus, the time required to decelerate upwards is the same as the time required to accelerate downwards. The ball returns to the surface with the same velocity with which it was thrown (neglecting friction). Therefore:

$\bar{a} = \dfrac{v_f - v_i}{t}$

$\bar{a}t = v_f - v_i$

$v_f = \bar{a}t + v_i$

$= \left(9.80 \dfrac{\text{m}}{\text{sec}^2}\right)(3.0 \text{ sec})$

$= (9.80)(3.0) \dfrac{\text{m}}{\text{sec}^2} \times \text{sec}$

$= 29 \dfrac{\text{m} \cdot \text{sec}}{\text{sec} \cdot \text{sec}}$

$= \boxed{29 \text{ m/sec}}$

(or 96 ft/sec in English units)

(The velocity of the ball—29 m/sec or 96 ft/sec—is about 65 mph.)

13. These three questions are easily answered by using the three sets of relationships, or equations, that were presented in this chapter:

(a) $v_f = \bar{a}t + v$ and when v_i is zero,

$v_f = \bar{a}t$

$v_f = \left(9.80 \dfrac{\text{m}}{\text{sec}^2}\right)(4.00 \text{ sec})$

$= 9.80 \times 4.00 \dfrac{\text{m}}{\text{sec}^2} \times \text{sec}$

$= 39.2 \dfrac{\text{m} \cdot \text{sec}}{\text{sec} \cdot \text{sec}}$

$= \boxed{39.2 \text{ m/sec}}$

(b) $\bar{v} = \dfrac{v_f + v_i}{2} = \dfrac{39.2 \text{ m/sec} + 0}{2} = \boxed{19.6 \text{ m/sec}}$

(c) $\bar{v} = \dfrac{d}{t} \quad \therefore d = \bar{v}t = \left(19.6 \dfrac{\text{m}}{\text{sec}}\right)(4.00 \text{ sec})$

$= 19.6 \times 4.00 \dfrac{\text{m}}{\text{sec}} \times \text{sec}$

$= 78.4 \dfrac{\text{m} \cdot \text{sec}}{\text{sec}}$

$= \boxed{78.4 \text{ m}}$

(A velocity of 39.2 m/sec is equivalent to about 88 mph.)

14. Note that this problem can be solved with a series of three steps as in the previous problem. It can also be solved by the equation that combines all the relationships into one step. Either method is acceptable, but the following example of a one-step solution reduces the possibilities of error since fewer calculations are involved:

$d = \dfrac{1}{2} gt^2 = \dfrac{1}{2}\left(9.80 \dfrac{\text{m}}{\text{sec}^2}\right)(5.00 \text{ sec})^2$

$= \dfrac{1}{2}\left(9.80 \dfrac{\text{m}}{\text{sec}^2}\right)(25.0 \text{ sec}^2)$

$= \left(\dfrac{1}{2}\right)(9.80)(25.0) \dfrac{\text{m}}{\text{sec}^2} \times \text{sec}^2$

$= 4.90 \times 25.0 \dfrac{\text{m} \cdot \text{sec}^2}{\text{sec}^2}$

$= \boxed{123 \text{ m}}$

15. Note that "how long must a car accelerate?" is asking for a unit of time and that the acceleration ($\bar{a}$), initial velocity (v_i), and the final velocity (v_f) are given. The *first step*, before anything else is done, is to write the equation expressing a relationship between these quantities and then solve the equation for the unknown (time in this problem). Since unit conversions are not necessary, quantities are then substituted for the symbols and mathematical operations are performed on both the numbers and units:

$\bar{a} = \dfrac{v_f - v_i}{t} \quad \therefore t = \dfrac{v_f - v_i}{\bar{a}} = \dfrac{50.0 \dfrac{\text{m}}{\text{sec}} - 10.0 \dfrac{\text{m}}{\text{sec}}}{4.00 \dfrac{\text{m}}{\text{sec}^2}}$

$= \dfrac{40.0}{4.00} \dfrac{\text{m}}{\text{sec}} \times \dfrac{\text{sec}^2}{\text{m}}$

$= 10.0 \dfrac{\text{m} \cdot \text{sec} \cdot \text{sec}}{\text{m} \cdot \text{sec}}$

$= \boxed{10.0 \text{ sec}}$

16. $\bar{a} = g \therefore g = \dfrac{v_f - v_i}{t} \therefore v_f = gt$ (when v_i is zero)

$$= \left(9.80 \,\dfrac{m}{sec^2}\right)(3.0 \ sec)$$

$$= (9.80)(3.0) \,\dfrac{m}{sec^2} \times sec$$

$$= 29.4 \,\dfrac{m \cdot sec}{sec \cdot sec}$$

$$= 29.4 \,\dfrac{m}{sec} = \boxed{29 \,\dfrac{m}{sec}}$$

Chapter 3

1. (a) Weight (W) is a downward force from the acceleration of gravity (g) on the mass (m) of an object. This relationship is the same as Newton's second law of motion, $F = ma$, and

$$\text{a) } W = mg = (1.25 \ kg)\left(9.80 \,\dfrac{m}{sec^2}\right)$$

$$= (1.25)(9.80) \ kg \times \dfrac{m}{sec^2}$$

$$= 12.25 \,\dfrac{kg \cdot m}{sec^2}$$

$$= \boxed{12.3 \ N}$$

(b) First, recall that a force (F) is measured in newtons (N) and a newton has units of $N = \dfrac{kg \cdot m}{sec^2}$. Second, the relationship between force (F), mass (m), and acceleration (a) is given by Newton's second law of motion, force = mass times acceleration, or $F = ma$. Thus,

$$\text{b) } F = ma \therefore a = \dfrac{F}{m} = \dfrac{10.0 \,\dfrac{kg \cdot m}{sec^2}}{1.25 \ kg}$$

$$= \dfrac{10.0}{1.25} \,\dfrac{kg \cdot m}{sec^2} \times \dfrac{1}{kg}$$

$$= 8.00 \,\dfrac{kg \cdot m}{kg \cdot sec^2}$$

$$= \boxed{8.00 \,\dfrac{m}{sec^2}}$$

(Note how the units were treated mathematically in this solution and why it is necessary to show the units for a newton of force. The resulting unit in the answer *is* a unit of acceleration, which provides a check that the problem was solved correctly. For your information, 8.00 m/sec² is equivalent to a change of velocity of about 29 km/hr each sec, or 18 mph each sec.)

2.

$$F = ma = (1.25 \ kg)\left(5.00 \,\dfrac{m}{sec^2}\right)$$

$$= (1.25)(5.00) \ kg \times \dfrac{m}{sec^2}$$

$$= 6.25 \,\dfrac{kg \cdot m}{sec^2}$$

$$= \boxed{6.25 \ N}$$

(Note that the solution is correctly reported in *newton* units of force rather than kg·m/sec².)

3. The bicycle tire exerts a backwards force on the road and the equal and opposite reaction force of the road on the bicycle produces the forward motion. (The motion is always in the direction of the applied force.) Therefore,

$$F = ma = (70.0 \ kg)\left(2.0 \,\dfrac{m}{sec^2}\right)$$

$$= (70.0)(2.0) \ kg \times \dfrac{m}{sec^2}$$

$$= 140 \,\dfrac{kg \cdot m}{sec^2}$$

$$= \boxed{140 \ N}$$

4. The question requires finding a force in the metric system, which is measured in newtons of force. Since newtons of force are defined in kg, m, and sec, unit conversions are necessary and these should be done first.

$$1 \,\dfrac{km}{hr} = \dfrac{1000 \ m}{3600 \ sec} = 0.2778 \,\dfrac{m}{sec}$$

Dividing both sides of this conversion factor by what you are converting *from* gives the conversion ratio of

$$\dfrac{0.2778 \,\dfrac{m}{sec}}{\dfrac{km}{hr}}$$

Multiplying this conversion ratio times the two velocities in km/hr will convert them to m/sec as follows:

$$\left(0.2778 \,\dfrac{\dfrac{m}{sec}}{\dfrac{km}{hr}}\right)\left(80.0 \,\dfrac{km}{hr}\right)$$

$$= [0.2778][80.0] \,\dfrac{m}{sec} \times \dfrac{hr}{km} \times \dfrac{km}{hr}$$

$$= 22.2 \,\dfrac{m}{sec}$$

$$\left(0.2778 \,\dfrac{\dfrac{m}{sec}}{\dfrac{km}{hr}}\right)\left(44.0 \,\dfrac{km}{hr}\right)$$

$$= [0.2778][44.0] \; \frac{m}{sec} \times \frac{hr}{km} \times \frac{km}{hr}$$

$$= 12.2 \; \frac{m}{sec}$$

Now you are ready to find the appropriate relationship between the quantities involved. This involves two separate equations: Newton's second law of motion and the relationship of quantities involved in acceleration. These may be combined as follows:

$$F = ma \text{ and } a = \frac{v_f - v_i}{t} \therefore F = m\left(\frac{v_f - v_i}{t}\right)$$

Now you are ready to substitute quantities for the symbols and perform the necessary mathematical operations:

$$= (1,500 \text{ kg})\left(\frac{22.2 \text{ m/sec} - 12.2 \text{ m/sec}}{10.0 \text{ sec}}\right)$$

$$= (1,500 \text{ kg})\left(\frac{10.0 \text{ m/sec}}{10.0 \text{ sec}}\right)$$

$$= 1,500 \times 1.00 \; \frac{kg \cdot \frac{m}{sec}}{sec}$$

$$= 1,500 \; \frac{kg \cdot m}{sec} \times \frac{1}{sec}$$

$$= 1,500 \; \frac{kg \cdot m}{sec \cdot sec}$$

$$= 1,500 \; \frac{kg \cdot m}{sec^2}$$

$$= 1,500 \text{ N} = \boxed{1.5 \times 10^3 \text{ N}}$$

5. A unit conversion is needed as in the previous problem:

$$\left(90.0 \; \frac{km}{hr}\right)\left(0.2778 \; \frac{\frac{m}{sec}}{\frac{km}{hr}}\right) = 25.0 \text{ m/sec}$$

(a) $F = ma \therefore m = \dfrac{F}{a}$ and $a = \dfrac{v_f - v_i}{t}$, so

$$m = \frac{F}{\frac{v_f - v_i}{t}} = \frac{5000.0 \; \frac{kg \cdot m}{sec^2}}{\frac{25.0 \text{ m/sec} - 0}{5.0 \text{ sec}}}$$

$$= \frac{5000.0 \; \frac{kg \cdot m}{sec^2}}{5.0 \; \frac{m}{sec^2}}$$

$$= \frac{5000.0}{5.0} \; \frac{kg \cdot m}{sec^2} \times \frac{sec^2}{m}$$

$$= 1000 \; \frac{kg \cdot m \cdot sec^2}{m \cdot sec^2}$$

$$= \boxed{1.0 \times 10^3 \text{ kg}}$$

(b)
$$W = mg$$

$$= (1.0 \times 10^3 \text{ kg})\left(9.80 \; \frac{m}{sec^2}\right)$$

$$= (1.0 \times 10^3)(9.80) \text{ kg} \times \frac{m}{sec^2}$$

$$= 9.8 \times 10^3 \; \frac{kg \cdot m}{sec^2}$$

$$= \boxed{9.8 \times 10^3 \text{ N}}$$

6.
$$W = mg$$

$$= (70.0 \text{ kg})\left(9.80 \; \frac{m}{sec^2}\right)$$

$$= 70.0 \times 9.80 \text{ kg} \times \frac{m}{sec^2}$$

$$= 686 \; \frac{kg \cdot m}{sec^2}$$

$$= \boxed{686 \text{ N}}$$

7. You were given a mass (m), a force (f), and a time (t) and asked to find a final speed (v_f). These relationships are found in Newton's second law of motion and in the definition of acceleration. The two equations can be combined and solved for v_f:

$$F = ma \text{ and } a = \frac{v_f - v_i}{t}$$

$$\therefore$$

$$F = m\left(\frac{v_f - v_i}{t}\right)$$

$$Ft = m(v_f - v_i)$$

$$\frac{Ft}{m} = v_f - v_i$$

$$v_f = \frac{Ft}{m} + v_i$$

Now the solution becomes a matter of substituting values and carrying out the mathematical operations:

$$v_f = \frac{\left(1,000.0 \; \frac{kg \cdot m}{sec^2}\right)(10.0 \text{ sec})}{1000.0 \text{ kg}} + 0 \; \frac{m}{sec}$$

$$= \frac{(1,000.0)(10.0)}{1,000.0} \; \frac{kg \cdot m}{sec^2} \times \frac{sec}{1} \times \frac{1}{kg}$$

$$= 10.0 \; \frac{kg \cdot m \cdot sec}{kg \cdot sec \cdot sec}$$

$$= \boxed{10.0 \; \frac{m}{sec}}$$

8.
$$p = mv$$

$$= (50 \text{ kg})\left(2 \frac{m}{sec}\right)$$

$$= \boxed{100 \; \frac{kg \cdot m}{sec}}$$

(Note the lower case "p" is the symbol used for momentum. This is one of the few cases where the English letter does not provide a clue about what it stands for. The units for momentum are also somewhat unusual for metric units since they do not have a name or single symbol to represent them.)

9.

$$F_c = \frac{mv^2}{r}$$

$$= \frac{(0.20 \text{ kg})\left(3.0 \dfrac{m}{sec}\right)^2}{1.5 \text{ m}}$$

$$= \frac{(0.20 \text{ kg})\left(9.0 \dfrac{m^2}{sec^2}\right)}{1.5 \text{ m}}$$

$$= \frac{0.20 \times 9.0 \text{ kg·m}^2}{1.5} \times \frac{1}{sec^2} \times \frac{1}{m}$$

$$= 1.2 \frac{\text{kg·m·}\cancel{m}}{sec^2·\cancel{m}}$$

$$= \boxed{1.2 \text{ N}}$$

10. Note that unit conversions are necessary since the definition of the newton unit of force involves different units:

$$F_c = \frac{mv^2}{r}$$

$$\frac{F_c r}{m} = v^2$$

$$v = \sqrt{\frac{F_c r}{m}}$$

$$v = \sqrt{\frac{\left(1.0\dfrac{\text{kg·m}}{sec^2}\right)(0.50 \text{ m})}{0.10 \text{ kg}}}$$

$$= \sqrt{\frac{(1.0)(0.50)}{0.10} \frac{\text{kg·m}}{sec^2} \times \frac{m}{1} \times \frac{1}{kg}}$$

$$= \sqrt{5.0 \text{ m}^2/sec^2}$$

$$= \boxed{2.2 \text{ m/sec}}$$

11. (a)

$$F_c = \frac{mv^2}{r} = \frac{(1000.0 \text{ kg})(10.0 \text{ m/sec})^2}{20.0 \text{ m}}$$

$$= \frac{(1000.0 \text{ kg})(100.0 \text{ m}^2/sec^2)}{20.0 \text{ m}}$$

$$= \frac{(1000.0)(100.0)}{20.0} \frac{\text{kg·m}^2}{sec^2} \times \frac{1}{m}$$

$$= 5000 \frac{\text{kg·}\cancel{m}·m}{sec^2·\cancel{m}}$$

$$= \boxed{5.00 \times 10^3 \text{ N}}$$

(b) The centripetal force that keeps a car on the road while moving around a curve is the frictional force between the tires and the road.

12. (a) Newton's laws of motion consider the resistance to a change of motion, or mass, and not weight. The astronaut's mass is

$$W = mg \therefore m = \frac{W}{g} = \frac{1960.0\dfrac{\text{kg·m}}{sec^2}}{9.80 \dfrac{m}{sec^2}}$$

$$= \frac{1960.0}{9.80} \frac{\text{kg·m}}{sec^2} \times \frac{sec^2}{m} = 200 \text{ kg}$$

(b) From Newton's second law of motion, you can see that the 100 N rocket gives the 200 kg astronaut an acceleration of:

$$F = ma \therefore a = \frac{F}{m} = \frac{100 \dfrac{\text{kg·m}}{sec^2}}{200 \text{ kg}}$$

$$= \frac{100 \text{ kg·m}}{200 \text{ sec}^2} \times \frac{1}{kg} = 0.5 \text{ m/sec}^2$$

(c) An acceleration of 0.5 m/sec² for 2.0 sec will result in a final velocity of

$$a = \frac{v_f - v_i}{t} \therefore v_f = at + v_i$$

$$= (0.5 \text{ m/sec}^2)(2.0 \text{ sec}) + 0 \text{ m/sec}$$

$$= \boxed{1 \text{ m/sec}}$$

Chapter 4

1. (a)

$$W = Fd$$

$$= (10 \text{ lb})(5 \text{ ft})$$

$$= (10)(5) \text{ ft} \times \text{lb}$$

$$= \boxed{50 \text{ ft·lb}}$$

(b) The distance of the bookcase from some horizontal reference level did not change so the gravitational potential energy does not change.

2. The force (F) needed to lift the book is equal to the weight (W) of the book, or F = W. Since W = mg, then F = mg. Work is defined as the product of a force moved through a distance, or W = Fd. The work done in lifting the book is therefore W = mgd, and:

(a)

$$W = mgd$$

$$= (2.0 \text{ kg})(9.80 \text{ m/sec}^2)(2.00 \text{ m})$$

$$= (2.0)(9.80)(2.00) \frac{\text{kg·m}}{sec^2} \times m$$

$$= 39.2 \frac{\text{kg·m}^2}{sec^2}$$

$$= 39.2 \text{ J} = \boxed{39 \text{ J}}$$

(b)

$$PE = Mgh = \boxed{39 \text{ J}}$$

(c)

$$PE_{lost} = KE_{gained} = mgh = \boxed{39 \text{ J}}$$

(or)

$$v = \sqrt{2gh} = \sqrt{(2)(9.80 \text{ m/sec}^2)(2.00 \text{ m})}$$
$$= \sqrt{39.2 \text{ m}^2/\text{sec}^2} \quad \text{(Note)}$$
$$= 6.26 \text{ m/sec}$$

$$KE = \frac{1}{2} mv^2 = \left(\frac{1}{2}\right)(2.0 \text{ kg})(6.26 \text{ m/sec})^2$$
$$= \left(\frac{1}{2}\right)(2.0 \text{ kg})(39.2 \text{ m}^2/\text{sec}^2)$$
$$= (1.0)(39.2) \frac{\text{kg} \cdot \text{m}^2}{\text{sec}^2}$$
$$= \boxed{39 \text{ J}}$$

3. Note that the gram unit must be converted to kg to be consistent with the definition of a newton-meter, or joule unit of energy:

$$KE = \frac{1}{2}mv^2 = \left(\frac{1}{2}\right)(0.15 \text{ kg})(30.0 \text{ m/sec})^2$$
$$= \left(\frac{1}{2}\right)(0.15 \text{ kg})(900 \text{ m}^2/\text{sec}^2)$$
$$= \left(\frac{1}{2}\right)(0.15)(900) \frac{\text{kg} \cdot \text{m}^2}{\text{sec}^2}$$
$$= 67.5 \text{ J} = \boxed{68 \text{ J}}$$

4. The km/hr unit must first be converted to m/sec before finding the kinetic energy. Note also that the work done to put an object in motion is equal to the energy of motion, or kinetic energy that it has as a result of the work. The work needed to bring the object to a stop is also equal to the kinetic energy of the moving object:

Unit conversion:

$$1 \frac{\text{km}}{\text{hr}} = 0.2778 \frac{\frac{\text{m}}{\text{sec}}}{\frac{\text{km}}{\text{hr}}} = \left(90.0 \frac{\text{km}}{\text{hr}}\right)\left(0.2778 \frac{\frac{\text{m}}{\text{sec}}}{\frac{\text{km}}{\text{hr}}}\right) = 25.0 \text{ m/sec}$$

(a)
$$KE = \frac{1}{2}mv^2 = \frac{1}{2}(1000.0 \text{ kg})\left(25.0 \frac{\text{m}}{\text{sec}}\right)^2$$
$$= \frac{1}{2}(1000.0 \text{ kg})\left(625 \frac{\text{m}^2}{\text{sec}^2}\right)$$
$$= \frac{1}{2}(1000.0)(625) \frac{\text{kg} \cdot \text{m}^2}{\text{sec}^2}$$
$$= 312.5 \text{ kJ} = \boxed{313 \text{ kJ}}$$

(b)
$$W = Fd = KE = \boxed{313 \text{ kJ}}$$

(c)
$$KE = W = Fd = \boxed{313 \text{ kJ}}$$

5.
$$KE = \frac{1}{2}mv^2$$
$$= \frac{1}{2}(60.0 \text{ kg})\left(2.0 \frac{\text{m}}{\text{sec}}\right)^2$$
$$= \frac{1}{2}(60.0 \text{ kg})\left(4.0 \frac{\text{m}^2}{\text{sec}^2}\right)$$
$$= 30.0 \times 4.0 \text{ kg} \times \left(\frac{\text{m}^2}{\text{sec}^2}\right)$$
$$= \boxed{120 \text{ J}}$$

$$KE = \frac{1}{2}mv^2$$
$$= \frac{1}{2}(60.0 \text{ kg})\left(4.0 \frac{\text{m}}{\text{sec}}\right)^2$$
$$= \frac{1}{2}(60.0 \text{ kg})\left(16 \frac{\text{m}^2}{\text{sec}^2}\right)$$
$$= 30.0 \times 16 \text{ kg} \times \left(\frac{\text{m}^2}{\text{sec}^2}\right)$$
$$= \boxed{480 \text{ J}}$$

Thus, doubling the speed results in a four-fold increase in kinetic energy.

6.
$$KE = \frac{1}{2}mv^2$$
$$= \frac{1}{2}(70.0 \text{ kg})(6.00 \text{ m/sec})^2$$
$$= (35.0 \text{ kg})(36.0 \text{ m}^2/\text{sec}^2)$$
$$= 35.0 \times 36.0 \text{ kg} \times \frac{\text{m}^2}{\text{sec}^2}$$
$$= \boxed{1260 \text{ J}}$$

$$KE = \frac{1}{2}mv^2$$
$$= \frac{1}{2}(140.0 \text{ kg})(6.00 \text{ m/sec})^2$$
$$= (70.0 \text{ kg})(36.0 \text{ m}^2/\text{sec}^2)$$
$$= 70.0 \times 36.0 \text{ kg} \times \frac{\text{m}^2}{\text{sec}^2}$$
$$= \boxed{2520 \text{ J}}$$

Thus, doubling the mass results in a doubling of the kinetic energy.

7. (a) The force needed is equal to the weight of the student. The English unit of a pound is a force unit, so

$$W = Fd$$
$$= (170.0 \text{ lb})(25.0 \text{ ft})$$
$$= \boxed{4250 \text{ ft} \cdot \text{lb}}$$

(b) Work (W) is defined as a force (F) moved through a distance (d), or $W = Fd$. Power (P) is defined as work (W) per unit of time (t), or $P = W/t$. Therefore,

$$P = \frac{Fd}{t}$$

$$= \frac{(170.0 \text{ lb})(25.0 \text{ ft})}{10.0 \text{ sec}}$$

$$= \frac{(170.0)(25.0)}{10.0} \frac{\text{ft} \cdot \text{lb}}{\text{sec}}$$

$$= 425 \frac{\text{ft} \cdot \text{lb}}{\text{sec}}$$

One hp is defined as $550 \dfrac{\text{ft} \cdot \text{lb}}{\text{sec}}$ and

$$\frac{425 \text{ ft} \cdot \text{lb/sec}}{550 \text{ ft} \cdot \text{lb/sec}} = \boxed{0.773 \text{ hp}}$$

(Note that the student's power rating (425 ft·lb/sec) is less than the power rating defined as one horsepower (550 ft·lb/sec). Thus, the student's horsepower must be *less* than one horsepower. A simple analysis such as this will let you know if you inverted the ratio or not.)

8. (a) The force (F) needed to lift the elevator is equal to the weight of the elevator. Since the work (W) is = to Fd and power (P) is = to W/t, then

$$P = \frac{Fd}{t} \therefore t = \frac{Fd}{P}$$

$$= \frac{[2000.0 \text{ lb}][20.0 \text{ ft}]}{\left(550 \dfrac{\dfrac{\text{ft} \cdot \text{lb}}{\text{sec}}}{\text{hp}}\right)[20.0 \text{ hp}]}$$

$$= \frac{40,000}{11,000} \frac{\text{ft} \cdot \text{lb}}{\dfrac{\text{ft} \cdot \text{lb}}{\text{sec}}} \times \frac{1}{\text{hp}} \times \text{hp}$$

$$= \frac{40,000}{11,000} \frac{\text{ft} \cdot \text{lb}}{1} \times \frac{\text{sec}}{\text{ft} \cdot \text{lb}}$$

$$= 3.64 \frac{\text{ft} \cdot \text{lb} \cdot \text{sec}}{\text{ft} \cdot \text{lb}}$$

$$= \boxed{3.64 \text{ sec}}$$

(b)
$$\bar{v} = \frac{d}{t}$$

$$= \frac{20.0 \text{ ft}}{3.64 \text{ sec}}$$

$$= \boxed{5.49 \text{ ft/sec}}$$

9. Since $PE_{lost} = KE_{gained}$ then $mgh = \dfrac{1}{2} mv^2$. Solving for v,

$$v = \sqrt{2gh} = \sqrt{(2)(32.0 \text{ ft/sec}^2)(9.80 \text{ ft})}$$

$$= \sqrt{(2)(32.0)(9.80) \text{ ft}^2/\text{sec}^2}$$

$$= \sqrt{627 \text{ ft}^2/\text{sec}^2}$$

$$= \boxed{25.0 \text{ ft/sec}}$$

10.
$$KE = \frac{1}{2}mv^2 \therefore v = \sqrt{\frac{2KE}{m}}$$

$$= \sqrt{\frac{(2)\left(200,000 \dfrac{\text{kg} \cdot \text{m}^2}{\text{sec}^2}\right)}{1000.0 \text{ kg}}}$$

$$= \sqrt{\frac{400,000}{1000.0} \frac{\text{kg} \cdot \text{m}^2}{\text{sec}^2} \times \frac{1}{\text{kg}}}$$

$$= \sqrt{\frac{400,000}{1000.0} \frac{\text{kg} \cdot \text{m}^2}{\text{kg} \cdot \text{sec}^2}}$$

$$= \sqrt{400 \text{ m}^2/\text{sec}^2}$$

$$= \boxed{20.0 \text{ m/sec}}$$

11. The maximum velocity occurs at the lowest point with a gain of kinetic energy equivalent to the loss of potential energy in falling 3.0 in (which is 0.25 ft), so

$$KE_{gained} = PE_{lost}$$

$$\frac{1}{2}mv^2 = mgh$$

$$v = \sqrt{2gh}$$

$$= \sqrt{(2)(32.0 \text{ ft/sec}^2)(0.25 \text{ ft})}$$

$$= \sqrt{(2)(32.0)(0.25) \text{ ft/sec}^2 \times \text{ft}}$$

$$= \sqrt{16 \text{ ft}^2/\text{sec}^2}$$

$$= \boxed{4.0 \text{ ft/sec}}$$

12. (a) $W = Fd$ and the force F that is needed to lift the load upward is mg, so $W = mgh$. Power is W/t, so

$$P = \frac{mgh}{t}$$

$$= \frac{(250.0 \text{ kg})(9.80 \text{ m/sec}^2)(80.0 \text{ m})}{39.2 \text{ sec}}$$

$$= \frac{(250.0)(9.80)(80.0)}{39.2} \frac{\text{kg}}{1} \times \frac{\text{m}}{\text{sec}^2} \times \frac{\text{m}}{1} \times \frac{1}{\text{sec}}$$

$$= \frac{196,000}{39.2} \frac{\text{kg} \cdot \text{m}^2}{\text{sec}^2} \times \frac{1}{\text{sec}}$$

$$= 5000 \frac{\text{J}}{\text{sec}}$$

$$= \boxed{5.00 \text{ kW}}$$

(b) There are 746 watts per horsepower, so

$$\frac{5000 \text{ W}}{746 \dfrac{\text{W}}{\text{hp}}} = \frac{5000}{746} \frac{\text{W}}{1} \times \frac{\text{hp}}{\text{W}}$$

$$= 6.70 \frac{\text{W} \cdot \text{hp}}{\text{W}}$$

$$= \boxed{6.70 \text{ hp}}$$

Chapter 5

1.
$$Q = mc\Delta T$$

$$= (221 \text{ g})\left(0.093 \, \frac{\text{cal}}{\text{g}^\circ\text{C}}\right)(38.0^\circ\text{C} - 20.0^\circ\text{C})$$

$$= (221)(0.093)(18.0) \text{ g} \times \frac{\text{cal}}{\text{g}^\circ\text{C}} \times {}^\circ\text{C}$$

$$= 370 \, \frac{\text{g} \cdot \text{cal} \cdot {}^\circ\cancel{C}}{\cancel{g}^\circ\cancel{C}}$$

$$= \boxed{370 \text{ cal}}$$

2. First, you need to know the energy of the moving bike and rider. Since the speed is given as 36 km/hr, convert to m/sec by multiplying times 0.2778 m/sec per km/hr:

$$\left(36 \, \frac{\text{km}}{\text{hr}}\right)\left(0.2278 \, \frac{\text{m/sec}}{\text{km/hr}}\right)$$

$$= (36)(0.2778) \, \frac{\text{km}}{\text{hr}} \times \frac{\text{hr}}{\text{km}} \times \frac{\text{m}}{\text{sec}}$$

$$= 10 \text{ m/sec}$$

Then,

$$KE = \frac{1}{2}mv^2$$

$$= \frac{1}{2}(100 \text{ kg})(10 \text{ m/sec})^2$$

$$= \frac{1}{2}(100 \text{ kg})(100 \text{ m}^2/\text{sec}^2)$$

$$= \frac{1}{2}(100)(100) \, \frac{\text{kg} \cdot \text{m}^2}{\text{sec}^2}$$

$$= 5,000 \text{ J}$$

Second, this energy is converted to the calorie heat unit through the mechanical equivalent of heat relationship, that 1.0 kcal = 4185 J, or that 1.0 cal = 4.185 J. Thus,

$$\frac{1.0 \text{ cal}}{4.185 \text{ J}} = 0.2389 \, \frac{\text{cal}}{\text{J}}$$

$$= (0.2389)(5,000) \, \frac{\text{cal}}{\text{J}} \times \text{J}$$

$$= 1,195 \, \frac{\text{cal} \cdot \cancel{J}}{\cancel{J}} = 1.195 \text{ kcal}$$

$$= \boxed{2 \text{ kcal}}$$

3. First, you need to find the energy of the falling bag. Since the potential energy lost equals the kinetic energy gained, the energy of the bag just as it hits the ground can be found from

$$PE = mgh$$

$$= (15.53 \text{ kg})(9.80 \text{ m/sec}^2)(5.50 \text{ m})$$

$$= (15.53)(9.80)(5.50) \, \frac{\text{kg} \cdot \text{m}}{\text{sec}^2} \times \text{m}$$

$$= 837 \text{ J}$$

In calories, this energy is equivalent to

$$(0.2389 \text{ cal/J})(837 \text{ J}) = 200 \text{ cal}$$

Second, the temperature change can be calculated from the equation giving the relationship between a quantity of heat (Q), mass (m), specific heat of the substance (c), and the change of temperature:

$$Q = mc\Delta T \therefore \Delta T = \frac{Q}{mc}$$

$$= \frac{0.200 \text{ kcal}}{(15.53 \text{ kg})\left(0.25 \, \frac{\text{kcal}}{\text{kg}^\circ\text{C}}\right)}$$

$$= \frac{0.200}{(15.53)(0.25)} \, \frac{\text{kcal}}{1} \times \frac{1}{\text{kg}} \times \frac{\text{kg}^\circ\text{C}}{\text{kcal}}$$

$$= 0.051 \, \frac{\text{kcal} \cdot \text{kg}^\circ\text{C}}{\text{kcal} \cdot \text{kg}}$$

$$= \boxed{0.051^\circ \text{ C}}$$

4. The Calorie used by dietitians is a kilocalorie; thus 250.0 Cal is 250.0 kcal. The mechanical energy equivalent is 1 kcal = 4185 J, so (250.0 kcal)(4185 J/kcal) = 1,046,250 J.

Since $W = $ Fd and the force needed is equal to the weight (mg) of the person, $W = mgh = (75.0 \text{ kg})(9.80 \text{ m/sec}^2)(10.0 \text{ m}) = $ 7350 J for each stairway climb.

A total of 1,046,250 J of energy from the French fries would require 1,046,250 J/7350 J per climb, or 142.3 trips up the stairs.

5. For unit consistency,

$$T_C = \frac{5}{9}(T_F - 32^\circ) = \frac{5}{9}(68^\circ - 32^\circ) = \frac{5}{9}(36^\circ) = 20^\circ \text{ C}$$

$$= \frac{5}{9}(32^\circ - 32^\circ) = \frac{5}{9}(0^\circ) = 0^\circ \text{ C}$$

Glass Bowl:

$$Q = mc\Delta T$$

$$= (0.5 \text{ kg})\left(0.2 \, \frac{\text{kcal}}{\text{kg}^\circ\text{C}}\right)(20^\circ \text{ C})$$

$$= (0.5)(0.2)(20) \, \frac{\text{kg}}{1} \times \frac{\text{kcal}}{\text{kg}^\circ\text{C}} \times \frac{{}^\circ\text{C}}{1}$$

$$= \boxed{2 \text{ kcal}}$$

Iron Pan:

$$Q = mc\Delta T$$

$$= (0.5 \text{ kg})\left(0.11 \, \frac{\text{kcal}}{\text{kg}^\circ\text{C}}\right)(20^\circ \text{ C})$$

$$= (0.5)(0.11)(20) \text{ kg} \times \frac{\text{kcal}}{\text{kg}^\circ\text{C}} \times {}^\circ \text{ C}$$

$$= \boxed{1 \text{ kcal}}$$

6. Note that a specific heat expressed in cal/g has the same numerical value as a specific heat expressed in kcal/kg because you can cancel the k units. You could convert 896 cal to 0.896 kcal, but one of the two conversion methods is needed for consistency with other units in the problem.

$$Q = mc\Delta T \therefore m = \frac{Q}{c\Delta T}$$

$$= \frac{896 \text{ cal}}{\left(0.056 \frac{\text{cal}}{\text{g}^\circ\text{C}}\right)(80.0^\circ \text{ C})}$$

$$= \frac{896}{(0.056)(80.0)} \frac{\text{cal}}{1} \times \frac{\text{g}^\circ\text{C}}{\text{cal}} \times \frac{1}{\text{C}}$$

$$= 200 \text{ g}$$

$$= \boxed{0.20 \text{ kg}}$$

7. Since a watt is defined as a joule/sec, finding the total energy in joules will tell the time:

$$Q = mc\Delta T$$

$$= (250.0 \text{ g})\left(1.00 \frac{\text{cal}}{\text{g}^\circ\text{C}}\right)(60.0^\circ \text{ C})$$

$$= (250.0)(1.00)(60.0) \text{ g} \times \frac{\text{cal}}{\text{g}^\circ\text{C}} \times {}^\circ \text{C}$$

$$= 1.50 \times 10^4 \text{ cal}$$

This energy in joules is $(1.50 \times 10^4 \text{ cal})\left(4.185 \frac{\text{J}}{\text{cal}}\right) = 62,800 \text{ J}$

A 300 watt heater uses energy at a rate of $300 \frac{\text{J}}{\text{sec}}$, so $\frac{62,800 \text{ J}}{300 \text{ J/sec}}$

$= 209$ sec is required, which is $\dfrac{209 \text{ sec}}{60 \dfrac{\text{sec}}{\text{min}}} = 3.48$ min, or

$$\boxed{\text{about } 3\frac{1}{2} \text{ min.}}$$

8. $$Q = mc\Delta T \therefore c = \frac{Q}{m\Delta T}$$

$$= \frac{60.0 \text{ cal}}{(100.0 \text{ g})(20.0^\circ \text{ C})}$$

$$= \frac{60.0}{(100.0)(20.0)} \frac{\text{cal}}{\text{g}^\circ\text{C}}$$

$$= \boxed{0.0300 \frac{\text{cal}}{\text{g}^\circ\text{C}}}$$

9. Since the problem specified a solid changing to a liquid without a temperature change, you should recognize that this is a question about a phase change only. The phase change from solid to liquid (or liquid to solid) is concerned with the latent heat of fusion. For water, the latent heat of fusion is given as 80.0 cal/g, and

$$m = 250.0 \text{ g} \quad Q = mL_f$$

$$L_{f \text{ (water)}} = 80.0 \text{ cal/g} \quad = (250.0 \text{ g})\left(80.0 \frac{\text{cal}}{\text{g}}\right)$$

$$Q = ? \qquad = 250.0 \times 80.0 \frac{\text{g}\cdot\text{cal}}{\text{g}}$$

$$= 20,000 \text{ cal} = \boxed{20.0 \text{ kcal}}$$

10. To change water at 80.0° C to steam at 100.0° C requires two separate quantities of heat that can be called Q_1 and Q_2. The quantity Q_1 is the amount of heat needed to warm the water from 80.0° C to the boiling point, which is 100.0° C at sea level pressure ($\Delta T = 20.0^\circ$ C). The relationship between the variable involved is $Q = mc\Delta T$. The quantity Q_2 is the amount of heat needed to take 100.0° water through the phase change to steam (water vapor) at 100.0° C. The phase change from a liquid to a gas (or gas to liquid) is concerned with the latent heat of vaporization. For water, the latent heat of vaporization is given as 540.0 cal/g.

$$m = 250.0 \text{ g} \qquad Q_1 = mc\Delta T$$

$$L_{v \text{ (water)}} = 540.0 \text{ cal/g} \quad = (250.0 \text{ g})\left(1.00 \frac{\text{cal}}{\text{g}\cdot^\circ\text{C}}\right)(20.0^\circ \text{ C})$$

$$Q = ? \qquad = (250.0)(1.00)(20.0) \text{ g} \times \frac{\text{cal}}{\text{g}^\circ\text{C}} \times {}^\circ\text{C}$$

$$= 5000 \frac{\text{g}\cdot\text{cal}\cdot^\circ\text{C}}{\text{g}\cdot^\circ\text{C}}$$

$$= 5000 \text{ cal}$$

$$= 5.00 \text{ kcal}$$

$$Q_2 = mL_v$$

$$= (250.0 \text{ g})\left(540.0 \frac{\text{cal}}{\text{g}}\right)$$

$$= 250.0 \times 540.0 \frac{\text{g}\cdot\text{cal}}{\text{g}}$$

$$= 135,000 \text{ cal}$$

$$= 135.0 \text{ kcal}$$

$$Q_{\text{Total}} = Q_1 + Q_2$$

$$= 5.00 \text{ kcal} + 135.0 \text{ kcal}$$

$$= \boxed{140.0 \text{ kcal}}$$

11. To change 20.0° C water to steam at 125.0° C requires three separate quantities of heat. First, the quantity Q_1 is the amount of heat needed to warm the water from 20.0° C to 100.0° C ($\Delta T = 80.0^\circ$ C). The quantity Q_2 is the amount of heat needed to take 100.0° C water to steam at 100.0° C. Finally, the quantity Q_3 is amount of heat needed to warm the steam from 100.0° to 125.0° C. According to table 5.1, the c for steam is 0.48 cal/g°C.

$$m = 100.0 \text{ g} \qquad Q_1 = mc\Delta T$$

$$\Delta T_{\text{water}} = 80.0^\circ \text{ C} \quad = (100.0 \text{ g})\left(1.00 \frac{\text{cal}}{\text{g}^\circ\text{C}}\right)(80.0^\circ \text{ C})$$

$$\Delta T_{\text{steam}} = 25.0^\circ \text{ C}$$

$$L_{v \text{ (water)}} = 540.0 \text{ cal/g} \quad = (100.0)(1.00)(80.0) \text{ g} \times \frac{\text{cal}}{\text{g}^\circ\text{C}} \times {}^\circ\text{C}$$

$$c_{\text{steam}} = 0.48 \text{ cal/g}^\circ\text{C} \quad = 8000 \frac{\text{g}\cdot\text{cal}\cdot^\circ\text{C}}{\text{g}\cdot^\circ\text{C}}$$

$$= 8000 \text{ cal}$$

$$= 8.00 \text{ kcal}$$

$$Q_2 = mL_v$$

$$= (100.0 \text{ g})\left(540.0 \ \frac{\text{cal}}{\text{g}}\right)$$

$$= 100.0 \times 540.0 \ \frac{\text{g} \cdot \text{cal}}{\text{g}}$$

$$= 54{,}000 \text{ cal}$$

$$= 54.00 \text{ kcal}$$

$$Q_3 = mc\Delta T$$

$$= (100.0 \text{ g})\left(0.48 \ \frac{\text{cal}}{\text{g}^\circ\text{C}}\right)(25.0^\circ \text{ C})$$

$$= (100.0)(0.48)(25.0) \ \text{g} \times \frac{\text{cal}}{\text{g}^\circ\text{C}} \times {}^\circ\text{C}$$

$$= 1200 \ \frac{\text{g} \cdot \text{cal} \cdot {}^\circ\text{C}}{\text{g} \cdot {}^\circ\text{C}}$$

$$= 1200 \text{ cal}$$

$$= 1.2 \text{ kcal}$$

$$Q_{total} = Q_1 + Q_2 + Q_3$$

$$= 8.00 \text{ kcal} + 54.00 \text{ kcal} + 1.2 \text{ kcal}$$

$$= \boxed{63.2 \text{ kcal}}$$

12. (a) **Step 1:** Cool water from 18.0° C to 0° C.

$$Q_1 = mc\Delta T$$

$$= (400.0 \text{ g})\left(1.00 \ \frac{\text{cal}}{\text{g}^\circ\text{C}}\right)(18.0^\circ \text{ C})$$

$$= (400.0)(1.00)(18.0) \ \text{g} \times \frac{\text{cal}}{\text{g}^\circ\text{C}} \times {}^\circ\text{C}$$

$$= 7200 \ \frac{\text{g} \cdot \text{cal} \cdot {}^\circ\text{C}}{\text{g} \cdot {}^\circ\text{C}}$$

$$= 7200 \text{ cal}$$

$$= 7.20 \text{ kcal}$$

Step 2: Find the energy needed for the phase change of water at 0° C to ice at 0° C.

$$Q_2 = mL_f$$

$$= (400.0 \text{ g})\left(80.0 \ \frac{\text{cal}}{\text{g}}\right)$$

$$= 400.0 \times 80.0 \ \frac{\text{g} \cdot \text{cal}}{\text{g}}$$

$$= 32{,}000 \text{ cal}$$

$$= 32.0 \text{ kcal}$$

Step 3: Cool the ice from 0° C to ice at −5.00° C.

$$Q_3 = mc\Delta T$$

$$= (400.0 \text{ g})\left(0.50 \ \frac{\text{cal}}{\text{g}^\circ\text{C}}\right)(5.00^\circ \text{ C})$$

$$= 400.0 \times 0.50 \times 5.00 \ \text{g} \times \frac{\text{cal}}{\text{g}^\circ\text{C}} \times {}^\circ\text{C}$$

$$= 1000 \ \frac{\text{g} \cdot \text{cal} \cdot {}^\circ\text{C}}{\text{g}^\circ\text{C}}$$

$$= 1000 \text{ cal}$$

$$= 1.0 \text{ kcal}$$

$$Q_{total} = Q_1 + Q_2 + Q_3$$

$$= 7.20 \text{ kcal} + 32.0 \text{ kcal} + 1.0 \text{ kcal}$$

$$= \boxed{40.2 \text{ kcal}}$$

(b)
$$Q = mL_v \ \therefore \ m = \frac{Q}{L_v}$$

$$= \frac{39{,}400 \text{ cal}}{40.0 \ \frac{\text{cal}}{\text{g}}}$$

$$= \frac{39{,}400}{40.0} \ \frac{\text{cal}}{1} \times \frac{\text{g}}{\text{cal}}$$

$$= 985 \ \frac{\text{cal} \cdot \text{g}}{\text{cal}}$$

$$= \boxed{985 \text{ g}}$$

Chapter 6

1.
$$v = f\lambda$$

$$= \left(10 \ \frac{1}{\text{sec}}\right)(0.50 \text{ m})$$

$$= 5 \ \frac{\text{m}}{\text{sec}}$$

2. The distance between two *consecutive* condensations (or rarefactions) is one wavelength, so $\lambda = 3.00$ m and

$$v = f\lambda$$

$$= \left(112.0 \ \frac{1}{\text{sec}}\right)(3.00 \text{ m})$$

$$= \boxed{336 \ \frac{\text{m}}{\text{sec}}}$$

3. (a) One complete wave every four sec means that T = 4.00 sec. (Note that the symbol for the *time of a cycle* is T. Do not confuse this symbol with the symbol for temperature.)

(b)
$$f = \frac{1}{T}$$

$$= \frac{1}{4.0 \text{ sec}}$$

$$= \frac{1}{4.0} \ \frac{1}{\text{sec}}$$

$$= 0.25 \ \frac{1}{\text{sec}}$$

$$= \boxed{0.25 \text{ Hz}}$$

4. The distance from one condensation to the next is one wavelength, so

$$v = f\lambda \therefore \lambda = \frac{v}{f}$$

$$= \frac{330 \dfrac{m}{sec}}{260 \dfrac{1}{sec}}$$

$$= \frac{330}{260} \dfrac{m}{sec} \times \dfrac{sec}{1}$$

$$= \boxed{1.3 \text{ m}}$$

5. (a) $v = f\lambda = \left(256 \dfrac{1}{sec}\right)(1.34 \text{ m}) = \boxed{343 \text{ m/sec}}$

(b) $= \left(440 \dfrac{1}{sec}\right)(0.780 \text{ m}) = \boxed{343 \text{ m/sec}}$

(c) $= \left(750 \dfrac{1}{sec}\right)(0.457 \text{ m}) = \boxed{343 \text{ m/sec}}$

(d) $= \left(2{,}500 \dfrac{1}{sec}\right)(0.1372 \text{ m}) = \boxed{343 \text{ m/sec}}$

6. The speed of sound at 0.0° C is 1087 ft/sec, and

(a) $V_{T_F} = V_0 + \left[\dfrac{2.0 \text{ ft/sec}}{^\circ C}\right][T_P]$

$$= 1087 \text{ ft/sec} + \left[\dfrac{2.0 \text{ ft/sec}}{^\circ C}\right][0.0^\circ \text{ C}]$$

$$= 1087 + (2.0)(0.0) \text{ ft/sec} + \dfrac{\text{ft/sec}}{^\circ\!\!\!\!/\,C} \times {^\circ\!\!\!\!/\,C}$$

$$= 1087 \text{ ft/sec} + 0.0 \text{ ft/sec}$$

$$= \boxed{1087 \text{ ft/sec}}$$

(b) $V_{20^\circ} = 1087 \text{ ft/sec} + \left[\dfrac{2.0 \text{ ft/sec}}{^\circ C}\right][20.0^\circ \text{ C}]$

$$= 1087 \text{ ft/sec} + 40.0 \text{ ft/sec}$$

$$= \boxed{1127 \text{ ft/sec}}$$

(c) $V_{40^\circ} = 1087 \text{ ft/sec} + \left[\dfrac{2.0 \text{ ft/sec}}{^\circ C}\right][40.0^\circ \text{ C}]$

$$= 1087 \text{ ft/sec} + 80.0 \text{ ft/sec}$$

$$= \boxed{1167 \text{ ft/sec}}$$

(d) $V_{80^\circ} = 1087 \text{ ft/sec} + \left[\dfrac{2.0 \text{ ft/sec}}{^\circ C}\right][80.0^\circ \text{ C}]$

$$= 1087 \text{ ft/sec} + 160 \text{ ft/sec}$$

$$= \boxed{1247 \text{ ft/sec}}$$

7. For consistency with the units of the equation given, 43.7° F is first converted to 6.50° C. The velocity of sound in this air is:

$$V_{T_F} = V_0 + \left[\dfrac{2.0 \text{ ft/sec}}{^\circ C}\right][T_P]$$

$$= 1087 \text{ ft/sec} + \left[\dfrac{2.0 \text{ ft/sec}}{^\circ C}\right][6.50^\circ \text{ C}]$$

$$= 1087 \text{ ft/sec} + 13 \text{ ft/sec}$$

$$= 1100 \text{ ft/sec}$$

The distance that a sound with this velocity travels in the given time is

$$v = \dfrac{d}{t} \therefore d = vt$$

$$= (1100 \text{ ft/sec})(4.80 \text{ sec})$$

$$= (1100)(4.80) \dfrac{\text{ft}\cdot\text{sec}}{\text{sec}}$$

$$= 5280 \text{ ft}$$

$$\dfrac{5280 \text{ ft}}{2}$$

$$= \boxed{2640 \text{ ft}}$$

Since the sound traveled from the rifle to the cliff and then back, the cliff must be 5280 feet/2 = 2640 feet, or one-half mile away.

8. This problem requires three steps, (1) conversion of the °F temperature value to °C, (2) calculating the velocity of sound in air at this temperature, and (3) calculating the distance from the calculated velocity and the given time:

$$V_{T_F} = V_0 + \left[\dfrac{2.0 \text{ ft/sec}}{^\circ C}\right][T_P]$$

$$= 1087 \text{ ft/sec} + \left[\dfrac{2.0 \text{ ft/sec}}{^\circ C}\right][26.67^\circ \text{ C}]$$

$$= 1087 \text{ ft/sec} + 53 \text{ ft/sec} = 1140 \text{ ft/sec}$$

$$v = \dfrac{d}{t} \therefore d = vt$$

$$= (1140 \text{ ft/sec})(4.63 \text{ sec})$$

$$= \boxed{5280 \text{ ft (one mile)}}$$

9. A tube closed at one end must have a minimum of 1/4 wavelength to have a node at the closed end and a resonate antinode at the open end. Thus,

$$L = \frac{1}{4}\lambda \therefore \lambda = 4 \text{ L}$$

$$= (4)(0.250 \text{ m})$$

$$= 1.00 \text{ m}$$

$$v = f\lambda$$

$$= \left(340 \ \frac{1}{\text{sec}}\right)(1.00 \ \text{m})$$

$$= (340)(1.00) \ \frac{1}{\text{sec}} \times \text{m}$$

$$= \boxed{340 \ \frac{\text{m}}{\text{sec}}}$$

10. (a)

$$v = f\lambda \ \therefore \ \lambda = \frac{v}{f}$$

$$= \frac{1125 \ \frac{\text{ft}}{\text{sec}}}{440 \ \frac{1}{\text{sec}}}$$

$$= \frac{1125}{440} \ \frac{\text{ft}}{\text{sec}} \times \frac{\text{sec}}{1}$$

$$= 2.56 \ \frac{\text{ft} \cdot \text{sec}}{\text{sec}}$$

$$= \boxed{2.6 \ \text{ft}}$$

(b)

$$v = f\lambda \ \therefore \ \lambda = \frac{v}{f}$$

$$= \frac{5020}{440} \ \frac{\text{ft}}{\text{sec}} \times \frac{\text{sec}}{1}$$

$$= 11.4 \ \text{ft} = \boxed{11 \ \text{ft}}$$

11. The wavelength can be found from the relationship between the length of the air column and λ (see the solution to problem number 9):

$$\lambda = 4L = 4(0.240 \ \text{m}) = 0.960 \ \text{m}.$$

The velocity can be determined from the given air temperature:

$$v_{T_F} = v_0 + \left[\frac{0.60 \ \text{m/sec}}{°C}\right][T_P]$$

$$= 331 \ \text{m/sec} + \left[\frac{0.60 \ \text{m/sec}}{°C}\right][20.0° \ C]$$

$$= 331 \ \text{m/sec} + 12 \ \text{m/sec} = 343 \ \text{m/sec}$$

From the wave equation,

$$v = f\lambda \ \therefore \ f = \frac{v}{\lambda}$$

$$= \frac{343 \ \text{m/sec}}{0.960 \ \text{m}}$$

$$= \frac{343}{0.960} \ \frac{\text{m}}{\text{sec}} \times \frac{1}{\text{m}}$$

$$= 357 \ \frac{1}{\text{sec}}$$

$$= \boxed{357 \ \text{Hz}}$$

12. The fundamental frequency is found from the relationship:

$$f_n = \frac{nv}{4L} \quad \text{where } n = 1 \text{ is the fundamental and } n = 3, 5, 7, \text{ and}$$

so on are the possible harmonics.

Assuming room temperature since you have to assume some temperature to find the velocity (room temperature is defined to be 20.0° C), the velocity of sound is 343 m/sec. Thus:

Fundamental frequency (n = 1):

$$f = \frac{nv}{4L} = \frac{(1)(343 \ \text{m/sec})}{(4)(0.700 \ \text{m})}$$

$$= \frac{343}{2.80} \ \frac{\text{m}}{\text{sec}} \times \frac{1}{\text{m}}$$

$$= 122.5 \ \frac{1}{\text{sec}}$$

$$= \boxed{123 \ \text{Hz}}$$

First overtone above the fundamental (n = 3):

$$f = \frac{nv}{4L} = \frac{(3)(343 \ \text{m/sec})}{(4)(0.700)}$$

$$= \frac{1029}{2.80} \ \frac{\text{m}}{\text{sec}} \times \frac{1}{\text{m}}$$

$$= 367.5 \ \frac{1}{\text{sec}}$$

$$= \boxed{368 \ \text{Hz}}$$

Chapter 7

1. First, recall that a negative charge means an excess of electrons. Second, the relationship between the total charge (q), the number of electrons (n), and the charge of a single electron (e) is $q = ne$. The fundamental charge of a single (n = 1) electron (e) is 1.60×10^{-19} C. Thus

$$q = ne, \ \therefore \ n = \frac{q}{e}$$

$$= \frac{1.00 \times 10^{-14} \ \text{C}}{1.60 \times 10^{-19} \ \frac{\text{C}}{\text{electron}}}$$

$$= \frac{1.00 \times 10^{-14}}{1.60 \times 10^{-19}} \ \frac{\text{C}}{1} \times \frac{\text{electron}}{\text{C}}$$

$$= 6.25 \times 10^4 \ \frac{\text{electron}}{\text{C}}$$

$$= \boxed{6.25 \times 10^4 \ \text{electron}}$$

2. (a) Both balloons have negative charges so the force is repulsive, pushing the balloons away from each other.

(b) The magnitude of the force can be found from Coulomb's law:

$$F = \frac{kq_1q_2}{d^2}$$

$$= \frac{(9.00 \times 10^9 \ \text{N} \cdot \text{m}^2/\text{C}^2)(3.00 \times 10^{-14} \ \text{C})(2.00 \times 10^{-12} \ \text{C})}{(2.00 \times 10^{-2} \ \text{m})^2}$$

$$= \frac{(9.00 \times 10^9)(3.00 \times 10^{-14})(2.00 \times 10^{-12})}{4.00 \times 10^{-4}} \frac{\frac{\text{N} \cdot \text{m}^2}{\text{C}^2} \times \text{C} \times \text{C}}{\text{m}^2}$$

$$= \frac{5.40 \times 10^{-16}}{4.00 \times 10^{-4}} \frac{\text{N} \cdot \cancel{\text{m}^2}}{\cancel{\text{C}^2}} \times \cancel{\text{C}^2} \times \frac{1}{\cancel{\text{m}^2}}$$

$$= \boxed{1.35 \times 10^{-12} \ \text{N}}$$

3.

$$\frac{\text{potential}}{\text{difference}} = \frac{\text{work}}{\text{charge}}$$

or

$$V = \frac{W}{q}$$

$$= \frac{7.50 \ \text{J}}{5.00 \ \text{C}}$$

$$= 1.50 \ \frac{\text{J}}{\text{C}}$$

$$= \boxed{1.50 \ \text{V}}$$

4.

$$\frac{\text{electric}}{\text{current}} = \frac{\text{charge}}{\text{time}}$$

or

$$I = \frac{q}{t}$$

$$= \frac{6.00 \ \text{C}}{2.00 \ \text{sec}}$$

$$= 3.00 \ \frac{\text{C}}{\text{sec}}$$

$$= \boxed{3.00 \ \text{A}}$$

5. A current of 1.00 amp is defined as 1.00 coulomb/sec. Since the fundamental charge of the electron is 1.60×10^{-19} C/electron,

$$\frac{1.00 \ \dfrac{\text{C}}{\text{sec}}}{1.60 \times 10^{-19} \ \dfrac{\text{C}}{\text{electron}}}$$

$$= 6.25 \times 10^{18} \ \frac{\cancel{\text{C}}}{\text{sec}} \times \frac{\text{electron}}{\cancel{\text{C}}}$$

$$= \boxed{6.25 \times 10^{18} \ \frac{\text{electrons}}{\text{sec}}}$$

6.

$$R = \frac{V}{I}$$

$$= \frac{120.0 \ \text{V}}{4.00 \ \text{A}}$$

$$= 30.0 \ \frac{\text{V}}{\text{A}}$$

$$= \boxed{30.0 \ \Omega}$$

7.

$$R = \frac{V}{I} \quad \therefore I = \frac{V}{R}$$

$$= \frac{120.0 \ \text{V}}{60.0 \ \dfrac{\text{V}}{\text{A}}}$$

$$= \frac{120.0}{60.0} \ \cancel{\text{V}} \times \frac{\text{A}}{\cancel{\text{V}}}$$

$$= \boxed{2.00 \ \text{A}}$$

8. (a)

$$R = \frac{V}{I} \quad \therefore V = IR$$

$$= (1.20 \ \text{A})\left(10.0 \ \frac{\text{V}}{\text{A}}\right)$$

$$= \boxed{12.0 \ \text{V}}$$

(b)

$$\text{Power} = (\text{current})(\text{potential difference})$$

or

$$P = IV$$

$$= \left(1.20 \ \frac{\text{C}}{\text{sec}}\right)\left(12.0 \ \frac{\text{J}}{\text{C}}\right)$$

$$= (1.20)(12.0) \ \frac{\cancel{\text{C}}}{\text{sec}} \times \frac{\text{J}}{\cancel{\text{C}}}$$

$$= 14.4 \ \frac{\text{J}}{\text{sec}}$$

$$= \boxed{14.4 \ \text{watt}}$$

9. Note that there are two separate electrical units that are rates: (1) the amp (coulomb/sec), and (2) the watt (joule/sec). The question asked for a rate of using energy. Energy is measured in joules, so you are looking for the power of the radio in watts. To find watts (P = IV), you will need to calculate the current (I) since it is not given. The current can be obtained from the relationship of Ohm's law:

$$I = \frac{V}{R}$$

$$= \frac{3.00 \ \text{V}}{15.0 \ \dfrac{\text{V}}{\text{A}}}$$

$$= 0.200 \ \text{A}$$

$$P = IV$$

$$= (0.200 \ \text{C/sec})(3.00 \ \text{J/C})$$

$$= \boxed{0.600 \ \text{watt}}$$

10.

$$\frac{(1{,}200 \text{ W})(0.25 \text{ hr})}{1{,}000 \frac{W}{kW}} = \frac{(1{,}200)(0.25)}{1{,}000} \frac{W \cdot hr}{1} \times \frac{kW}{W}$$

$$= 0.30 \text{ kWhr}$$

$$(0.30 \text{ kWhr})\left(\frac{\$0.10}{kWhr}\right) = \boxed{\$0.03} \quad (3 \text{ cents})$$

11. The relationship between power (P), current (I), and volts (V) will provide a solution. Since the relationship considers power in watts the first step is to convert horsepower to watts. One horsepower is equivalent to 746 watts, so:

$$(746 \text{ W/hp})(2.00 \text{ hp}) = 1492 \text{ W}$$

$$P = IV \therefore I = \frac{P}{V}$$

$$= \frac{1492 \frac{J}{sec}}{12.0 \frac{J}{C}}$$

$$= \frac{1492}{12.0} \frac{J}{sec} \times \frac{C}{J}$$

$$= 124.3 \frac{C}{sec}$$

$$= \boxed{124 \text{ A}}$$

12. (a) The rate of using energy is joules/sec, or the watt. Since 1.00 hp = 746 W,

inside motor: $(746 \text{ W/hp})(1/3 \text{ hp}) = 249 \text{ W}$

outside motor: $(746 \text{ W/hp})(1/3 \text{ hp}) = 249 \text{ W}$

compressor motor: $(746 \text{ W/hp})(3.70 \text{ hp}) - 2760 \text{ W}$

$$249 \text{ W} + 249 \text{ W} + 2760 \text{ W} = \boxed{3258 \text{ W}}$$

(b)
$$\frac{(3258 \text{ W})(1.00 \text{ hr})}{1000 \text{ W/kW}} = 3.26 \text{ kWhr}$$

$$(3.26 \text{ kWhr})(\$0.10/\text{kWhr}) = \boxed{\$0.33 \text{ per hour}}$$

(c) $(\$0.33/\text{hr})(12 \text{ hr/day})(30 \text{ day/mo})$
$$= \$118.80 = \boxed{\$120}$$

13. The solution is to find how much current each device draws and then to see if the total current is less or greater than the breaker rating:

Toaster: $I = \dfrac{V}{R} = \dfrac{120 \text{ V}}{15 \text{ V/A}} = 8.0 \text{ A}$

Motor: $(0.20 \text{ hp})(746 \text{ W/hp}) = 150 \text{ W}$

$$I = \frac{P}{V} = \frac{150 \text{ J/sec}}{120 \text{ J/C}} = 1.3 \text{ A}$$

Three 100 W bulbs: $3 \times 100 \text{ W} = 300 \text{ W}$

$$I = \frac{P}{V} = \frac{300 \text{ J/sec}}{120 \text{ J/C}} = 2.5 \text{ A}$$

Iron: $I = \dfrac{P}{V} = \dfrac{600 \text{ J/sec}}{120 \text{ J/C}} = 5.0 \text{ A}$

The sum of the currents is $8.0A + 1.3A + 2.5A + 5.0A = 16.8A$, so the total current is greater than 15.0 amp and the circuit breaker will trip.

14. (a) $V_P = 1200 \text{ V}$

$N_P = 1 \text{ loop}$

$N_s = 200 \text{ loops}$

$V_s = ?$

$$\frac{V_P}{N_P} - \frac{V_s}{N_s} \therefore V_s = \frac{V_P N_s}{N_P}$$

$$V_s = \frac{(1200 \text{ V})(200 \text{ loop})}{1 \text{ loop}}$$

$$= \boxed{240{,}000 \text{ V}}$$

(b) $I_P = 40 \text{ A}$

$I_s = ?$

$$V_P I_P = V_s I_s \therefore I_s = \frac{V_P I_P}{V_s}$$

$$I_s = \frac{1200 \text{ V} \times 40 \text{ A}}{240{,}000 \text{ V}}$$

$$= \frac{1200 \times 40}{240{,}000} \frac{V \cdot A}{V}$$

$$= \boxed{0.2 \text{ A}}$$

15. (a) $V_s = 12 \text{ V}$

$I_s = 0.5 \text{ A}$

$V_P = 120 \text{ V}$

$\dfrac{N_P}{N_s} = ?$

$$\frac{V_P}{N_P} = \frac{V_s}{N_s} \therefore \frac{N_P}{N_s} = \frac{V_P}{V_s}$$

$$\frac{N_P}{N_s} = \frac{120 \text{ V}}{12 \text{ V}} = \frac{10}{1}$$

or

$$\boxed{10 \text{ primary to 1 secondary}}$$

(b) $I_P = ?$

$$V_P I_P = V_s I_s \therefore I_P = \frac{V_s I_s}{V_P}$$

$$I_P = \frac{(12 \text{ V})(0.5 \text{ A})}{120 \text{ V}}$$

$$= \frac{12 \times 0.5}{120} \frac{V \cdot A}{V}$$

$$= \boxed{0.05 \text{ A}}$$

(c) $P_s = ?$

$$P_s = I_s V_s$$

$$= (0.5 \text{ A})(12 \text{ V})$$

$$= 0.5 \times 12 \frac{C}{sec} \times \frac{J}{C}$$

$$= 6 \frac{J}{sec}$$

$$= \boxed{6 \text{ W}}$$

16. (a) $V_P = 120 \text{ V}$

$N_P = 50 \text{ loops}$

$N_s = 150 \text{ loops}$

$I_P = 5.0 \text{ A}$

$V_s = ?$

$$\frac{V_P}{N_P} = \frac{V_s}{N_s} \therefore V_s = \frac{V_P N_s}{N_P}$$

$$V_s = \frac{120 \text{ V} \times 150 \text{ loops}}{50 \text{ loops}}$$

$$= \frac{120 \times 150}{50} \frac{V \cdot loops}{loops}$$

$$= \boxed{360 \text{ V}}$$

(b) $I_s = ?$ $\quad V_P I_P = V_s I_s \therefore I_s = \dfrac{V_P I_P}{V_s}$

$$I_s = \dfrac{(120 \text{ V})(5.0 \text{ A})}{360 \text{ V}}$$

$$= \dfrac{120 \times 5.0}{360} \dfrac{\text{V} \cdot \text{A}}{\text{V}}$$

$$= \boxed{1.7 \text{ A}}$$

(c) $P_s = ?$ $\quad P_s = I_s V_s$

$$= \left(1.7 \dfrac{\text{C}}{\text{sec}}\right)\left(360 \dfrac{\text{J}}{\text{C}}\right)$$

$$= 1.7 \times 360 \dfrac{\text{C}}{\text{sec}} \times \dfrac{\text{J}}{\text{C}}$$

$$= 612 \dfrac{\text{J}}{\text{sec}}$$

$$= \boxed{600 \text{ W}}$$

Chapter 8

1. The relationship between the speed of light in a transparent material (v), the speed of light in a vacuum ($c = 3.00 \times 10^8$ m/sec) and the index of refraction (n) is $n = c/v$. According to table 8.1, the index of refraction for water is $n = 1.33$ and ice is $n = 1.31$.

(a) $c = 3.00 \times 10^8$ m/sec

$n = 1.33$

$v = ?$

$n = \dfrac{c}{v} \therefore v = \dfrac{c}{n}$

$$v = \dfrac{3.00 \times 10^8 \text{ m/sec}}{1.33}$$

$$= \boxed{2.26 \times 10^8 \text{ m/sec}}$$

(b) $c = 3.00 \times 10^8$ m/sec

$n = 1.31$

$v = ?$

$$v = \dfrac{3.00 \times 10^8 \text{ m/sec}}{1.31}$$

$$= \boxed{2.29 \times 10^8 \text{ m/sec}}$$

2. $d = 1.50 \times 10^8$ km

$\quad = 1.50 \times 10^{11}$ m

$c = 3.00 \times 10^8$ m/sec

$t = ?$

$v = \dfrac{d}{t} \therefore t = \dfrac{d}{v}$

$$t = \dfrac{1.50 \times 10^{11} \text{ m}}{3.00 \times 10^8 \dfrac{\text{m}}{\text{sec}}}$$

$$= \dfrac{1.50 \times 10^{11}}{3.00 \times 10^8} \text{ m} \times \dfrac{\text{sec}}{\text{m}}$$

$$= 5.00 \times 10^2 \dfrac{\text{m} \cdot \text{sec}}{\text{m}}$$

$$= \dfrac{5.00 \times 10^2 \text{ sec}}{60.0 \dfrac{\text{sec}}{\text{min}}}$$

$$= \dfrac{5.00 \times 10^2}{60.0} \text{ sec} \times \dfrac{\text{min}}{\text{sec}}$$

$$= \boxed{8.33 \text{ min}}$$

3. $d = 6.00 \times 10^9$ km

$\quad = 6.00 \times 10^{12}$ m

$c = 3.00 \times 10^8$ m/sec

$t = ?$

$v = \dfrac{d}{t} \therefore t = \dfrac{d}{v}$

$$t = \dfrac{6.00 \times 10^{12} \text{ m}}{3.00 \times 10^8 \dfrac{\text{m}}{\text{sec}}}$$

$$= \dfrac{6.00 \times 10^{12}}{3.00 \times 10^8} \text{ m} \times \dfrac{\text{sec}}{\text{m}}$$

$$= 2.00 \times 10^4 \text{ sec}$$

$$= \dfrac{2.00 \times 10^4 \text{ sec}}{3,600 \dfrac{\text{sec}}{\text{hr}}}$$

$$= \dfrac{2.00 \times 10^4}{3.600 \times 10^3} \text{ sec} \times \dfrac{\text{hr}}{\text{sec}}$$

$$= \boxed{5.56 \text{ hr}}$$

4. From equation 8.1, note that both angles are measured from the normal and that the angle of incidence (θ_i) equals the angle of reflection (θ_r), or

$$\theta_i = \theta_r \therefore \boxed{\theta_i = 10°}$$

5. $v = 2.20 \times 10^8$ m/sec

$c = 3.00 \times 10^8$ m/sec

$n = ?$

$n = \dfrac{c}{v}$

$$= \dfrac{3.00 \times 10^8 \dfrac{\text{m}}{\text{sec}}}{2.20 \times 10^8 \dfrac{\text{m}}{\text{sec}}}$$

$$= 1.36$$

According to table 8.1, the substance with an index of refraction of 1.36 is $\boxed{\text{ethyl alcohol.}}$

6. (a) From equation 8.3:

$\lambda = 6.00 \times 10^{-7}$ m $\qquad c = \lambda f \therefore f = \dfrac{c}{\lambda}$

$c = 3.00 \times 10^8$ m/sec

$f = ?$

$f = \dfrac{3.00 \times 10^8 \frac{m}{sec}}{6.00 \times 10^{-7} \text{ m}}$

$= \dfrac{3.00 \times 10^8}{6.00 \times 10^{-7}} \dfrac{\cancel{m}}{sec} \times \dfrac{1}{\cancel{m}}$

$= 5.00 \times 10^{14} \dfrac{1}{sec}$

$= \boxed{5.00 \times 10^{14} \text{ Hz}}$

(b) From equation 8.5:

$f = 5.00 \times 10^{14}$ Hz

$h = 6.63 \times 10^{-34}$ J·sec

$E = ?$

$E = hf$

$= (6.63 \times 10^{-34} \text{ J·sec})\left(5.00 \times 10^{14} \dfrac{1}{sec}\right)$

$= (6.63 \times 10^{-34})(5.00 \times 10^{14}) \text{ J·}\cancel{sec} \times \dfrac{1}{\cancel{sec}}$

$= \boxed{3.32 \times 10^{-19} \text{ J}}$

7. First, you can find the energy of one photon of the peak intensity wavelength (5.60×10^{-7} m) by using equation 8.3 to find the frequency, then equation 8.5 to find the energy:

Step 1: $c = \lambda f \therefore f = \dfrac{c}{\lambda}$

$= \dfrac{3.00 \times 10^8 \frac{m}{sec}}{5.60 \times 10^{-7} \text{ m}}$

$= 5.36 \times 10^{14}$ Hz

Step 2: $E = hf$

$= (6.63 \times 10^{-34} \text{ J·sec})(5.36 \times 10^{14} \text{ Hz})$

$= 3.55 \times 10^{-19}$ J

Step 3: Since one photon carries an energy of 3.55×10^{-19} J and the overall intensity is 1,000.0 W, each square meter must receive an average of

$\dfrac{1000.0 \frac{J}{sec}}{3.55 \times 10^{-19} \frac{J}{photon}}$

$\dfrac{1.000 \times 10^3}{3.55 \times 10^{-19}} \dfrac{\cancel{J}}{sec} \times \dfrac{photon}{\cancel{J}}$

$\boxed{2.82 \times 10^{21} \dfrac{photon}{sec}}$

8. (a) $f = 4.90 \times 10^{14}$ Hz $\qquad c = \lambda f \therefore \lambda = \dfrac{c}{f}$

$c = 3.00 \times 10^8$ m/sec

$\lambda = ?$

$\lambda = \dfrac{3.00 \times 10^8 \frac{m}{sec}}{4.90 \times 10^{14} \frac{1}{sec}}$

$= \dfrac{3.00 \times 10^8}{4.90 \times 10^{14}} \dfrac{m}{\cancel{sec}} \times \dfrac{\cancel{sec}}{1}$

$= \boxed{6.12 \times 10^{-7} \text{ m}}$

(b) According to table 8.2, this is the frequency and wavelength of orange light.

9. $f = 5.00 \times 10^{20}$ Hz

$h = 6.63 \times 10^{-34}$ J·sec

$E = ?$

$E = hf$

$= (6.63 \times 10^{-34} \text{ J·sec})\left(5.00 \times 10^{20} \dfrac{1}{sec}\right)$

$= (6.63 \times 10^{-34})(5.00 \times 10^{20}) \text{ J·}\cancel{sec} \times \dfrac{1}{\cancel{sec}}$

$= \boxed{3.32 \times 10^{-13} \text{ J}}$

10. $\lambda = 1.00$ mm

$= 0.001$ m

$f = ?$

$c = 3.00 \times 10^8$ m/sec

$h = 6.63 \times 10^{-34}$ J·sec

$E = ?$

Step 1: $c = \lambda f \therefore f = \dfrac{v}{\lambda}$

$f = \dfrac{3.00 \times 10^8 \frac{m}{sec}}{1.00 \times 10^{-3} \text{ m}}$

$= \dfrac{3.00 \times 10^8}{1.00 \times 10^{-3}} \dfrac{\cancel{m}}{sec} \times \dfrac{1}{\cancel{m}}$

$= 3.00 \times 10^{11}$ Hz

Step 2:

$E = hf$

$= (6.63 \times 10^{-34} \text{ J·sec})\left(3.00 \times 10^{11} \dfrac{1}{sec}\right)$

$= (6.63 \times 10^{-34})(3.00 \times 10^{11}) \text{ J·}\cancel{sec} \times \dfrac{1}{\cancel{sec}}$

$= \boxed{1.99 \times 10^{-22} \text{ J}}$

Chapter 9

1. $m = 1.68 \times 10^{-27}$ kg

$v = 3.22 \times 10^3$ m/sec

$h = 6.63 \times 10^{-34}$ J·sec

$\lambda = ?$

$$\lambda = \frac{h}{mv}$$

$$= \frac{6.63 \times 10^{-34} \text{ J·sec}}{(1.68 \times 10^{-27} \text{ kg})\left(3.22 \times 10^3 \dfrac{m}{sec}\right)}$$

$$= \frac{6.63 \times 10^{-34}}{(1.68 \times 10^{-27})(3.22 \times 10^3)} \frac{\text{J·sec}}{\text{kg} \times \dfrac{m}{sec}}$$

$$= \frac{6.63 \times 10^{-34}}{5.41 \times 10^{-24}} \frac{\dfrac{\text{kg·m}^2}{\text{sec·sec}} \times \text{sec}}{\text{kg} \times \dfrac{m}{sec}}$$

$$= 1.23 \times 10^{-10} \frac{\text{kg·m·m}}{\text{sec}} \times \frac{1}{\text{kg}} \times \frac{\text{sec}}{m}$$

$$= \boxed{1.23 \times 10^{-10} \text{ m}}$$

2. (a)

$$n = 6$$

$$E_1 = -13.6 \text{ eV}$$

$$E_6 = ?$$

$$E_n = \frac{E_1}{n^2}$$

$$E_6 = \frac{-13.6 \text{ eV}}{6^2}$$

$$= \frac{-13.6 \text{ eV}}{36}$$

$$= \boxed{-0.378 \text{ eV}}$$

(b)

$$= (-0.378 \text{ eV})\left(1.60 \times 10^{-19} \frac{J}{eV}\right)$$

$$= (-0.378)(1.60 \times 10^{-19}) \text{ eV} \times \frac{J}{eV}$$

$$= \boxed{-6.05 \times 10^{-20} \text{ J}}$$

3. (a) Energy is related to the frequency and Planck's constant in equation 9.1, $E = hf$. From equation 9.6,

$$hf = E_h - E_l \therefore E = E_h - E_l$$

$$E_h = (n = 2) = -5.44 \times 10^{-19} \text{ J}$$

$$E_l = (n = 6) = -6.05 \times 10^{-20} \text{ J}$$

$$E = ? \text{ J}$$

$$E = E_h - E_l$$

$$= (-6.05 \times 10^{-20} \text{ J}) - (-5.44 \times 10^{-19} \text{ J})$$

$$= \boxed{4.84 \times 10^{-19} \text{ J}}$$

(b) $E_h = -0.377$ eV*

$E_l = -3.40$ eV*

$E = ?$ eV

$$E = E_h - E_l$$

$$= (-0.377 \text{ eV}) - (-3.40 \text{ eV})$$

$$= \boxed{3.02 \text{ eV}}$$

*From figure 9.14

4. $(n = 6) = -6.05 \times 10^{-20}$ J

$(n = 2) = -5.44 \times 10^{-19}$ J

$h = 6.63 \times 10^{-34}$ J·sec

$f = ?$

$$hf = E_h - E_l \therefore f = \frac{E_h - E_l}{h}$$

$$f = \frac{(-6.05 \times 10^{-20} \text{ J}) - (-5.44 \times 10^{-19} \text{ J})}{6.63 \times 10^{-34} \text{ J·sec}}$$

$$= \frac{4.84 \times 10^{-19} \text{ J}}{6.63 \times 10^{-34} \text{ J·sec}}$$

$$= 7.29 \times 10^{14} \frac{1}{\text{sec}}$$

$$= \boxed{7.29 \times 10^{14} \text{ Hz}}$$

5. $(n = 1) = -13.6$ eV $E_n = \dfrac{E_1}{n^2}$

$E = ?$

$$= \frac{-13.6 \text{ eV}}{1^2}$$

$$= -13.6 \text{ eV}$$

Since the energy of the electron is -13.6 eV, it will require 13.6 eV (or 2.18×10^{-19} J) to remove the electron.

6. $q/m = -1.76 \times 10^{11}$ C/kg

$q = -1.60 \times 10^{-19}$ C

$m = ?$

$$\text{mass} = \frac{\text{charge}}{\text{charge/mass}}$$

$$= \frac{-1.60 \times 10^{-19} \text{ C}}{-1.76 \times 10^{11} \dfrac{C}{kg}}$$

$$= \frac{-1.60 \times 10^{-19}}{-1.76 \times 10^{11}} \text{ C} \times \frac{kg}{C}$$

$$= \boxed{9.09 \times 10^{-31} \text{ kg}}$$

7. $\lambda = -1.67 \times 10^{-10}$ m

$m = 9.11 \times 10^{-31}$ kg

$v = ?$

$$\lambda = \frac{h}{mv} \quad \therefore \quad v = \frac{h}{m\lambda}$$

$$v = \frac{6.63 \times 10^{-34} \text{ J·sec}}{(9.11 \times 10^{-31} \text{ kg})(1.67 \times 10^{-10} \text{ m})}$$

$$= \frac{6.63 \times 10^{-34}}{(9.11 \times 10^{-31})(1.67 \times 10^{-10})} \frac{\text{J·sec}}{\text{kg·m}}$$

$$= \frac{6.63 \times 10^{-34}}{1.52 \times 10^{-40}} \frac{\frac{\text{kg·m}^2}{\text{sec·sec}} \times \text{sec}}{\text{kg·m}}$$

$$= 4.36 \times 10^{6} \frac{\text{kg·m·m}}{\text{sec}} \times \frac{1}{\text{kg}} \times \frac{1}{\text{m}}$$

$$= \boxed{4.36 \times 10^{6} \frac{\text{m}}{\text{sec}}}$$

8. (a) Boron: $1s^2 2s^2 2p^1$
   (b) Aluminum: $1s^2 2s^2 2p^6 3s^2 3p^1$
   (c) Potassium: $1s^2 2s^2 2p^6 3s^2 3p^6 4s^1$

9. (a) Boron is atomic number 5 and there are 5 electrons.
   (b) Aluminum is atomic number 13 and there are 13 electrons.
   (c) Potassium is atomic number 19 and there are 19 electrons.

10. (a) Argon: $1s^2 2s^2 2p^6 3s^2 3p^6$
    (b) Zinc: $1s^2 2s^2 2p^6 3s^2 3p^6 4s^2 3d^{10}$
    (c) Bromine: $1s^2 2s^2 2p^6 3s^2 3p^6 4s^2 3d^{10} 4p^5$

Chapter 10

1. The answers to this question are found in the list of elements and their symbols, which is located on the inside back cover of this text.

   (a) Silicon: Si
   (b) Silver: Ag
   (c) Helium: He
   (d) Potassium: K
   (e) Magnesium: Mg
   (f) Iron: Fe

2. Atomic weight is the weighted average of the isotopes as they occur in nature. Thus

   Lithium-6: $6.01512 \text{ u} \times 0.0742 = 0.446 \text{ u}$

   Lithium-7: $7.016 \text{ u} \times 0.9258 = 6.4054 \text{ u}$

   Lithium-6 contributes 0.446 u of the weighted average and lithium-7 contributes 6.4954 u. The atomic weight of lithium is therefore

   $$\begin{array}{r} 0.446 \text{ u} \\ + 6.4954 \text{ u} \\ \hline 6.941 \text{ u} \end{array}$$

3. Recall that the subscript is the atomic number, which identifies the number of protons. In a neutral atom, the number of protons equals the number of electrons so the atomic number tells you the number of electrons, too. The superscript is the mass number, which identifies the number of neutrons and the number of protons in the nucleus. The number of neutrons is therefore the mass number minus the atomic number.

	Protons	Neutrons	Electrons
(a)	6	6	6
(b)	1	0	1
(c)	18	22	18
(d)	1	1	1
(e)	79	118	79
(f)	92	143	92

4.

		Period	Family
(a)	Radon (Rn)	6	VIIIA
(b)	Sodium (Na)	3	IA
(c)	Copper (Cu)	4	IB
(d)	Neon (Ne)	2	VIIIA
(e)	Iodine (I)	5	VIIA
(f)	Lead (Pb)	6	IVA

5. Recall that the number of outer shell electrons is the same as the family number for the representative elements:

   (a) Li: 1   (d) Cl: 7
   (b) N: 5   (e) Ra: 2
   (c) F: 7   (f) Be: 2

6. The same information that was used in question 5 can be used to draw the dot notation:

   (a) $\overset{\Large\cdot}{B}\cdot$ (c) Ca: (e) $\cdot\overset{\Large\cdot\cdot}{\underset{\cdot\cdot}{O}}\cdot$

   (b) $\cdot\overset{\cdot\cdot}{\underset{\cdot\cdot}{B}r}:$ (d) K·   (f) $\cdot\overset{\cdot\cdot}{\underset{\cdot\cdot}{S}}\cdot$

7. The charge is found by identifying how many electrons are lost or gained in achieving the noble gas structure:

   (a) Boron $3+$
   (b) Bromine $1-$
   (c) Calcium $2+$
   (d) Potassium $1+$
   (e) Oxygen $2-$
   (f) Nitrogen $3-$

8. Metals have one, two, or three outer electrons and are located in the left two-thirds of the periodic table. Semiconductors are adjacent to the line that separates the metals and nonmetals. Look at the periodic table on the inside back cover and you will see:

   (a) Krypton—nonmetal
   (b) Cesium—metal
   (c) Silicon—semiconductor
   (d) Sulfur—nonmetal
   (e) Molybdenum—metal
   (f) Plutonium—metal

9. (a) Bromine gained an electron to acquire a $1-$ charge, so it must be in family VIIA (the members of this family have seven electrons and need one more to acquire the noble gas structure).
   (b) Potassium must have lost one electron, so it is in IA.
   (c) Aluminum lost three electrons, so it is in IIIA.
   (d) Sulfur gained two electrons, so it is in VIA.
   (e) Barium lost two electrons, so it is in IIA.
   (f) Oxygen gained two electrons, so it is in VIA.

10. (a) $^{16}_{8}\text{O}$ (c) $^{3}_{1}\text{H}$

    (b) $^{23}_{11}\text{Na}$ (d) $^{35}_{17}\text{Cl}$

Chapter 11

1.

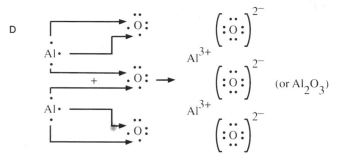

A $K\cdot$ + $\cdot\ddot{\underset{\cdot\cdot}{I}}:$ $\longrightarrow$ $K^+\left(:\ddot{\underset{\cdot\cdot}{I}}:\right)^-$ (or KI)

B $Sr:$ + $\cdot\ddot{\underset{\cdot\cdot}{S}}:$ $\longrightarrow$ $Sr^{2+}\left(:\ddot{\underset{\cdot\cdot}{S}}:\right)^{2-}$ (or SrS)

C $\begin{matrix} Na\cdot \\ Na\cdot \end{matrix}$ + $\cdot\ddot{\underset{\cdot\cdot}{O}}:$ $\longrightarrow$ $\begin{matrix} Na^+ \\ Na^+ \end{matrix}\left(:\ddot{\underset{\cdot\cdot}{O}}:\right)^{2-}$ (or Na_2O)

D $\begin{matrix} Al\cdot \\ Al\cdot \end{matrix}$ + $\begin{matrix} \cdot\ddot{\underset{\cdot\cdot}{O}}: \\ \cdot\ddot{\underset{\cdot\cdot}{O}}: \\ \cdot\ddot{\underset{\cdot\cdot}{O}}: \end{matrix}$ $\longrightarrow$ $\begin{matrix} Al^{3+} \\ Al^{3+} \end{matrix}$ $\begin{matrix} \left(:\ddot{\underset{\cdot\cdot}{O}}:\right)^{2-} \\ \left(:\ddot{\underset{\cdot\cdot}{O}}:\right)^{2-} \\ \left(:\ddot{\underset{\cdot\cdot}{O}}:\right)^{2-} \end{matrix}$ (or Al_2O_3)

2. (a) Sulfur is in family VIA, so sulfur has six valence electrons and will need two more to achieve a stable outer structure like the noble gases. Two more outer shell electrons will give the sulfur atom a charge of 2−. Copper²⁺ will balance the 2− charge of sulfur, so the name is copper(II) sulfide. Note the "-ide" ending for compounds that have only two different elements.

(b) Oxygen is in family VIA, so oxygen has six valence electrons and will have a charge of 2−. Using the crossover technique in reverse, you can see that the charge on the oxygen is 2− and the charge on the iron is 3−. Therefore the name is iron(III) oxide.

(c) From information in (a) and (b), you know that oxygen has a charge of 2−. The chromium ion must have the same charge to make a neutral compound as it must be, so the name is chromium(II) oxide. Again, note the "-ide" ending for a compound with two different elements.

(d) Sulfur has a charge of 2−, so the lead ion must have the same positive charge to make a neutral compound. The name is lead(II) sulfide.

3. The name of some common polyatomic ions are in table 11.4. Using this table as a reference, the names are

(a) hydroxide
(b) sulfite
(c) hypochlorite
(d) nitrate
(e) carbonate
(f) perchlorate

4. The Roman numeral tells you the charge on the variable-charge elements. The charges for the polyatomic ions are found in table 11.4. The charges for metallic elements can be found in tables 11.1 and 11.2. Using these resources and the crossover technique, the formulas are as follows:

(a) $Fe(OH)_3$
(b) $Pb_3(PO_4)_2$
(c) $ZnCO_3$
(d) NH_4NO_3
(e) $KHCO_3$
(f) K_2SO_3

5. Table 11.6 has information about the meaning of prefixes and stem names used in naming covalent compounds. (a), for example, asks for the formula of carbon tetrachloride. Carbon has no prefixes, so there is one carbon atom, and it comes first in the formula because it comes first in the name. The "tetra-" prefix means four, so there are four chlorine atoms. The name ends in "-ide" so you know there are only two elements in the compound. The symbols can be obtained from the list of elements on the inside back cover of this text. Using all this information from the name, you can think out the formula for carbon tetrachloride. The same process is used for the other compounds and formulas:

(a) CCl_4
(b) H_2O
(c) MnO_2
(d) SO_3
(e) N_2O_5
(f) As_2S_5

6. Again using information from table 11.6, this question requires you to reverse the thinking procedure you learned in question 5.

(a) carbon monoxide
(b) carbon dioxide
(c) carbon disulfide
(d) dinitrogen monoxide
(e) tetraphosphorus trisulfide
(f) dinitrogen trioxide

7. The types of bonds formed are predicted by using the electronegativity scale in figure 11.7 and finding the absolute difference. On this basis:

(a) Difference = 1.7, which means ionic bond
(b) Difference = 0, which means nonpolar covalent
(c) Difference = 0, which means nonpolar covalent
(d) Difference = 0.4, which means nonpolar covalent
(e) Difference = 3.0, which means ionic
(f) Difference = 1.6, which means polar covalent and almost ionic

Chapter 12

1. (a) $MgCl_2$ is an ionic compound, so the formula has to be empirical.

(b) C_2H_2 is a covalent compound, so the formula might be molecular. Since it is not the simplest whole number ratio (which would be CH), then the formula is molecular.

(c) BaF_2 is ionic; the formula is empirical.

(d) C_8H_{18} is not the simplest whole number ratio of a covalent compound, so the formula is molecular.

(e) CH_4 is covalent, but the formula might or might not be molecular (?).

(f) S_8 is a nonmetal bonded to a nonmetal (itself); this is a molecular formula.

2. (a) $CuSO_4$

   1 of Cu $= 1 \times 63.5$ u $=$ 63.5 u
   1 of S $\ = 1 \times 32.1$ u $=$ 32.1 u
   4 of O $\ = 4 \times 16.0$ u $=$ 64.0 u
   $\underline{}$
   159.6 u

   (b) CS_2

   1 of C $= 1 \times 12.0$ u $= 12.0$ u
   2 of S $= 2 \times 32.0$ u $= 64.0$ u
   $\underline{}$
   76.0 u

   (c) $CaSO_4$

   1 of Ca $= 1 \times 40.1$ u $=$ 40.1 u
   1 of S $\ = 1 \times 32.0$ u $=$ 32.0 u
   4 of O $\ = 4 \times 16.0$ u $=$ 64.0 u
   $\underline{}$
   136.1 u

   (d) Na_2CO_3

   2 of Na $= 2 \times 23.0$ u $-$ 46.0 u
   1 of C $\ = 1 \times 12.0$ u $=$ 12.0 u
   3 of O $\ = 3 \times 16.0$ u $=$ 48.0 u
   $\underline{}$
   106.0 u

3. (a) FeS_2

   For Fe: $\dfrac{(55.9 \text{ u Fe})(1)}{119.9 \text{ u } FeS_2} \times 100\% \ FeS_2 = 46.6\%$ Fe

   For S: $\dfrac{(32.0 \text{ u S})(2)}{119.9 \text{ u } FeS_2} \times 100\% \ FeS_2 = 53.4\%$ S

   or $(100\% \ FeS_2) - (46.6\% \text{ Fe}) = 53.4\%$ S

   (b) H_3BO_3

   For H: $\dfrac{(1.0 \text{ u H})(3)}{61.8 \text{ u } H_3BO_3} \times 100\% \ H_3BO_3 = 4.85\%$ H

   For B: $\dfrac{(10.8 \text{ u B})(1)}{61.8 \text{ u } H_3BO_3} \times 100\% \ H_3BO_3 = 17.5\%$ B

   For O: $\dfrac{(16 \text{ u O})(3)}{61.8 \text{ u } H_3BO_3} \times 100\% \ H_3BO_3 = 77.7\%$ O

   (c) $NaHCO_3$

   For Na: $\dfrac{(23.0 \text{ u Na})(1)}{84.0 \text{ u } NaHCO_3} \times 100\% \ NaHCO_3 = 27.4\%$ Na

   For H: $\dfrac{(1.0 \text{ u H})(1)}{84.0 \text{ u } NaHCO_3} \times 100\% \ NaHCO_3 = 1.2\%$ H

   For C: $\dfrac{(12.0 \text{ u C})(1)}{84.0 \text{ u } NaHCO_3} \times 100\% \ NaHCO_3 = 14.3\%$ C

   For O: $\dfrac{(16.0 \text{ u O})(3)}{84.0 \text{ u } NaHCO_3} \times 100\% \ NaHCO_3 = 57.1\%$ O

   (d) $C_9H_8O_4$

   For C: $\dfrac{(12.0 \text{ u C})(9)}{180.0 \text{ u } C_9H_8O_4} \times 100\% \ C_9H_8O_4 = 60.0\%$ C

   For H: $\dfrac{(1.0 \text{ u H})(8)}{180.0 \text{ u } C_9H_8O_4} \times 100\% \ C_9H_8O_4 = 4.4\%$ H

   For O: $\dfrac{(16.0 \text{ u O})(4)}{180.0 \text{ u } C_9H_8O_4} \times 100\% \ C_9H_8O_4 = 35.6\%$ O

4. (a) $2 SO_2 + O_2 \rightarrow 2 SO_3$
   (b) $4 P + 5 O_2 \rightarrow 2 P_2O_5$
   (c) $2 Al + 6 HCl \rightarrow 2 AlCl_3 + 3 H_2$
   (d) $2 NaOH + 6 H_2SO_4 \rightarrow Na_2SO_4 + 2 H_2O$
   (e) $Fe_2O_3 + 3 CO \rightarrow 2 Fe + 3 CO_2$
   (f) $3 Mg(OH)_2 + 2 H_3PO_4 \rightarrow Mg_3(PO_4)_2 + 6 H_2O$

5. (a) General form of $XY + AZ \rightarrow XZ + AY$ with precipitate formed: Ion exchange reaction.
   (b) General form of $X + Y \rightarrow XY$: Combination reaction.
   (c) General form of $XY \rightarrow X + Y + \ldots$: Decomposition reaction.
   (d) General form of $X + Y \rightarrow XY$: Combination reaction.
   (e) General form of $XY + A \rightarrow AY + X$: Replacement reaction.
   (f) General form of $X + Y \rightarrow XY$: Combination reaction.

6. (a) $C_5H_{12}(g) + 8 O_2(g) \rightarrow 5 CO_2(g) + 6 H_2O(g)$
   (b) $HCl(aq) + NaOH(aq) \rightarrow NaCl(aq) + H_2O(l)$
   (c) $2 Al(s) + Fe_2O_3(s) \rightarrow Al_2O_3(s) + 2 Fe(l)$
   (d) $Fe(s) + CuSO_4(aq) \rightarrow FeSO_4(aq) + Cu(s)$
   (e) $MgCl(aq) + Fe(NO_3)_2(aq) \rightarrow$ No reaction (all possible compounds are soluble and no gas or water was formed).
   (f) $C_6H_{10}O_5(s) + 6 O_2(g) \rightarrow 6 CO_2(g) + 5 H_2O(g)$

7. (a) $2 KClO_3(s) \overset{\Delta}{\rightarrow} 2 KCl(s) + 3 O_2\uparrow$
   (b) $2 Al_2O_3(l) \overset{elec}{\rightarrow} 4 Al(s) + 3 O_2\uparrow$
   (c) $CaCO_3(s) \overset{\Delta}{\rightarrow} CaO(s) + CO_2\uparrow$

8. (a) $2 Na(s) + 2 H_2O(l) \rightarrow 2 NaOH(aq) + H_2\uparrow$
   (b) $Au(s) + HCl(aq) \rightarrow$ No reaction (gold is below hydrogen in the activity series)
   (c) $Al(s) + FeCl_3(aq) \rightarrow AlCl_3(aq) + Fe(s)$
   (d) $Zn(s) + CuCl_2(aq) \rightarrow ZnCl_2(aq) + Cu(s)$

9. (a) $NaOH(aq) + HNO_3(aq) \rightarrow NaNO_3(aq) + H_2O(l)$
   (b) $CaCl_2(aq) + KNO_3(aq) \rightarrow$ No reaction
   (c) $3 Ba(NO_3)_2(aq) + 2 Na_3PO_4(aq) \rightarrow 6 NaNO_3(aq) + Ba_3(PO_4)_2\downarrow$
   (d) $2 KOH(aq) + ZnSO_4(aq) \rightarrow K_2SO_4(aq) + Zn(OH)_2\downarrow$

10. One mole of oxygen combines with 2 moles of acetylene, so 0.5 mole of oxygen would be needed for 1 mole of acetylene. Therefore, 1 L of C_2H_2 requires 0.5 L of O_2.

Chapter 13

1. $m_{solute} = 1.75$ g

   $m_{solution} = 50.0$ g

   $\%_{weight} = ?$

   $\% \text{ solute} = \dfrac{m_{solute}}{m_{solution}} \times 100\% \text{ solution}$

   $= \dfrac{1.75 \text{ g NaCl}}{50.0 \text{ g solution}} \times 100\% \text{ solution}$

   $= \boxed{3.50\% \text{ NaCl}}$

2. $m_{solution} = 103.5$ g

$m_{solute} = 3.50$ g

$\%_{weight} = ?$

$$\% \text{ solute} = \frac{m_{solute}}{m_{solution}} \times 100\% \text{ solution}$$

$$= \frac{3.50 \text{ g NaCl}}{103.5 \text{ g solution}} \times 100\% \text{ solution}$$

$$= \boxed{3.38\% \text{ NaCl}}$$

3. Since ppm is defined as the weight unit of solute in 1,000,000 weight units of solution, the percent by weight can be calculated just like any other percent. The weight of the dissolved sodium and chlorine ions is the part, and the weight of the solution is the whole, so

$$\% = \frac{\text{part}}{\text{whole}} \times 100\%$$

$$= \frac{30,113 \text{ g NaCl ions}}{1,000,000 \text{ g seawater}} \times 100\% \text{ seawater}$$

$$= \boxed{3.00\% \text{ NaCl ions}}$$

4. $m_{solution} = 250$ g

$\%$ solute $= 3.0\%$

$m_{solute} = ?$

$$\% \text{ solute} = \frac{m_{solute}}{m_{solution}} \times 100\% \text{ solution}$$

$$\therefore$$

$$m_{solute} = \frac{(m_{solution})(\% \text{ solute})}{100\% \text{ solution}}$$

$$= \frac{(250 \text{ g})(3.0\%)}{100\%}$$

$$= \boxed{7.5 \text{ g}}$$

5. $\%$ solution $= 12\%$ solution

$V_{solution} = 200$ mL

$V_{solute} = ?$

$$\% \text{ solution} = \frac{V_{solute}}{V_{solution}} \times 100\% \text{ solution}$$

$$\therefore$$

$$V_{solute} = \frac{(\% \text{ solution})(V_{solution})}{100\% \text{ solution}}$$

$$= \frac{(12\% \text{ solution})(200 \text{ mL})}{100\% \text{ solution}}$$

$$= \boxed{24 \text{ mL alcohol}}$$

6. $\%$ solution $= 40\%$

$V_{solution} = 50$ mL

$V_{solute} = ?$

$$\% \text{ solution} = \frac{V_{solute}}{V_{solution}} \times 100\% \text{ solution}$$

$$\therefore$$

$$V_{solute} = \frac{(\% \text{ solution})(V_{solution})}{100\% \text{ solution}}$$

$$= \frac{(40\% \text{ solution})(50 \text{ mL})}{100\% \text{ solution}}$$

$$= \boxed{20 \text{ mL alcohol}}$$

7. (a)

$$\% \text{ concentration} = \frac{\text{ppm}}{1 \times 10^4}$$

$$= \frac{5}{1 \times 10^4}$$

$$= \boxed{0.0005\% \text{ DDT}}$$

(b)

$$\% \text{ part} = \frac{\text{part}}{\text{whole}} \times 100\% \text{ whole}$$

$$\therefore$$

$$\text{whole} = \frac{(100\%)(\text{part})}{\% \text{ part}}$$

$$= \frac{(100\%)(17.0 \text{ g})}{0.0005\%} = \boxed{3,400,000 \text{ g or } 3,400 \text{ kg}}$$

8. (a) $HC_2H_3O_2(aq) + H_2O(l) \rightleftarrows H_3O^+(aq) + C_2H_3O_2^-(aq)$
     acid   base

(b) $C_6H_5NH_2(l) + H_2O(l) \rightleftarrows C_6H_5NH_3^+(aq) + OH^-(aq)$
     base   acid

(c) $HClO_4(aq) + HC_2H_3O_2(aq) \rightleftarrows H_2C_2H_3O_2^+(aq)$
     acid   base
    $+ ClO_4^-(aq)$

(d) $H_2O(l) + H_2O(l) \rightleftarrows H_3O^+(aq) + OH^-(aq)$
     base   acid

Chapter 14

1.

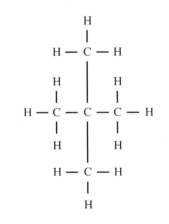

(a)

(b)

(c) 2,2-dimethylpropane

2. *n*-hexane

n-**hexane**

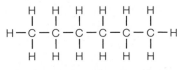

3-methylpentane

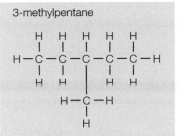

2-methylpentane

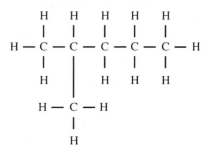

2,2-dimethylbutane

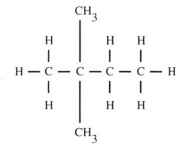

2,3-dimethylbutane

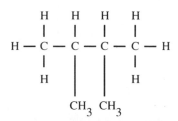

3.

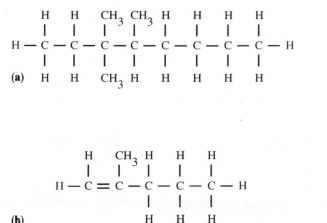

(a)

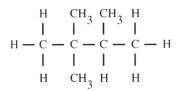

(b)

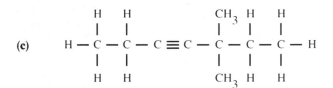

(c)

4. (a) 2-chloro-4-methylpentane
(b) 2-methyl-1-pentene
(c) 4-methyl-3-ethyl-2-pentene

5. The 2,2,3-trimethylbutane is more highly branched, so it will have the higher octane rating.

2,2,3 trimethylbutane

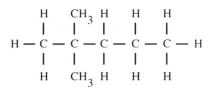

2,2-dimethylpentane

6. (a) alcohol
   (b) amide
   (c) ether
   (d) ester
   (e) organic acid

Chapter 15

1. (a) cobalt-60: 27 protons, 33 neutrons
   (b) potassium-40: 19 protons, 21 neutrons
   (c) neon-24: 10 protons, 14 neutrons
   (d) lead-204: 82 protons, 122 neutrons

2. (a) $^{60}_{27}Co$ (c) $^{24}_{10}Ne$
   (b) $^{40}_{19}K$ (d) $^{204}_{82}Pb$

3. (a) cobalt-60: Radioactive because odd numbers of protons (27) and odd numbers of neutrons (33) are usually unstable.
   (b) potassium-40: Radioactive, again having an odd number of protons (19) and an odd number of neutrons (21).
   (c) neon-24: Stable, because even numbers of protons and neutrons are usually stable.
   (d) lead-204: Stable, because even numbers of protons and neutrons *and* because 82 is a particularly stable number of nucleons.

4. (a) $^{56}_{26}Fe \rightarrow {}^{0}_{-1}e + {}^{56}_{27}Co$
   (b) $^{7}_{4}Be \rightarrow {}^{0}_{-1}e + {}^{7}_{5}B$
   (c) $^{64}_{29}Cu \rightarrow {}^{0}_{-1}e + {}^{64}_{30}Zn$
   (d) $^{24}_{11}Na \rightarrow {}^{0}_{-1}e + {}^{24}_{12}Mg$
   (e) $^{214}_{82}Pb \rightarrow {}^{0}_{-1}e + {}^{214}_{83}Bi$
   (f) $^{32}_{15}P \rightarrow {}^{0}_{-1}e + {}^{32}_{16}S$

5. (a) $^{235}_{92}U \rightarrow {}^{4}_{2}He + {}^{231}_{90}Th$
   (b) $^{226}_{88}Ra \rightarrow {}^{4}_{2}He + {}^{222}_{86}Rn$
   (c) $^{239}_{94}Pu \rightarrow {}^{4}_{2}He + {}^{235}_{92}U$
   (d) $^{214}_{83}Bi \rightarrow {}^{4}_{2}He + {}^{210}_{81}Tl$
   (e) $^{230}_{90}Th \rightarrow {}^{4}_{2}He + {}^{226}_{88}Ra$
   (f) $^{210}_{84}Po \rightarrow {}^{4}_{2}He + {}^{206}_{82}Pb$

6. Thirty-two days is four half-lives. After the first half-life (8 days), 1/2 oz will remain. After the second half-life (8 + 8, or 16 days), 1/4 oz will remain. After the third half-life (8 + 8 + 8, or 24 days), 1/8 oz will remain. After the fourth half-life (8 + 8 + 8 + 8, or 32 days), 1/16 oz will remain, or 6.3×10^{-2} oz.

7. The relationship between half-life and the radioactive decay constant is found in equation 17.2, which is first solved for the decay constant:

$$t_{1/2} = \frac{0.693}{k} \therefore k = \frac{0.693}{t_{1/2}}$$

A decay constant calculation requires units of seconds, so 27.6 years is next converted to seconds:

$$(27.6 \text{ yr})(3.15 \times 10^7 \text{ sec/yr}) = 8.69 \times 10^8 \text{ sec}$$

This value in seconds is now used in equation 17.2 solved for k:

$$k = \frac{0.693}{8.69 \times 10^8 \text{ sec}} = 7.97 \times 10^{-10}/\text{sec}$$

8. The relationship between the radioactive decay constant, the decay rate, and the number of nuclei is found in equation 17.1, which can be solved for the rate:

$$k = \frac{\text{rate}}{n} \therefore \text{rate} = kn$$

The number of nuclei in a molar mass of strontium-90 (87.6 g) is Avogadro's number, 6.02×10^{23}, so

$$\text{rate} = (7.97 \times 10^{-10}/\text{sec})(6.02 \times 10^{23} \text{ nuclei})$$
$$= 4.80 \times 10^{14} \text{ nuclei/sec}$$

9. A curie (Ci) is defined as 3.70×10^{10} disintegrations/sec, so a molar mass that is decaying at a rate of 4.80×10^{14} nuclei/sec has a curie activity of

$$\frac{4.80 \times 10^{14}}{3.70 \times 10^{10}} = 1.30 \times 10^4 \text{ Ci}$$

10. The Fe-56 nucleus has a mass of 55.9206 u, but the individual masses of the nucleons are

$$26 \text{ protons} \times 1.00728 \text{ u} = 26.1893 \text{ u}$$
$$30 \text{ neutrons} \times 1.00867 \text{ u} = \underline{30.2601 \text{ u}}$$
$$56.4494 \text{ u}$$

The mass defect is thus

$$\begin{array}{r} 56.4494 \text{ u} \\ - 55.9206 \text{ u} \\ \hline 0.5288 \text{ u} \end{array}$$

The atomic mass unit (u) is equal to the mass of a mole (g), therefore 0.5288 u = 0.5288 g. The mass defect is equivalent to the binding energy according to $E = mc^2$. For a molar mass of Fe-56, the mass defect is

$$E = (5.29 \times 10^{-4} \text{ kg})\left(3.00 \times 10^8 \frac{m}{\text{sec}}\right)^2$$
$$= (5.29 \times 10^{-4} \text{ kg})\left(9.00 \times 10^{16} \frac{m^2}{\text{sec}^2}\right)$$
$$= 4.76 \times 10^{13} \frac{\text{kg} \cdot m^2}{\text{sec}^2}$$
$$= 4.76 \times 10^{13} \text{ J}$$

For a single nucleus,

$$\frac{4.76 \times 10^{13} \text{ J}}{6.02 \times 10^{23} \text{ nuclei}} = 7.90 \times 10^{-11} \text{ J/nuclei}$$

Glossary

absolute humidity a measure of the actual amount of water vapor in the air at a given time—for example, in grams per cubic meter

absolute magnitude a classification scheme to compensate for the distance differences to stars; calculations of the brightness that stars would appear to have if they were all at a defined, standard distance of 10 parsec

absolute scale temperature scale set so that zero is at the theoretical lowest temperature possible, which would occur when all random motion of molecules has ceased

absolute zero the theoretical lowest temperature possible, which occurs when all random motion of molecules has ceased

abyssal plain the practically level plain of the ocean floor

acceleration a change in velocity per change in time; by definition, this change in velocity can result from a change in speed, a change in direction, or a combination of changes in speed and direction

accretion disk fat bulging disk of gas and dust from the remains of the gas cloud that forms around a protostar

achondrites homogeneously textured stony meteorites

acid any substance that is a proton donor when dissolved in water; generally considered a solution of hydronium ions in water that can neutralize a base, forming a salt and water

acid-base indicator a vegetable dye used to distinguish acid and base solutions by a color change

adiabatic cooling the decrease in temperature of an expanding gas that involves no additional heat flowing out of the gas; cooling from the energy lost by expansion

adiabatic heating the increase in temperature of compressed gas that involves no additional heat flowing into the gas; heating from the energy gained by compression

air mass a large, more or less uniform body of air with nearly the same temperature and moisture conditions throughout

air mass weather the weather experienced within a given air mass; characterized by slow gradual changes from day to day

alcohol an organic compound with a general formula of ROH, where R is one of the hydrocarbon groups—for example, methyl or ethyl

aldehyde an organic molecule with the general formula RCHO, where R is one of the hydrocarbon groups—for example, methyl or ethyl

alkali metals members of family IA of the periodic table, having common properties of shiny, low density metals that can be cut with a knife and that react violently with water to form an alkaline solution

alkaline earth metals members of family IIA of the periodic table, having common properties of soft, reactive metals that are less reactive than alkali metals

alkanes hydrocarbons with single covalent bonds between the carbon atoms

alkenes hydrocarbons with a double covalent carbon-carbon bond

alkyne hydrocarbon with a carbon-carbon triple bond

allotropic forms elements that can have several different structures with different physical properties—for example, graphite and diamond are two allotropic forms of carbon

alpha particle the nucleus of a helium atom (two protons and two neutrons) emitted as radiation from a decaying heavy nucleus; also known as an alpha ray

alpine glaciers glaciers that form at high elevations in mountainous regions

alternating current an electric current that first moves one direction, then the opposite direction with a regular frequency

amino acids organic functional groups that form polypeptides and proteins

amp unit of electric current; equivalent to C/sec

ampere full name of the unit amp

amplitude the extent of displacement from the equilibrium condition; the size of a wave from the rest (equilibrium) position

angle of incidence angle of an incident (arriving) ray or particle to a surface; measured from a line perpendicular to the surface (the normal)

angle of reflection angle of a reflected ray or particle from a surface; measured from a line perpendicular to the surface (the normal)

angular momentum quantum number from quantum mechanics model of the atom, one of four descriptions of the energy state of an electron wave; this quantum number describes the energy sublevels of electrons within the main energy levels of an atom

angular unconformity a boundary in rock where the bedding planes above and below the time interruption unconformity are not parallel, meaning probable tilting or folding followed by a significant period of erosion, which in turn was followed by a period of deposition

annular eclipse occurs when the penumbra reaches the surface of the earth; as seen from earth, the sun forms a bright ring around the disk of the new moon

antarctic circle parallel identifying the limit toward the equator where the sun appears above the horizon all day for six months during the summer; located at 66.5° S latitude

anticline an arch-shaped fold in layered bed rock

anticyclone a high pressure center with winds flowing away from the center; associated with clear, fair weather

antinode region of maximum amplitude between adjacent nodes in a standing wave

aphelion the point at which an orbit is farthest from the sun

apogee the point at which the moon's elliptical orbit takes the moon farthest from the earth

apparent local noon the instant of the sun crossing the celestial meridian at any particular longitude

apparent local solar time the time found from the position of the sun in the sky; the shadow of the gnomon on a sundial

apparent magnitude a classification scheme for different levels of brightness of stars that you see; brightness values range from one to six with the number one (first magnitude) assigned to the brightest star and the number six (sixth magnitude) assigned to the faintest star that can be seen

apparent solar day the interval between two consecutive crossings of the celestial meridian by the sun

aquifer a layer of sand, gravel, or other highly permeable material beneath the surface that is saturated with water and is capable of producing water

arctic circle parallel identifying the limit toward the equator where the sun appears above the horizon all day for six months during the summer; located at 66.5° N latitude

area the extent of a surface; the surface bounded by three or more lines

arid dry climate classification; receives less than 25 cm (10 in) precipitation per year

aromatic hydrocarbon organic compound with at least one benzene ring structure; cyclic hydrocarbons and their derivatives

artesian term describing the condition where confining pressure forces groundwater from a well to rise above the aquifer

asbestos The common name for any one of several incombustible fibrous minerals that will not melt or ignite and can be woven into a fireproof cloth or used directly in fireproof insulation; about six different commercial varieties of asbestos are used, one of which has been linked to cancer under heavy exposure

asteroids small rocky bodies left over from the formation of the solar system; most are accumulated in a zone between the orbits of Mars and Jupiter

asthenosphere a plastic, mobile layer of the earth's structure that extends unbroken around the entire earth below the lithosphere; ranges in thickness from a depth of 130 km to 160 km

astronomical unit the radius of the earth's orbit is defined as one astronomical unit (A.U.)

atmospheric stability the condition of a parcel of air found by relating the parcel's actual lapse rate to the dry adiabatic lapse rate

atom the smallest unit of an element that can exist alone or in combination with other elements

atomic mass unit relative mass unit (u) of an isotope based on the standard of the carbon-12 isotope, which is defined as a mass of exactly 12.00 u; one atomic mass unit (1 u) is 1/12 the mass of a carbon-12 atom

atomic number the number of protons in the nucleus of an atom

atomic weight weighted average of the masses of stable isotopes of an element as they occur in nature, based on the abundance of each isotope of the element and the atomic mass of the isotope compared to C-12

autumnal equinox one of two times a year that daylight and night are of equal length; occurs on or about September 23 and identifies the beginning of the fall season

avalanche a mass movement of a wide variety of materials such as rocks, snow, trees, soils, and so forth in a single chaotic flow; also called debris avalanche

Avogadro's number the number of C-12 atoms in exactly 12.00 g of C; 6.02×10^{23} atoms or other chemical units; the number of chemical units in one mole of a substance

axis the imaginary line about which a planet or other object rotates

background radiation ionizing radiation (alpha, beta, gamma, etc.) from natural sources; between 100 and 500 millirems/yr of exposure to natural radioactivity from the environment

Balmer series a set of four line spectra, narrow lines of color emitted by hydrogen atom electrons as they drop from excited states to the ground state

band of stability a region of a graph of the number of neutrons versus the number of protons in nuclei; nuclei that have the neutron to proton

ratios located in this band do not undergo radioactive decay

barometer an instrument that measures atmospheric pressure, used in weather forecasting and in determining elevation above sea level

base any substance that is a proton acceptor when dissolved in water; generally considered a solution that forms hydroxide ions in water that can neutralize an acid, forming a salt and water

basin a large, bowl-shaped fold in the land into which streams drain; also a small enclosed or partly enclosed body of water

batholith a large magma that has cooled and solidified below the surface, forming a large mass of intrusive rock

beat rhythmic increases and decreases of volume from constructive and destructive interference between two sound waves of slightly different frequencies

beta particle high energy electron emitted as ionizing radiation from a decaying nucleus; also known as a beta ray

big bang theory current model of galactic evolution in which the universe was created from an intense and brilliant explosion from a primeval fireball

binding energy the energy required to break a nucleus into its constituent protons and neutrons; also the energy equivalent released when a nucleus is formed

black hole the theoretical remaining core of a supernova that is so dense that even light cannot escape

blackbody radiation electromagnetic radiation emitted by an ideal material (the blackbody) that perfectly absorbs and perfectly emits radiation

body wave a seismic wave that travels through the earth's interior, spreading outward from a disturbance in all directions

Bohr model model of the structure of the atom that attempted to correct the deficiencies of the solar system model and account for the Balmer series

boiling point the temperature at which a phase change of liquid to gas takes place through boiling; the same temperature as the condensation point

boundary the division between two regions of differing physical properties

Bowen's reaction series crystallization series that occurs as a result of the different freezing point temperatures of various minerals present in magma

breaker a wave whose front has become so steep that the top part has broken forward of the wave, breaking into foam, especially against a shoreline

British thermal unit the amount of energy or heat needed to increase the temperature of one pound of water one degree Fahrenheit (abbreviated Btu)

buffer solution a solution consisting of a weak acid and a salt that has the same negative ion as the acid; has the ability to resist changes in the pH when small amounts of an acid or a base are added to the solution

calorie the amount of energy (or heat) needed to increase the temperature of one gram of water one degree Celsius

Calorie the dieter's "calorie"; equivalent to one kilocalorie

carbohydrates organic compounds that include sugars, starches, and cellulose; carbohydrates are used by plants and animals for structure, protection, and food

carbon film a type of fossil formed when the volatile and gaseous constituents of a buried organic structure are distilled away, leaving a carbon film as a record

carbonation in chemical weathering a reaction that occurs naturally between carbonic acid (H_2CO_3) and rock minerals

cast sediments deposited by groundwater in a mold, taking the shape and external features of the organism that was removed to form the mold, then gradually changing to sedimentary rock

cathode rays negatively charged particles (electrons) that are emitted from a negative terminal in an evacuated glass tube

celestial equator line of the equator of the earth directly above the earth; the equator of the earth projected on the celestial sphere

celestial meridian an imaginary line in the sky directly above you that runs north through the north celestial pole, south through the south celestial pole, and back around the other side to make a big circle around the earth

celestial sphere a coordinate system of lines used to locate objects in the sky by imagining a huge turning sphere surrounding the earth with the stars and other objects attached to the sphere; the latitude and longitude lines of the earth's surface are projected to the celestial sphere

cellulose a polysaccharide abundant in plants that forms the fibers in cell walls that preserves the structure of plant materials

Celsius scale referent scale that defines numerical values for measuring hotness or coldness, defined as degrees of temperature; based on the reference points of the freezing point of water and the boiling point of water at sea level pressure, with 100 degrees between the two points

cementation process by which spaces between buried sediment particles under compaction are filled with binding chemical deposits, binding the particles into a rigid, cohesive mass of a sedimentary rock

Cenozoic one of four geologic eras; the time of recent life, meaning the fossils of this era are identical to the life found on earth today

centigrade alternate name for the Celsius scale

centrifugal force an apparent outward force on an object following a circular path that is a consequence of the third law of motion

centripetal force the force required to pull an object out of its natural straight-line path and into a circular path; centripetal means "center seeking"

cepheid variables stars that have a regular variation in brightness over a period of time

chain reaction a self-sustaining reaction where some of the products are able to produce more reactions of the same kind; in a nuclear chain reaction neutrons are the products that produce more nuclear reactions in a self-sustaining series

chemical bond an attractive force that holds atoms together in a compound

chemical change a change in which the identity of matter is altered and new substances are formed

chemical energy a form of energy involved in chemical reactions associated with changes in internal potential energy; a kind of potential energy that is stored and later released during a chemical reaction

chemical equation concise way of describing what happens in a chemical reaction

chemical equilibrium occurs when two opposing reactions happen at the same time and at the same rate

chemical reaction a change in matter where different chemical substances are created by forming or breaking chemical bonds

chemical sediments ions from rock materials that have removed from solution, for example, carbonate ions removed by crystallization or organisms to form calcium carbonate chemical sediments

chemical weathering the breakdown of minerals in rocks by chemical reactions with water, gases of the atmosphere, or solutions

chemistry the science concerned with the study of the composition, structure, and properties of substances and the transformations they undergo

Chinook a warm wind that has been warmed by compression; also called Santa Ana

chondrites subdivision of stony meteorites containing small spherical lumps of silicate minerals or glass

chondrules small spherical lumps of silicate minerals or glass found in some meteorites

cinder cone volcano a volcanic cone that formed from cinders, sharp-edged rock fragments that cooled from frothy blobs of lava as they were thrown into the air

cirque a bowl-like depression in the side of a mountain, usually at the upper end of a mountain valley, formed by glacial erosion

clastic sediments weathered rock fragments that are in various states of being broken down from solid bedrock; boulders, gravel, and silt

climate the general pattern of weather that occurs in a region over a number of years

coalescence process (meteorology) the process by which large raindrops form from the merging and uniting of millions of tiny water droplets

cold front the front that is formed as a cold air mass moves into warmer air

combination chemical reaction a synthesis reaction in which two or more substances combine to form a single compound

comets celestial objects originating from the outer edges of the solar system that move about the sun in highly elliptical orbits; solar heating and pressure from the solar wind form a tail on the comet that points away from the sun

compaction the process of pressure from a depth of overlying sediments squeezing the deeper sediments together and squeezing water out

composite volcano a volcanic cone that formed from a build-up of alternating layers of cinders, ash, and lava flows

compound a pure chemical substance that can be decomposed by a chemical change into simpler substances with a fixed mass ratio

compressive stress a force that tends to compress the surface as the earth's plates move into each other

concentration an arbitrary description of the relative amounts of solute and solvent in a solution; a larger amount of solute makes a concentrated solution and a small amount of solute makes a dilute concentration

condensation (sound) a compression of gas molecules; a pulse of increased density and pressure that moves through the air at the speed of sound

condensation (water vapor) where more vapor or gas molecules are returning to the liquid state than are evaporating

condensation nuclei tiny particles such as tiny dust, smoke, soot, and salt crystals that are suspended in the air on which water condenses

condensation point the temperature at which a gas or vapor changes back to a liquid

conduction the transfer of heat from a region of higher temperature to a region of lower temperature by increased kinetic energy moving from molecule to molecule

constellations patterns of 88 groups of stars imagined to resemble various mythological characters, inanimate objects, and animals

constructive interference the condition in which two waves arriving at the same place, at the same time and in phase, add amplitudes to create a new wave

continental air mass dry air masses that form over large land areas

continental climate a climate influenced by air masses from large land areas; hot summers and cold winters

continental drift a concept that continents shift positions on the earth's surface, moving across the surface rather than being fixed, stationary land masses

continental glaciers glaciers that cover a large area of a continent, e.g., Greenland and the Antarctic

continental shelf a feature of the ocean floor; the flooded margins of the continents that form a zone of relatively shallow water adjacent to the continents

continental slope a feature of the ocean floor; a steep slope forming the transition between the continental shelf and the deep ocean basin

control rods rods inserted between fuel rods in a nuclear reactor to absorb neutrons and thus control the rate of the nuclear chain reaction

convection transfer of heat from a region of higher temperature to a region of lower temperature by the displacement of high energy molecules—for example, the displacement of warmer, less dense air (higher kinetic energy) by cooler, more dense air (lower kinetic energy)

convection cell complete convective circulation pattern; also, slowly turning regions in the plastic asthenosphere that might drive the motion of plate tectonics

convection zone (of a star) part of the interior of a star according to a model; the region directly above the radiation zone where gases are heated

by the radiation zone below, move upward by convection to the surface, where they emit energy in the form of visible light, ultraviolet radiation, and infrared radiation

conventional current opposite to electron current—that is, considers an electric current to consist of a drift of positive charges that flow from the positive terminal to the negative terminal of a battery

convergent boundaries boundaries that occur between two plates moving toward each other

coordinate covalent bond a "hole and plug" kind of covalent bond, formed when the shared electron pair is contributed by one atom

Copernican system heliocentric, or sun-centered solar system, model developed by Nicholas Copernicus in 1543

core (of earth) the center part of the earth, which consists of a solid inner part and liquid outer part, makes up about 15 percent of the earth's total volume and about one third of its mass.

core (of a star) dense, very hot region of a star where nuclear fusion reactions release gamma and X-ray radiation

Coriolis effect the apparent deflection due to the rotation of the earth; it is to the right in the Northern Hemisphere

correlation the determination of the equivalence in geologic age by comparing the rocks in two separate locations

coulomb unit used to measure quantity of electric charge; equivalent to the charge resulting from the transfer of 6.24 billion particles such as the electron

Coulomb's law relationship between charge, distance, and magnitude of the electrical force between two bodies

covalent bond a chemical bond formed by the sharing of a pair of electrons

covalent compound chemical compound held together by a covalent bond or bonds

creep the slow downhill movement of soil down a steep slope

crest the high mound of water that is part of a wave; also refers to the condensation, or high pressure part, of a sound wave

critical angle limit to the angle of incidence when all light rays are reflected internally

critical mass mass of fissionable material needed to sustain a chain reaction

crude oil petroleum pumped from the ground that has not yet been refined into usable products

crust the outermost part of the earth's interior structure; the thin, solid layer of rock that rests on top of the Mohorovicic discontinuity

curie unit of nuclear activity defined as 3.70×10^{10} nuclear disintegrations per second

cycle a complete vibration

cyclone a low-pressure center where the winds move into the low-pressure center and are forced upward; a low-pressure center with clouds, precipitation, and stormy conditions

data measurement information used to describe something

data points points that may be plotted on a graph to represent simultaneous measurements of two related variables

daylight saving time setting clocks ahead one hour during the summer to more effectively utilize the longer days of summer, then setting the clock back in the fall

decibel scale a nonlinear scale of loudness based on the ratio of the intensity level of a sound to the intensity at the threshold of hearing

decomposition chemical reaction a chemical reaction in which a compound is broken down into the elements that make up the compound, into simpler compounds, or into elements and simpler compounds

deep focus earthquakes earthquakes that occur in the lower part of the upper mantle, between 350 to 700 km below the surface of the earth

deflation the widespread removal of base materials from the surface by the wind

degassing process where gases and water vapor were released from rocks heated to melting during the early stages of the formation of a planet

delta a somewhat triangular deposit at the mouth of a river formed where a stream flowing into a body of water slowed and lost its sediment-carrying ability

density the compactness of matter described by a ratio of mass (or weight) per unit volume

density current an ocean current that flows because of density differences in seawater

destructive interference the condition in which two waves arriving at the same point at the same time out of phase add amplitudes to create zero total disturbance

dew condensation of water vapor into droplets of liquid on surfaces

dew point temperature the temperature at which condensation begins

diastrophism all inclusive term that means any and all possible movements of the earth's plates, including drift, isostatic adjustment, and any other process that deforms or changes the earth's surface by movement

diffraction the bending of light around the edge of an opaque object

diffuse reflection light rays reflected in many random directions, as opposed to the parallel rays reflected from a perfectly smooth surface such as a mirror

dike a tabular shaped intrusive rock that formed when magma moved into joints or faults that cut across other rock bodies

direct current an electrical current that always moves in one direction

direct proportion when two variables increase or decrease together in the same ratio (at the same rate)

disaccharides two monosaccharides joined together with the loss of a water molecule; examples of disaccharides are sucrose (table sugar), lactose, and maltose

dispersion the effect of spreading colors of light into a spectrum with a material that has an index of refraction that varies with wavelength

divergent boundaries boundaries that occur between two plates moving away from each other

divide line separating two adjacent watersheds

dome a large upwardly bulging, symmetrical fold that resembles a dome

Doppler effect an apparent shift in the frequency of sound or light due to relative motion between the source of the sound or light and the observer

double bond covalent bond formed when two pairs of electrons are shared by two atoms

dry adiabatic lapse rate the rate of adiabatic cooling or warming of air in the absence of condensation or evaporation in a parcel of air that is descending or ascending

dune a hill, low mound, or ridge of wind-blown sand or other sediments

earthflow a mass movement of a variety of materials such as soil, rocks, and water with a thick, fluid-like flow

earthquake a quaking, shaking, vibrating, or upheaval of the earth's surface

earthquake epicenter point on the earth's surface directly above an earthquake focus

earthquake focus place where seismic waves originate beneath the surface of the earth

echo a reflected sound that can be distinguished from the original sound, which usually arrives 0.1 sec or more after the original sound

eclipse when the shadow of a celestial body falls on the surface of another celestial body

El Niño changes in the Pacific Ocean involving complex interacting conditions of atmospheric pressure systems, ocean currents, water temperatures, and wind patterns that seem to be linked to worldwide changes in the weather

elastic rebound the sudden snap of stressed rock into new positions; the recovery from elastic strain that results in an earthquake

elastic strain an adjustment to stress in which materials recover their original shape after a stress is released

electric circuit consists of a voltage source that maintains an electrical potential, a continuous conducting path for a current to follow, and a device where work is done by the electrical potential; a switch in the circuit is used to complete or interrupt the conducting path

electric current the flow of electric charge

electric field force field produced by an electrical charge

electric field lines a map of an electric field representing the direction of the force that a test charge would experience; the direction of an electric field shown by lines of force

electric generator a mechanical device that uses wire loops rotating in a magnetic field to produce electromagnetic induction in order to generate electricity

electric potential energy potential energy due to the position of a charge near other charges

electrical conductors materials that have electrons that are free to move throughout the material; for example, metals

electrical energy a form of energy from electromagnetic interactions; one of five forms of energy—mechanical, chemical, radiant, electrical, and nuclear

electrical force a fundamental force that results from the interaction of electrical charge and is billions and billions of times stronger than the gravitational force; sometimes called the

"electromagnetic force" because of the strong association between electricity and magnetism

electrical insulators electrical nonconductors, or materials that obstruct the flow of electric current

electrical nonconductors materials that have electrons that are not moved easily within the material—for example, rubber; electrical nonconductors are also called electrical insulators

electrical resistance the property of opposing or reducing electric current

electrolyte water solution of ionic substances that conducts an electric current

electromagnet a magnet formed by a solenoid that can be turned on and off by turning the current on and off

electromagnetic force one of four fundamental forces; the force of attraction or repulsion between two charged particles

electromagnetic induction process in which current is induced by moving a loop of wire in a magnetic field or by changing the magnetic field

electron subatomic particle that has the smallest negative charge possible and usually found in an orbital of an atom, but gained or lost when atoms become ions

electron configuration the arrangement of electrons in orbitals and suborbitals about the nucleus of an atom

electron current opposite to conventional current; that is, considers electric current to consist of a drift of negative charges that flows from the negative terminal to the positive terminal of a battery

electron dot notation notation made by writing the chemical symbol of an element with dots around the symbol to indicate the number of outer shell electrons

electron pair a pair of electrons with different spin quantum numbers that may occupy an orbital

electron volt the energy gained by an electron moving across a potential difference of one volt; equivalent to 1.60×10^{-19} J

electronegativity the comparative ability of atoms of an element to attract bonding electrons

electrostatic charge an accumulated electric charge on an object from a surplus or deficiency of electrons; also called "static electricity"

element a pure chemical substance that cannot be broken down into anything simpler by chemical or physical means; there are over 100 known elements, the fundamental materials of which all matter is made

empirical formula identifies the elements present in a compound and describes the simplest whole number ratio of atoms of these elements with subscripts

energy the ability to do work

English system a system of measurement that originally used sizes of parts of the human body as referents

epicycle small secondary circular orbit in the geocentric model that was invented to explain the occasional retrograde motion of the planets

epochs subdivisions of geologic periods

equation a statement that describes a relationship in which quantities on one side of the equal sign are identical to quantities on the other side

equation of time the cumulative variation between the apparent local solar time and the mean solar time

equinoxes Latin meaning "equal nights"; time when daylight and night are of equal length, which occurs during the spring equinox and the autumnal equinox

eras the major blocks of time in the earth's geologic history; the Cenozoic, Mesozoic, Paleozoic, and Precambrian

erosion the process of physically removing weathered materials, for example, rock fragments are physically picked up by an erosion agent such as a stream or a glacier

esters esters class of organic compounds with the general structure of RCOOR', where R is one of the hydrocarbon groups—for example, methyl or ethyl; esters make up fats, oils, and waxes and some give fruit and flowers their taste and odor

ether class of organic compounds with the general formula ROR', where R is one of the hydrocarbon groups—for example, methyl or ethyl; mostly used as industrial and laboratory solvents

evaporation process of more molecules leaving a liquid for the gaseous state than returning from the gas to the liquid; can occur at any given temperature from the surface of a liquid

excited states as applied to an atom, describes the energy state of an atom that has electrons in a state above the minimum energy state for that atom; as applied to a nucleus, describes the energy state of a nucleus that has particles in a state above the minimum energy state for that nuclear configuration

exfoliation the fracturing and breaking away of curved, sheetlike plates from bare rock surfaces via physical or chemical weathering, resulting in dome-shaped hills and rounded boulders

exosphere the outermost layer of the atmosphere where gas molecules merge with the diffuse vacuum of space

extrusive igneous rocks fine-grained igneous rocks formed as lava cools rapidly on the surface

Fahrenheit scale referent scale that defines numerical values for measuring hotness or coldness, defined as degrees of temperature; based on the reference points of the freezing point of water and the boiling point of water at sea level pressure, with 180 degrees between the two points

family vertical columns of the periodic table consisting of elements that have similar properties

fats organic compounds of esters formed from glycerol and three long chain carboxylic acids that are also called triglycerides; called fats in animals and oils in plants

fault a break in the continuity of a rock formation along which relative movement has occurred between the rocks on either side

fault plane the surface along which relative movement has occurred between the rocks on either side; the surface of the break in continuity of a rock formation

ferromagnesian silicates silicates that contain iron and magnesium; examples include the dark colored minerals olivine, augite, hornblende, and biotite

first law of motion every object remains at rest or in a state of uniform straight-line motion unless acted on by an unbalanced force

first quarter the moon phase between the new phase and the full phase when the moon is perpendicular to a line drawn through the earth and the sun; one half of the lighted moon can be seen from earth so this phase is called the first quarter

floodplain the wide, level floor of a valley built by a stream; the river valley where a stream floods when it spills out of its channel

fluids matter that has the ability to flow or be poured; the individual molecules of a fluid are able to move, rolling over or by one another

folds bends in layered bed rock as a result of stress or stresses that occurred when the rock layers were in a ductile condition, probably under considerable confining pressure from deep burial

foliation the alignment of flat crystal flakes of a rock into parallel sheets

force a push or pull capable of changing the state of motion of an object; since a force has magnitude (strength) as well as direction, it is a vector quantity

force field a model describing action at a distance by giving the magnitude and direction of force on a unit particle; considers a charge or a mass to alter the space surrounding it and a second charge or mass to interact with the altered space with a force

formula describes what elements are in a compound and in what proportions

formula weight the sum of the atomic weights of all the atoms in a chemical formula

fossil any evidence of former prehistoric life

fossil fuels organic fuels that contain the stored radiant energy of the sun converted to chemical energy by plants or animals that lived millions of years ago; coal, petroleum, and natural gas are the common fossil fuels

Foucault pendulum a heavy mass swinging from a long wire that can be used to provide evidence about the rotation of the earth

fracture strain an adjustment to stress in which materials crack or break as a result of the stress

free fall when objects fall toward the earth with no forces acting upward; air resistance is neglected when considering an object to be in free fall

freezing point the temperature at which a phase change of liquid to solid takes place; the same temperature as the melting point for a given substance

frequency the number of cycles of a vibration or of a wave occurring in one second, measured in units of cycles per second (hertz)

fresh water water that is not saline and is fit for human consumption

front the boundary, or thin transition zone, between air masses of different temperatures

frost ice crystals formed by water vapor condensing directly from the vapor phase; frozen water vapor that forms on objects

frost wedging the process of freezing and thawing water in small rock pores and cracks that become larger and larger, eventually forcing pieces of rock to break off

fuel rod long zirconium alloy tubes containing fissionable material for use in a nuclear reactor

full moon the moon phase when the earth is between the sun and the moon and the entire side of the moon facing the earth is illuminated by sunlight

functional group the atom or group of atoms in an organic molecule that is responsible for the chemical properties of a particular class or group of organic chemicals

fundamental charge smallest common charge known; the magnitude of the charge of an electron and a proton, which is 1.60×10^{-19} coulomb

fundamental frequency the lowest frequency (longest wavelength) that can set up standing waves in an air column or on a string

fundamental properties a property that cannot be defined in simpler terms other than to describe how it is measured; the fundamental properties are length, mass, time, and charge

g symbol representing the acceleration of an object in free fall due to the force of gravity; its magnitude is 9.80 m/sec² (32.0 ft/sec²)

galactic clusters gravitationally bound subgroups of as many as 1,000 stars that move together within the Milky Way galaxy

galaxy group of billions and billions of stars that form the basic unit of the universe, for example, the Earth is part of the Solar System that is located in the Milky Way galaxy

gamma ray very short wavelength electromagnetic radiation emitted by decaying nuclei

gases a phase of matter composed of molecules that are relatively far apart moving freely in a constant, random motion and have weak cohesive forces acting between them, resulting in the characteristic indefinite shape and indefinite volume of a gas

gasohol solution of ethanol and gasoline

Geiger counter a device that indirectly measures ionizing radiation (beta and/or gamma) by detecting "avalanches" of electrons that are able to move because of the ions produced by the passage of ionizing radiation

geocentric the idea that the earth is the center of the universe

geologic time scale a "calendar" of geologic history based on the appearance and disappearance of particular fossils in the sedimentary rock record

geomagnetic time scale time scale established from the number and duration of magnetic field reversals during the past 6 million years

geosyncline a large trough-shaped fold, or syncline, in the earth; a syncline of great extent

giant planets the large outer planets of Jupiter, Saturn, Uranus, and Neptune that all have similar densities and compositions

glacier a large mass of ice on land that formed from compacted snow and slowly moves under its own weight

globular clusters symmetrical and tightly packed clusters of as many as a million stars that move together as a subgroup within the Milky Way galaxy

glycerol an alcohol with three hydroxyl groups per molecule for example, glycerin (1,2,3-propanetriol)

glycogen a highly branched polysaccharide synthesized by the human body and stored in the muscles and liver; serves as a direct reserve source of energy

glycol an alcohol with two hydroxyl groups per molecule; for example, ethylene glycol that is used as an antifreeze

gram-atomic weight the mass in grams of one mole of an element that is numerically equal to its atomic weight

gram-formula weight the mass in grams of one mole of a compound that is numerically equal to its formula weight

gram-molecular weight the gram-formula weight of a molecular compound

granite light colored, coarse-grained igneous rock common on continents; igneous rocks formed by blends of quartz and feldspars, with small amounts of micas, hornblende, and other minerals

greenhouse effect the process of increasing the temperature of the lower parts of the atmosphere through redirecting energy back toward the surface; the absorption and reemission of infrared radiation by carbon dioxide, water vapor, and a few other gases in the atmosphere

ground state energy state of an atom with electrons at the lowest energy state possible for that atom

groundwater water from a saturated zone beneath the surface; water from beneath the surface that supplies wells and springs

gyre the great circular systems of moving water in each ocean

hail a frozen form of precipitation, sometimes with alternating layers of clear and opaque, cloudy ice

hair hygrometer a device that measures relative humidity from changes in the length of hair

half-life the time required for one-half of the unstable nuclei in a radioactive substance to decay into a new element

halogen member of family VIIA of the periodic table, having common properties of very reactive nonmetallic elements common in salt compounds

hard water water that contains relatively high concentrations of dissolved salts of calcium and magnesium

heat total internal energy of molecules, which is increased by gaining energy from a temperature difference (conduction, convection, radiation) or by gaining energy from a form conversion (mechanic, chemical, radiant, electrical, nuclear)

heat of formation energy released in a chemical reaction

Heisenberg uncertainty principle you cannot measure both the exact momentum and the exact position of a subatomic particle at the same time—when the more exact of the two is known, the less certain you are of the value of the other

hertz unit of frequency; equivalent to one cycle per second

Hertzsprung-Russell diagram diagram to classify stars with a temperature-luminosity graph

high short for high-pressure center (anticyclone), which is associated with clear, fair weather

high latitudes latitudes close to the poles; those that sometime receive no solar radiation at noon

high pressure center another term for anticyclone

horsepower measurement of power defined as a power rating of 550 ft·Ib/sec

hot spots sites on the earth's surface where plumes of hot rock materials rise from deep within the mantle

humid moist climate classification; receives more than 50 cm (20 in) precipitation per year

humidity the amount of water vapor in the air; see relative humidity

hurricane a tropical cyclone with heavy rains and winds exceeding 120 km/hr

hydration the attraction of water molecules for ions; a reaction that occurs between water and minerals that make up rocks

hydrocarbon an organic compound consisting of only the two elements hydrogen and carbon

hydrocarbon derivatives organic compounds that can be thought of as forming when one or more hydrogen atoms on a hydrocarbon have been replaced by an element or a group of elements other than hydrogen

hydrogen bond a weak to moderate bond between the hydrogen end ($+$) of a polar molecule and the negative end ($-$) of a second polar molecule

hydrologic cycle water vapor cycling into and out of the atmosphere through continuous evaporation of liquid water from the surface and precipitation of water back to the surface

hydronium ion a molecule of water with an attached hydrogen ion, H_3O^+

hypothesis a tentative explanation of a phenomenon that is compatible with the data and provides a framework for understanding and describing that phenomenon

ice-crystal process a precipitation-forming process that brings water droplets of a cloud together through the formation of ice crystals

ice-forming nuclei small, solid particles suspended in air; ice can form on the suspended particles

igneous rocks rocks that formed from a magma, which is a hot, molten mass of melted rock materials

incandescent matter emitting visible light as a result of high temperature for example, a light bulb, a flame from any burning source, and the sun are all incandescent sources because of high temperature

incident ray line representing the direction of motion of incoming light approaching a boundary

inclination of earth axis tilt of the earth's axis measured from the plane of the ecliptic (23.5°); considered to be the same throughout the year

index fossils distinctive fossils of organisms that lived only a brief time; used to compare the age of rocks exposed in two different locations

index of refraction the ratio of the speed of light in a vacuum to the speed of light in a material

inertia a property of matter describing the tendency of an object to resist a change in its state of motion; an object will remain in unchanging motion or at rest in the absence of an unbalanced force

infrasonic sound waves having too low a frequency to be heard by the human ear; sound having a frequency of less than 20 Hz

inorganic chemistry the study of all compounds and elements in which carbon is not the principal element

insulators materials that are poor conductors of heat—for example, heat flows slowly through materials with air pockets because the molecules making up air are far apart; also, materials that are poor conductors of electricity, for example, glass or wood

intensity a measure of the energy carried by a wave

interference phenomenon of light where the relative phase difference between two light waves produces light or dark spots, a result of light's wavelike nature

intermediate focus earthquakes earthquakes that occur in the upper part of the mantle, between 70 to 350 km below the surface of the earth

intermolecular forces forces of interaction between molecules

internal energy sum of all the potential energy and all the kinetic energy of all the molecules of an object

international date line the 180° meridian is arbitrarily called the international date line; used to compensate for cumulative time zone changes by adding or subtracting a day when the line is crossed

intertropical convergence zone a part of the lower troposphere in belt from 10° N to 10° S of the equator where air is heated, expands, and becomes less dense and rises around the belt

intrusive igneous rocks coarse-grained igneous rocks formed as magma cools slowly deep below the surface

inverse proportion the relationship in which the value of one variable increases while the value of the second variable decreases at the same rate (in the same ratio)

inversion a condition of the troposphere when temperature increases with height rather than decreasing with height; a cap of cold air over warmer air that results in increased air pollution

ion an atom or a particle that has a net charge because of the gain or loss of electrons; polyatomic ions are groups of banded atoms that have a net charge

ion exchange reaction a reaction that takes place when the ions of one compound interact with the ions of another, forming a solid that comes out of solution, a gas, or water

ionic bond chemical bond of electrostatic attraction between negative and positive ions

ionic compounds chemical compounds that are held together by ionic bonds—that is, bonds of electrostatic attraction between negative and positive ions

ionization process of forming ions from molecules

ionization counter a device that measures ionizing radiation (alpha, beta, gamma, etc.) by indirectly counting the ions produced by the radiation

ionized an atom or a particle that has a net charge because it has gained or lost electrons

ionosphere refers to that part of the atmosphere—parts of the thermosphere and upper mesosphere—where free electrons and ions reflect radio waves around the earth and where the northern lights occur

iron meteorites meteorite classification group whose members are composed mainly of iron

island arcs curving chains of volcanic islands that occur over belts of deep-seated earthquakes; for example, the Japanese and Indonesian islands

isomers chemical compounds with the same molecular formula but different molecular structure; compounds that are made from the same numbers of the same elements but have different molecular arrangements

isostasy a balance or equilibrium between adjacent blocks of crust "floating" on the asthenosphere; the concept of less dense rock floating on more dense rock

isostatic adjustment a concept of crustal rocks thought of as tending to rise or sink gradually until they are balanced by the weight of displaced mantle rocks

isotope atoms of an element with identical chemical properties but with different masses; isotopes are atoms of the same element with different numbers of neutrons

jet stream a powerful, winding belt of wind near the top of the troposphere that tends to extend all the way around the earth, moving generally from the west in both hemispheres at speeds of 160 km/hr or more

joint a break in the continuity of a rock formation without a relative movement of the rock on either side of the break

joule metric unit used to measure work and energy; can also be used to measure heat; equivalent to newton-meter

Kelvin scale a temperature scale that does not have arbitrarily assigned referent points and zero means nothing; the zero point on the Kelvin scale (also called absolute scale) is the lowest limit of temperature, where all random kinetic energy of molecules ceases

Kepler's first law relationship in planetary motion that each planet moves in an elliptical orbit, with the sun located at one focus

Kepler's laws of planetary motion the three laws describing the motion of the planets

Kepler's second law relationship in planetary motion that an imaginary line between the sun and a planet moves over equal areas of the ellipse during equal time intervals

Kepler's third law relationship in planetary motion that the square of the period of an orbit is directly proportional to the cube of the radius of the major axis of the orbit

ketone an organic compound with the general formula RCOR', where R is one of the hydrocarbon groups; for example, methyl or ethyl

kilocalorie the amount of energy required to increase the temperature of one kilogram of water one degree Celsius: equivalent to 1,000 calories

kilogram the fundamental unit of mass in the metric system of measurement

kinetic energy the energy of motion; can be measured from the work done to put an object in motion, from the mass and velocity of the object while in motion, or from the amount of work the object can do because of its motion

kinetic molecular theory the collection of assumptions that all matter is made up of tiny atoms and molecules that interact physically, that explain the various states of matter, and that have an average kinetic energy that defines the temperature of a substance

L-wave seismic waves that move on the solid surface of the earth much as water waves move across the surface of a body of water

laccolith an intrusive rock feature that formed when magma flowed into the plane of contact between sedimentary rock layers, then raised the overlying rock into a blister-like uplift

lake a large inland body of standing water

landforms the features of the surface of the earth such as mountains, valleys, and plains

landslide general term for rapid movement of any type or mass of materials

last quarter the moon phase between the full phase and the new phase when the moon is perpendicular to a line drawn through the earth and the sun; one half of the lighted moon can be seen from earth so this phase is called the last quarter

latent heat refers to the heat "hidden" in phase changes

latent heat of fusion the heat absorbed when one gram of a substance changes from the solid to the liquid phase, or the heat released by one gram of a substance when changing from the liquid phase to the solid phase

latent heat of vaporization the heat absorbed when one gram of a substance changes from the liquid phase to the gaseous phase, or the heat released when one gram of gas changes from the gaseous phase to the liquid phase

laterites highly leached soils of tropical climates; usually red with high iron and aluminum oxide content

latitude the angular distance from the equator to a point on a parallel that tell you how far north or south of the equator the point is located

lava magma, or molten rock, that is forced to the surface from a volcano or a crack in the earth's surface

law of conservation of energy energy is never created or destroyed; it can only be converted from one form to another as the total energy remains constant

law of conservation of mass same as law of conservation of matter; mass, including single atoms, is neither created nor destroyed in a chemical reaction

law of conservation of matter matter is neither created nor destroyed in a chemical reaction

law of conservation of momentum the total momentum of a group of interacting objects remains constant in the absence of external forces

light ray model using lines to show the direction of motion of light to describe the travels of light

light-year the distance that light travels through empty space in one year, approximately 9.5×10^{12} km

linear scale a scale, generally on a graph, where equal intervals represent equal changes in the value of a variable

line spectrum narrow lines of color in an otherwise dark spectrum; these lines can be used as "fingerprints" to identify gases

lines of force lines drawn to make an electric field strength map, with each line originating on a positive charge and ending on a negative charge; each line represents a path on which a charge would experience a constant force and lines closer together mean a stronger electric field

liquids a phase of matter composed of molecules that have interactions stronger than those found in a gas but not strong enough to keep the molecules near the equilibrium positions of a solid, resulting in the characteristic definite volume but indefinite shape of a liquid

liter a metric system unit of volume usually used for liquids

lithosphere solid layer of the earth's structure that is above the asthenosphere and includes the entire crust, the Moho, and the upper part of the mantle

loess very fine dust or silt that has been deposited by the wind over a large area

longitude angular distance of a point east or west from the prime meridian on a parallel

longitudinal wave a mechanical disturbance that causes particles to move closer together and farther apart in the same direction that the wave is traveling

longshore current a current that moves parallel to the shore, pushed along by waves that move accumulated water from breakers

loudness a subjective interpretation of a sound that is related to the energy of the vibrating source, related to the condition of the transmitting medium, and related to the distance involved

low latitudes latitudes close to the equator; those that sometimes receive vertical solar radiation at noon

luminosity the total amount of energy radiated into space each second from the surface of a star

luminous an object or objects that produce visible light; for example, the sun, stars, light bulbs, and burning materials are all luminous

lunar eclipse occurs when the moon is full and the sun, moon, and earth are lined up so the shadow of the earth falls on the moon

lunar highlands light colored mountainous regions of the moon

macromolecule very large molecule, with a molecular weight of thousands or millions of atomic mass units, that is made up of a combination of many smaller, similar molecules

magma a mass of molten rock material either below or on the earth's crust from which igneous rock is formed by cooling and hardening

magnetic domain tiny physical regions in permanent magnets, approximately 0.01 to 1 mm, that have magnetically aligned atoms, giving the domain an overall polarity

magnetic field model used to describe how magnetic forces on moving charges act at a distance

magnetic poles the ends, or sides, of a magnet about which the force of magnetic attraction seems to be concentrated

magnetic quantum number from quantum mechanics model of the atom, one of four descriptions of the energy state of an electron wave; this quantum number describes the energy of an electron orbital as the orbital is oriented in space by an external magnetic field, a kind of energy sub-sublevel

magnetic reversal the flipping of polarity of the earth's magnetic field as the north magnetic pole and the south magnetic pole exchange positions

magnitude the size of a measurement of a vector; scalar quantities that consist of a number and unit only, no direction, for example

main sequence stars normal, mature stars that use their nuclear fuel at a steady rate; stars on the Hertzsprung-Russell diagram in a narrow band that runs from the top left to the lower right

manipulated variable in an experiment, a quantity that can be controlled or manipulated; also known as the independent variable

mantle middle part of the earth's interior; a 2,870 km (about 1,780 mile) thick shell between the core and the crust

maria smooth dark areas on the moon

marine climate a climate influenced by air masses from over an ocean, with mild winters and cool summers compared to areas farther inland

maritime air mass a moist air mass that forms over the ocean

mass a measure of inertia, which means a resistance to a change of motion

mass defect the difference between the sum of the masses of the the individual nucleons forming a nucleus and the actual mass of that nucleus

mass movement erosion caused by the direct action of gravity

mass number the sum of the number of protons and neutrons in a nucleus defines the mass number of an atom; used to identify isotopes; for example, uranium 238

matter anything that occupies space and has mass

matter waves any moving object has wave properties, but at ordinary velocities these properties are observed only for objects with a tiny mass; term for the wavelike properties of subatomic particles

mean solar day is 24 hours long and is averaged from the mean solar time

mean solar time a uniform time averaged from the apparent solar time

meanders winding, circuitous turns or bends of a stream

measurement the process of comparing a property of an object to a well-defined and agreed-upon referent

mechanical energy the form of energy associated with machines, objects in motion, and objects having potential energy that results from gravity

mechanical weathering the physical breaking up of rocks without any changes in their chemical composition

melting point the temperature at which a phase change of solid to liquid takes place; the same temperature as the freezing point for a given substance

Mercalli scale expresses the relative intensity of an earthquake in terms of effects on people and building using Roman numerals that range from I to XII

meridians north-south running arcs that intersect at both poles and are perpendicular to the parallels

mesosphere the term means "middle layer"—the solid, dense layer of the earth's structure below the asthenosphere but above the core; also the layer of the atmosphere below the thermosphere and above the stratosphere

Mesozoic one of four geologic eras; the time of middle life, meaning some of the fossils for this time period are similar to the life found on earth today but many are different from anything living today

metal matter having the physical properties of conductivity, malleability, ductility, and luster

metamorphic rocks previously existing rocks that have been changed into a distinctly different rock by heat, pressure, or hot solutions

meteor the streak of light and smoke that appears in the sky when a meteoroid is made incandescent by friction with the earth's atmosphere

meteor shower event when many meteorites fall in a short period of time

meteorite the solid iron or stony material of a meteoroid that survives passage through the earth's atmosphere and reaches the surface

meteoroids remnants of comets and asteroids in space

meteorology the science of understanding and predicting weather

meter the fundamental metric unit of length

metric system a system of referent units based on invariable referents of nature that have been defined as standards

microclimate a local, small-scale pattern of climate; for example, the north side of a house has a different microclimate than the south side

middle latitudes latitudes equally far from the poles and equator; between the high and low latitudes

millibar a measure of atmospheric pressure equivalent to 1.000 dynes per cm²

mineral a naturally occurring, inorganic solid element or chemical compound with a crystalline structure

miscible fluids fluids that can mix in any proportion

mixture matter made of unlike parts that have a variable composition and can be separated into their component parts by physical means

model a mental or physical representation of something that cannot be observed directly that is usually used as an aid to understanding

moderator a substance in a nuclear reactor that slows fast neutrons so the neutrons can participate in nuclear reactions

Mohorovicic discontinuity boundary between the crust and mantle that is marked by a sharp increase in the velocity of seismic waves as they pass from the crust to the mantle

mold the preservation of the shape of an organism by the dissolution of the remains of a buried organism, leaving an empty space where the remains were

mole an amount of a substance that contains Avogadro's number of atoms, ions, molecules, or any other chemical unit; a mole is thus 6.02×10^{23} atoms, ions, or other chemical units

molecular formula a chemical formula that identifies the actual numbers of atoms in a molecule

molecular weight the formula weight of a molecular substance

molecule from the chemical point of view: a particle composed of two or more atoms held together by an attractive force called a chemical bond; from the kinetic theory point of view: smallest particle of a compound or gaseous element that can exist and still retain the characteristic properties of a substance

momentum the product of the mass of an object times its velocity

monadnocks hills of resistant rock that are found on peneplains

monosaccharides simple sugars that are mostly 6-carbon molecules such as glucose and fructose

moraines deposits of bulldozed rocks and other mounded maerials left behind by a melted glacier

mountain a natural elevation of the earth's crust that rises above the surrounding surface

mudflow a mass movement of a slurry of debris and water with the consistency of a thick milkshake

natural frequency the frequency of vibration of an elastic object that depends on the size, composition, and shape of the object

neap tide period of less pronounced high and low tides, occurs when the sun and moon are at right angles to one another

nebulae a diffuse mass of interstellar clouds of hydrogen gas or dust thay may develop into a star

negative electric charge one of the two types of electric charge; repels other negative charges and attracts positive charges

negative ion atom or particle that has a surplus, or imbalance, of electrons and, thus, a negative charge

net force the resulting force after all vector forces have been added; if a net force is zero, all the forces have canceled each other and there is not an unbalanced force

neutralized acid or base properties have been lost through a chemical reaction

neutron neutral subatomic particle usually found in the nucleus of an atom

neutron star very small super-dense remains of a supernova with a center core of pure neutrons

new crust zone zone of a divergent boundary where new crust is formed by magma upwelling at the boundary

new moon the moon phase when moon is between the earth and the sun and the entire side of the moon facing the earth is dark

newton a unit of force defined as $kg \cdot m/sec^2$; that is, a 1 newton force is needed to accelerate a 1 kg mass 1 m/sec^2

noble gas members of family Viii of the periodic table, having common properties of colorless, odorless, chemically inert gases; also known as rare gases or inert gases

node regions on a standing wave that do not oscillate

noise sounds made up of groups of waves of random frequency and intensity

nonelectrolytes water solutions that do not conduct an electric current; covalent compounds that form molecular solutions and cannot conduct an electric current

nonferromagnesian silicates silicates that do not contain iron or magnesium ions; examples include the minerals of muscovite (white mica), the feldspars, and quartz

nonmetal an element that is brittle (when a solid), does not have a metallic luster, is a poor conductor of heat and electricity, and is not malleable or ductile

nonsilicates minerals that do not have the silicon-oxygen tetrahedra in their crystal structure

noon the event of time when the sun moves across the celestial meridian

normal a line perpendicular to the surface of a boundary

normal fault a fault where the hanging wall has moved downward with respect to the foot wall

north celestial pole a point directly above the north pole of the earth; the point above the north pole on the celestial sphere

north pole the north pole of a magnet or lodestone is "north seeking," meaning that the pole of a magnet points northward when the magnet is free to turn

nova irregularly flaring star

nuclear energy the form of energy from reactions involving the nucleus, the innermost part of an atom

nuclear fission nuclear reaction of splitting a massive nucleus into more stable, less massive nuclei with an accompanying release of energy

nuclear force one of four fundamental forces, a strong force of attraction that operates over very short distances between subatomic particles; this force overcomes the electric repulsion of protons in a nucleus and binds the nucleus together

nuclear fusion nuclear reaction of low mass nuclei fusing together to form more stable and more massive nuclei with an accompanying release of energy

nuclear reactor steel vessel in which a controlled chain reaction of fissionable materials releases energy

nucleons name used to refer to both the protons and neutrons in the nucleus of an atom

nucleus tiny, relatively massive and positively charged center of an atom containing protons and neutrons; the small, dense center of an atom

numerical constant a constant without units; a number

oblate spheroid the shape of the earth—a somewhat squashed spherical shape

observed lapse rate the rate of change in temperature compared to change in altitude

occluded front a front that has been lifted completely off the ground into the atmosphere, forming a cyclonic storm

ocean the single, continuous body of salt water on the surface of the earth

ocean basin the deep bottom of the ocean floor, which starts beyond the continental slope

ocean currents streams of water within the ocean that stay in about the same path as they move over large distances; steady and continuous onward movement of a channel of water in the ocean

ocean wave a moving disturbance that travels across the surface of the ocean

oceanic ridges long, high, continuous, suboceanic mountain chains; for example, the Mid-Atlantic Ridge in the center of the Atlantic Ocean Basin.

oceanic trenches long, narrow, deep troughs with steep sides that run parallel to the edge of continents

octet rule a generalization that helps keep track of the valence electrons in most representative elements; atoms of the representative elements (A families) attempt to acquire an outer orbital with eight electrons through chemical reactions

ohm unit of resistance; equivalent to volts/amps

Ohm's law the electric potential difference is directly proportional to the product of the current times the resistance

oil field petroleum accumulated and trapped in extensive porous rock structure or structures

oils organic compounds of esters formed from glycerol and three long-chain carboxylic acids that are also called triglycerides; called fats in animals and oils in plants

Oort cloud the cloud of icy, dusty aggregates in the outer reaches of the solar system from which comets are thought to originate

opaque materials that do not allow the transmission of any light

orbital the region of space around the nucleus of an atom where an electron is likely to be found

ore mineral mineral deposits with an economic value

organic acids acids derived from organisms; organic compounds with a general formula of RCOOH, where R is one of the hydrocarbon groups; for example, methyl or ethyl

organic chemistry the study of compounds in which carbon is the principal element

orientation of earth's axis direction that the earth's axis points; considered to be the same throughout the year

origin the only point on a graph where both the x and y variables have a value of zero at the same time

oscillating theory model of the universe based on the observation that symmetry is observed in nature, so the universe will end in another big bang after contracting and collapsing

overtones higher resonant frequencies that occur at the same time as the fundamental frequency, giving a musical instrument its characteristic sound quality

oxbow lake a small body of water, or lake, that formed when two bends of a stream came together and cut off a meander

oxidation the process of a substance losing electrons during a chemical reaction; a reaction between oxygen and the minerals making up rocks

oxidation-reduction reaction a chemical reaction in which electrons are transferred from one atom to another; sometimes called "redox" for short

oxidizing agents substances that take electrons from other substances

ozone shield concentration of ozone in the lower portions of the stratosphere that absorbs potentially damaging ultraviolet radiation, preventing it from reaching the surface of the earth

P-wave a pressure, or compressional wave in which a disturbance vibrates materials back and forth in the same direction as the direction of wave movement

P-wave shadow zone a region on the earth between 103° and 142° of arc from an earthquake where no P-waves are received; believed to be explained by P-waves being refracted by the core

Paleozoic one of four geologic eras; time of ancient life, meaning the fossils from this time period are very different from anything living on the earth today

parallels reference lines on the earth used to identify where in the world you are northward or southward from the equator; east and west running circles that are parallel to the equator on a globe with the distance from the equator called the latitude

parsec astronomical unit of distance where the distance at which the angle made from 1 A.U. baseline is 1 arc second

parts per billion concentration ratio of parts of solute in every one billion parts of solution (ppb); could be expressed as ppb by volume or as ppb by weight

parts per million concentration ratio of parts of solute in every one million parts of solution (ppm); could be expressed as ppm by volume or as ppm by weight

Pauli exclusion principle no two electrons in an atom can have the same four quantum numbers; thus, a maximum of two electrons can occupy a given orbital

pedalfers soils formed in wet, humid climates that are characterized by strong leaching and percolation and tend to be acid from the decay of abundant humus

pedocals soils formed in arid climates where evaporation exceeds precipitation and water moves calcium carbonate upward beneath the land surface where it is precipitated, forming alkali soil

peneplain a nearly flat landform that is the end result of the weathering and erosion of the land surface

penumbra the zone of partial darkness in a shadow

percent by volume the volume of solute in 100 volumes of solution

percent by weight the weight of solute in 100 weight units of solution

perigee when the moon's elliptical orbit brings the moon closest to the earth

perihelion the point at which an orbit comes closest to the sun

period (geologic time) subdivisions of geologic eras

period (periodic table) horizontal rows of elements with increasing atomic numbers; runs from left to right on the element table

period (wave) the time required for one complete cycle of a wave

periodic law similar physical and chemical properties recur periodically when the elements are listed in order of increasing atomic number

permeability the ability to transmit fluids through openings, small passageways, or gaps

permineralization the process that forms a fossil by alteration of an organism's buried remains by circulating ground water depositing calcium carbonate, silica, or pyrite

petroleum oil that comes from oil-bearing rock, a mixture of hydrocarbons that is believed to have formed from ancient accumulations of buried organic materials such as remains of algae

phase change the action of a substance changing from one state of matter to another; a phase change always absorbs or releases internal potential energy that is not associated with a temperature change

phases of matter the different physical forms that matter can take as a result of different molecular arrangements, resulting in characteristics of the common phases of a solid, liquid, or gas

photoelectric effect the movement of electrons in some materials as a result of energy acquired from absorbed light

photons a quanta of energy in light wave; the particle associated with light

pH scale scale that measures the acidity of a solution with numbers below seven representing acids, seven representing neutral, and numbers above seven representing bases

physical change a change of the state of a substance but not the identity of the substance

pitch the frequency of a sound wave

Planck's constant proportionality constant in the relationship between the energy of vibrating molecules and their frequency of vibration; a value of 6.63×10^{-34} J sec

plasma a phase of matter; a very hot gas consisting of electrons and atoms that have been stripped of their electrons because of high kinetic energies

plastic strain an adjustment to stress in which materials become molded or bent out of shape under stress and do not return to their original shape after the stress is released

plate tectonics the theory that the earth's crust is made of rigid plates that float on the asthenosphere

plunging folds synclines and anticlines that are not parallel to the surface of the earth

polar air mass cold air mass that forms in cold regions

polar climate zone climate zone of the high latitudes; average monthly temperatures stay below 10° C (50° F), even during the warmest month of the year

polar covalent bond a covalent bond in which there is an unequal sharing of bonding electrons

polarized light whose constituent transverse waves are all vibrating in the same plane; also known as planepolarized light

Polaroid a film that transmits only polarized light

polyatomic ion ion made up of many atoms

polymer long chain of repeating monomer molecules

polymers huge, chainlike molecules made of hundreds or thousands of smaller repeating molecular units called monomers

polysaccharides polymers consisting of monosaccharide units joined together in straight or branched chains; starches, glycogen, or cellulose

pond a small body of standing water, smaller than a lake

porosity the ratio of pore space to the total volume of a rock or soil sample, expressed as a percentage; freely admitting the passage of fluids through pores or small spaces between parts of the rock or soil

positive electric charge one of the two types of electric charge; repels other positive charges and attracts negative charges

positive ion atom or particle that has a net positive charge due to an electron or electrons being torn away

potential energy energy due to position; energy associated with changes in position (e.g., gravitational potential energy) or changes in shape (e.g., compressed or stretched spring)

power the rate at which energy is transferred or the rate at which work is performed; defined as work per unit of time

Precambrian one of four geologic eras; the time before the time of ancient life, meaning the rocks for this time period contain very few fossils

precession the slow wobble of the axis of the earth similar to the wobble of a spinning top

precipitation water that falls to the surface of the earth in the solid or liquid form

pressure defined as force per unit area; for example, pounds per square inch (Ib/in^2)

primary coil part of a transformer; a coil of wire that is connected to a source of alternating current

primary loop part of the energy-converting system of a nuclear power plant; the closed pipe system that carries heated water from the nuclear reactor to a steam generator

prime meridian the referent meridian (0°) that passes through the Greenwich Observatory in England

principle of crosscutting relationships a frame of reference based on the understanding that any geologic feature that cuts across or is intruded into a rock mass must be younger than the rock mass

principle of faunal succession a frame of reference based on the understanding that life forms have changed through time as old life forms disappear from the fossil record and new ones appear, but the same form is never exactly duplicated independently at two different times in history

principle of original horizontality a frame of reference based on the understanding that on a large scale sediments are deposited in flat-lying layers, so any layers of sedimentary rocks that are not horizontal have been subjected to forces that have deformed the earth's surface

principle quantum number from quantum mechanics model of the atom, one of four descriptions of the energy state of an electron wave; this quantum number describes the main energy level of an electron in terms of its most probable distance from the nucleus

principle of superposition a frame of reference based on the understanding that an undisturbed sequence of horizontal rock layers is arranged in chronological order with the oldest layers at the bottom and each consecutive layer will be younger than the one below it

principle of uniformity a frame of reference of slow, uniform changes in the earth's history; the processes changing rocks today are the processes that changed them in the past, or "the present is the key to the past"

proof a measure of ethanol concentration of an alcoholic beverage; proof is double the concentration by volume; for example, 50 percent by volume is 100 proof.

properties qualities or attributes that, taken together, are usually unique to an object; for example, color, texture, and size

proportionality constant a constant applied to a proportionality statement that transforms the statement into an equation

proteins macromolecular polymers made of smaller molecules of amino acids, with molecular weight from about six thousand to fifty million; proteins are amino acid polymers with a role as a biological structure or function and without such a function are known as polypeptides

protogalaxy collection of gas, dust, and young stars in the process of forming a galaxy

proton subatomic particle that has the smallest possible positive charge, usually found in the nucleus of an atom

protoplanet nebular model a model of the formation of the solar system that states that the planets formed from gas and dust left over from the formation of the sun

protostar an accumulation of gases that will become a star

psychrometer a two-thermometer device used to measure the relative humidity

Ptolemaic system geocentric model of the structure of the solar system that uses epicycles to explain retrograde motion

pulsars the source of regular, equally spaced pulsating radio signals believed to be the result of the magnetic field of a rotating neutron star

pure substance materials that are the same throughout and have a fixed definite composition

pure tone sound made by very regular intensities and very regular frequencies from regular repeating vibrations

quad one quadrillion Btu (10^{15} Btu); used to describe very large amounts of energy

quanta nxcd amounts; usually referring to fixed amounts of energy absorbed or emitted by matter ("quanta" is plural and "quantum" is singular)

quantities measured properties; includes the numerical value of the measurement and the unit used in the measurement

quantum mechanics model of the atom based on the wave nature of subatomic particles, the mechanics of electron waves; also called wave mechanics

quantum numbers numbers that describe energy states of an electron; in the Bohr model of the atom, the orbit quantum numbers could be any whole number 1, 2, 3, and so on out from the nucleus; in the quantum mechanics model of the atom, four quantum numbers are used to describe the energy state of an electron wave

rad a measure of radiation received by a material (radiation absorbed dose)

radiant energy the form of energy that can travel through space; for example, visible light and other parts of the electromagnetic spectrum

radiation the transfer of heat from a region of higher temperature to a region of lower temperature by greater emission of radiant energy from the region of higher temperature

radiation zone part of the interior of a star according to a model; the region directly above the core where gamma and X-rays from the core are absorbed and reemitted, with the radiation slowly working its way outward

radioactive decay the natural spontaneous disintegration or decomposition of a nucleus

radioactive decay constant a specific constant for a particular isotope that is the ratio of the rate of nuclear disintegration per unit of time to the total number of radioactive nuclei

radioactive decay series series of decay reactions that begins with one radioactive nucleus that decays to a second nucleus that decays to a third nucleus and so on until a stable nucleus is reached

radioactivity spontaneous emission of particles or energy from an atomic nucleus as it disintegrates

radiometric age age of rocks determined by measuring the radioactive decay of unstable elements within the crystals of certain miners in the rocks

rarefaction a thinning or pulse of decreased density and pressure of gas molecules

ratio a relationship between two numbers, one divided by the other, such as the ratio of distance per time is speed

real image an image generated by a lens or mirror that can be projected onto a screen

red giant stars one of two groups of stars on the Hertzsprung-Russell diagram that have a different set of properties than the main sequence stars; bright, low temperature giant stars that are enormously bright for their temperature

redox reaction short name for oxidation-reduction reaction

reducing agents supplies electrons to the substance being reduced in a chemical reaction

referent referring to or thinking of a property in terms of another, more familiar object

reflected ray a line representing direction of motion of light reflected from a boundary

reflection the change when light, sound, or other waves bounce backwards off a boundary

refraction a change in the direction of travel of light, sound, or other waves crossing a boundary

rejuvenation process of uplifting land that renews the effectiveness of weathering and erosion processes

relative dating dating the age of a rock unit or geological event relative to some other unit or event

relative humidity ratio (times 100%) of how much water vapor is in the air to the maximum amount of water vapor that could be in the air at a given temperature

rem measure of radiation that considers the biological effects of different kinds of ionizing radiation

replacement (chemical reaction) reaction in which an atom or polyatomic ion is replaced in a compound by a different atom or polyatomic ion

replacement (fossil formation) process in which an organism's buried remains are altered by circulating ground waters carrying elements in solution; the removal of original materials by dissolutions and the replacement of new materials an atom or molecule at a time

representative elements name given to the members of the A group families of the periodic table; also called the main-group elements

reservoir a natural or artificial pond or lake used to store water, control floods, or generate electricity; a body of water stored for public use

resonance when the frequency of an external force matches the natural frequency and standing waves are set up

responding variable the variable that responds to changes in the manipulated variable; also known as the dependent variable because its value depends on the value of the manipulated variable

reverberation apparent increase in volume caused by reflections, usually arriving within 0.1 second after the original sound

reverse fault a fault where the hanging wall has moved upward with respect to the foot wall

revolution the motion of a planet as it orbits the sun

Richter Scale expresses the intensity of an earthquake in terms of a scale with each higher number indicating 10 times more ground movement and about 30 times more energy released than the preceding number

ridges long, rugged mountain chains rising thousands of meters above the abyssal plains of the ocean basin

rift a split or fracture in a rock formation, land formation, or in the crust of the earth

rip current strong, brief current that runs against the surf and out to sea

rock a solid aggregation of minerals or mineral materials that have been brought together into a cohesive solid

rock cycle understanding of igneous, sedimentary, or metamorphic rock as a temporary state in an ongoing transformation of rocks into new types; the process of rocks continually changing from one type to another

rock flour rock pulverized by a glacier into powdery, silt-sized sediment

rockfall the rapid tumbling, bouncing, or free fall of rock fragments from a cliff or steep slope

rockslide a sudden, rapid movement of a coherent unit of rock along a clearly defined surface or plane

rotation the spinning of a planet on its axis

runoff water moving across the surface of the earth as opposed to soaking into the ground

S-wave a sideways, or shear wave in which a disturbance vibrates materials from side to side, perpendicular to the direction of wave movement

S-wave shadow zone a region of the earth more than 103° of arc away from the epicenter of an earthquake where S-waves are not recorded; believed to be the result of core of the earth that is a liquid, or at least acts like a liquid

salinity a measure of dissolved salts in seawater, defined as the mass of salts dissolved in 1,000 g of solution

salt any ionic compound except one with hydroxide or oxide ions

San Andreas fault in California, the boundary between the North American Plate and the Pacific Plate that runs north-south for some 1300 km (800 miles) with the Pacific Plate moving northwest and the North American Plate moving southeast

saturated air air in which an equilibrium exists between evaporation and condensation; the relative humidity is 100 percent

saturated molecule an organic molecule that has the maximum number of hydrogen atoms possible

saturated solution the apparent limit to dissolving a given solid in a specified amount of water at a given temperature; a state of equilibrium that exists between dissolving solute and solute coming out of solution

scalars measurements that have magnitude only, defined by the numerical value and a unit such as 30 miles per hour

scientific law a relationship between quantities, usually described by an equation in the physical sciences; is more important and describes a wider range of phenomena than a scientific principle

scientific principle a relationship between quantities concerned with a specific, or narrow range of observations and behavior

scintillation counter a device that indirectly measures ionizing radiation (alpha, beta, gamma, etc.) by measuring the flashes of light produced when the radiation strikes a phosphor

sea a smaller part of the ocean with characteristics that distinguish it from the larger ocean

sea breeze cool, dense air from over water moving over land as part of convective circulation

seafloor spreading the process by which hot, molten rock moves up from the interior of the earth to emerge along mid-oceanic rifts, flowing out in both directions to create new rocks and spread the seafloor apart

seamounts steep, submerged volcanic peaks on the abyssal plain

second the standard unit of time in both the metric and English systems of measurement

second law of motion the acceleration of an object is directly proportional to the net force acting on that object and inversely proportional to the mass of the object

secondary coil part of a transformer, a coil of wire in which the voltage of the original alternating current in the primary coil is stepped up or down by way of electromagnetic induction

secondary loop part of nuclear power plant; the closed pipe system that carries steam from a steam generator to the turbines, then back to the steam generator as feedwater

sedimentary rocks rocks formed from particles or dissolved minerals from previously existing rocks

sediments accumulations of silt, sand, or gravel that settled out of the atmosphere or out of water

seismic waves vibrations that move as waves through any part of the earth, usually associated with earthquakes, volcanoes, or large explosions

seismograph an instrument that measures and records seismic wave data

semiarid climate classification between arid and humid; receives between 25 and 50 cm (10 and 20 in) precipitation per year

semiconductors elements that have properties between those of a metal and those of a nonmetal, sometimes conducting an electric current and sometimes acting like an electrical insulator depending on the conditions and their purity; also called metalloids

shallow focus earthquakes earthquakes that occur from the surface down to 70 km deep

shear stress produced when two plates slide past one another or by one plate sliding past another plate that is not moving

shell model of the nucleus model of the nucleus that has protons and neutrons moving in energy levels or shells in the nucleus (similar to the shell structure of electrons in an atom)

shells the layers that electrons occupy around the nucleus

shield volcano a broad, gently sloping volcanic cone constructed of solidified lava flows

shock wave a large, intense wave disturbance of very high pressure, the pressure wave created by an explosion, for example

sial term for rocks made primarily of minerals containing the elements silicon and aluminum, for example, granit; composition of crustal rocks

sidereal day the interval between two consecutive crossings of the celestial meridian by a particular star

sidereal month the time interval between two consecutive crossings of the moon across any star

sidereal year the time interval required for the earth to move around its orbit so that the sun is again in the same position against the stars

silicates minerals that contain silicon-oxygen tetrahedra either isolated or joined together in a crystal structure

sill a tabular-shaped intrusive rock that formed when magma moved into the plane of contact between sedimentary rock layers

sima refers to rocks such as basalt made primarily of minerals containing the elements silicon and magnesium

simple harmonic motion the vibratory motion that occurs when there is a restoring force opposite to and proportional to a displacement

single bond covalent bond in which a single pair of electrons is shared by two atoms

slope the ratio of changes in the y variable to changes in the x variable or how fast the y-value increases as the x-value increases

soil a mixture of unconsolidated weathered earth materials and humus, which is altered, decay resistant organic matter

soil horizons layers of soil in a soil profile, each with its own set of chemical and physical properties that develop as the profile matures with time

soil profile a distinct soil structure of layers with different physical and chemical properties designated as A, B, and C

solar constant the averaged solar power received by the outermost part of the earth's atmosphere when the sunlight is perpendicular to the outer edge and the earth is at an average distance from the sun; about 1,370 watts per square meter

solenoid a cylindrical coil of wire that becomes electromagnetic when a current runs through it

solids a phase of matter with molecules that remain close to fixed equilibrium positions due to strong interactions between the molecules, resulting in the characteristic definite shape and definite volume of a solid

solstices time when the sun is at its maximum or minimum altitude in the sky, known as the summer solstice and the winter solstice

solubility dissolving ability of a given solute in a specified amount of solvent, the concentration that is reached as a saturate solution is achieved at a particular temperature

solute the component of a solution that dissolves in the other component; the solvent

solution a homogeneous mixture of ions or molecules of two or more substances

solvent the component of a solution present in the larger amount; the solute dissolves in the solvent to make a solution

sonic boom sound waves that pile up into a shock wave when a source is traveling at or faster than the speed of sound

sound quality characteristic of the sound produced by a musical instrument; determined by the presence and relative strengths of the overtones produced by the instrument

south celestial pole a point directly above the south pole of the earth; the point above the south pole on the celestial sphere

southpole short for "south seeking"; the pole of a magnet that points southward when it is free to turn

specific heat each substance has its own specific heat, which is defined as the amount of energy

(or heat) needed to increase the temperature of one gram of a substance one degree Celsius

speed a measure of how fast an object is moving—the rate of change of position per change in time; speed has magnitude only and does not include the direction of change

spin quantum number from quantum mechanics model of the atom, one of four descriptions of the energy state of an electron wave; this quantum number describes the spin orientation of an electron relative to an external magnetic field

spring equinox one of two times a year that daylight and night are of equal length; occurs on or about March 21 and identifies the beginning of the spring season

spring tides unusually high and low tides that occur every two weeks because of the relative positions of the earth, moon, and sun

standard atmospheric pressure the average atmospheric pressure at sea level, which is also known as normal pressure; the standard pressure is 29.92 inches or 760.0 mm of mercury (1,013.25 millibar)

standard time zones 15° wide zones defined to have the same time throughout the zone, defined as the mean solar time at the middle of each zone

standard unit a measurement unit established as the standard upon which the value of the other referent units of the same type are based

standing waves condition where two waves of equal frequency traveling in opposite directions meet and form stationary regions of maximum displacement due to constructive interference and stationary regions of zero displacement due to destructive interference

starch group of complex carbohydrates (polysaccharides) that plants use as a stored food source and that serves as an important source of food for animals

stationary front occurs when the edge of a front is not advancing

steam generator part of nuclear power plant; the heat exchanger that heats feedwater from the secondary loop to steam with the very hot water from the primary loop

step-down transformer a transformer that decreases the voltage of a current

step-up transformer a transformer that increases the voltage of a current

stony meteorites rock-forming meteorites composed mostly of silicate minerals

stony-iron meteorites meteorites composed of silicate minerals and metallic iron

storm a rapid and violent weather change with strong winds, heavy rain, snow, or hail

strain adjustment to stress; a rock unit might respond to stress by changes in volume, changes in shape, or by breaking

stratopause the upper boundary of the stratosphere

stratosphere the layer of the atmosphere above the troposphere where temperature increases with height

stream a large or small body of running water

stress a force that tends to compress, pull apart, or deform rock; stress on rocks in the earth's solid outer crust results as the earth's plates move into, away from, or alongside each other

strong acid acid that ionizes completely in water, with all molecules dissociating into ions

strong base base that is completely ionic in solution and has hydroxide ions

subduction zone the region of a convergent boundary where the crust of one plate is forced under the crust of another plate into the interior of the earth

sublimation the phase change of a solid directly into a vapor or gas

submarine canyons a feature of the ocean basin; deep, steep-sided canyons that cut through the continental slopes

summer solstice in the Northern Hemisphere, the time when the sun reaches its maximum altitude in the sky, which occurs on or about June 22 and identifies the beginning of the summer season

superconductors some materials in which, under certain conditions, the electrical resistance approaches zero

supercooled water in the liquid phase when the temperature is below the freezing point

supernova a rare catastrophic explosion of a star into an extremely bright, but short-lived phenomenon

supersaturated containing more than the normal saturation amount of a solute at a given temperature

surf the zone where breakers occur; the water zone between the shoreline and the outermost boundary of the breakers

surface wave a seismic wave that moves across the earth's surface, spreading across the surface as water waves spread on the surface of a pond from a disturbance

swell regular groups of low profile, long wavelength waves that move continuously

syncline a trough-shaped fold in layered bed rock

synodic month the interval of time from new moon to new moon (or any two consecutive identical phases)

talus steep, conical or apron-like accumulations of rock fragments at the base of a slope

temperate climate zone climate zone of the middle latitudes; average monthly temperatures stay between 10° C and 18° C (50° F and 64° F) throughout the year

temperature how hot or how cold something is; a measure of the average kinetic energy of the molecules making up a substance

tensional stress the opposite of compressional stress; occurs when one part of a plate moves away from another part that does not move

terrestrial planets planets Mercury, Venus, Earth, and Mars that have similar densities and compositions as compared to the outer giant planets

theory a broad, detailed explanation that guides the development of hypotheses and interpretations of experiments in a field of study

thermometer a device used to measure the hotness or coldness of a substance

thermosphere thin, high outer atmospheric layer of the earth where the molecules are far apart and have a high kinetic energy

third law of motion whenever two objects interact, the force exerted on one object is equal in size and opposite in direction to the force exerted on the other object; forces always occur in matched pairs that are equal and opposite

thrust fault a reverse fault with a low-angle fault plane

thunderstorm a brief, intense electrical storm with rain, lightning, thunder, strong winds, and sometimes hail

tidal bore a strong tidal current, sometimes resembling a wave, produced in very long, very narrow bays as the tide rises

tidal currents a steady and continuous onward movement of water produced in narrow bays by the tides

tides periodic rise and fall of the level of the sea from the gravitational attraction of the moon and sun

tornado a long, narrow funnel-shaped column of violently whirling air from a thundercloud that moves destructively over a narrow path when it touches the ground

total internal reflection condition where all light is reflected back from a boundary between materials; occurs when light arrives at a boundary at the critical angle or beyond

total solar eclipse eclipse that occurs when the earth, moon, and the sun are lined up so the new moon completely covers the disk of the sun; the umbra of the moon's shadow falls on the surface of the earth

trace fossils indications of former life; for example, tracks, burrows, and coprolites

transform boundaries in plate tectonics, boundaries that occur between two plates sliding horizontally by each other along a long, vertical fault; sudden jerks along the boundary result in the vibrations of earthquakes

transformer a device consisting of a primary coil of wire connected to a source of alternating current and a secondary coil of wire in which electromagnetic induction increases or decreases the voltage of the source

transition elements members of the B group families of the periodic table

transparent term describing materials that allow the transmission of light; for example, glass and clear water are transparent materials

transportation the movement of eroded materials by agents such as rivers, glaciers, wind, or waves

transverse wave a mechanical disturbance that causes particles to move perpendicular to the direction that the wave is traveling

trenches a long, relatively narrow, steep-sided trough that occurs along the edges of the ocean basins

triglyceride organic compound of esters formed from glycerol and three long-chain carboxylic acids; also called fats in animals and oil in plants

triple bond covalent bond formed when three pairs of electrons are shared by two atoms

tropic of Cancer parallel identifying the Northern limit where the sun appears directly overhead; located at 23.5° N latitude

tropic of Capricorn parallel identifying the Southern limit where the sun appears directly overhead; located at 23.5° S latitude

tropical air mass a warm air mass from warm regions

tropical climate zone climate zone of the low latitudes; average monthly temperatures stay above 18° C (64° F), even during the coldest month of the year

tropical cyclone a large, violent circular storm that is born over the warm, tropical ocean near the equator; also called hurricane (Atlantic and eastern Pacific) and typhoon (in western Pacific)

tropical year the time interval between two consecutive spring equinoxes; used as standard for the common calendar year

tropopause the upper boundary of the troposphere, identified by the altitude where the temperature stops decreasing and remains constant with increasing altitude

troposphere layer of the atmosphere from the surface to where the temperature stops decreasing with height

trough the low mound of water that is part of a wave; also refers to the rarefaction, or low pressure part of a sound wave

tsunamis very large, fast, and destructive ocean wave created by an undersea earthquake, landslide, or volcanic explosion; a seismic sea wave

turbidity current a muddy curent produced by underwater landslides

typhoon the name for hurricanes in the western Pacific

ultrasonic sound waves too high in frequency to be heard by the human ear; frequencies above 20,000 Hz

unmbra the inner core of a complete shadow

unconformity a time break in the rock record

undertow a current beneath the surface of the water produced by the return of water from the shore to the sea

unit in measurement, a well-defined and agreed-upon referent

universal law of gravitation every object in the universe is attracted to every other object with a force directly proportional to the product of their masses and inversely proportional to the square of the distance between the centers of the two masses

unpolarized light light consisting of transverse waves vibrating in all conceivable random directions

unsaturated molecule an organic molecule that does not contain the maximum number of hydrogen atoms; a molecule that can add more hydrogen atoms because of the presence of double or triple bonds

valence the number of covalent bonds an atom can form

valence electrons electrons of the outermost shell; the electrons that determine the chemical properties of an atom and the electrons that participate in chemical bonding

Van Allen belts belts of radiation caused by cosmic-ray particles becoming trapped and following the earth's magnetic field lines between the poles

van der Walls force general term for weak attractive intermolecular forces

vapor the gaseous state of a substance that is normally in the liquid state

variables changing quantities usually represented by a letter or symbol

vectors quantities that require both a magnitude and direction to describe them

velocity describes both the speed and direction of a moving object; a change in velocity is a change in speed, in direction of travel, or both

ventifacts rocks sculpted by wind abrasion

vernal equinox another name for the spring equinox, which occurs on or about March 21 and marks the beginning of the spring season

vibration a back and forth motion that repeats itself

virtual image an image where light rays appear to originate from a mirror or lens; this image cannot be projected on a screen

volcanism volcanic activity; the movement of magma

volcano a hill or mountain formed by the extrusion of lava or rock fragments from a magma below

volt unit of potential difference equivalent to J/C

voltage drop the electric potential difference across a resistor or other part of a circuit that consumes power

voltage source source of electric power in an electric circuit that maintains a constant voltage supply to the circuit

volume how much space something occupies

vulcanism volcanic activity; the movement of magma

warm front the front that forms when a warm air mass advances against a cool air mass

watershed the region or land area drained by a stream; a stream drainage basin

water table the boundary below which the ground is saturated with water

watt metric unit for power; equivalent to J/sec

wave a disturbance or oscillation that moves through a medium

wave equation the relationship of the velocity of a wave to the product of the wavelength and frequency of the wave

wave front a region of maximum displacement in a wave; a condensation in a sound wave

wave height the vertical distance of an ocean wave between the top of the wave crest and the bottom of the next trough

wave mechanics alternate name for quantum mechanics derived from the wavelike properties of subatomic particles

wave period the time required for two successive crests or other successive parts of the wave to pass a given point

wavelength the horizontal distance between successive wave crests or other successive parts of the wave

weak acid acids that only partially ionize because of an equilibrium reaction with water

weak base a base only partially ionized because of an equilibrium reaction with water

weathering slow changes that result in the breaking up, crumbling, and destruction of any kind of solid rock

wet adiabatic lapse rate for a rising parcel of air adiabatic cooling rate of air with condensation of water vapor; adiabatic cooling plus the release of latent heat

white dwarf stars one of two groups of stars on the Hertzsprung-Russell diagram that have a different set of properties than the main sequence stars; faint, white-hot stars that are very small and dense

wind a horizontal movement of air that moves along or parallel to the ground, sometimes in currents or streams

wind abrasion the natural sand-blasting process that occurs when wind particles break off small particles of rock and polish the rock they strike

wind chill factor the cooling equivalent temperature since the wind makes the air temperature seem much lower; the cooling power of wind

winter solstice in the Northern Hemisphere, the time when the sun reaches its minimum altitude, which occurs on or about December 22 and identifies the beginning of the winter season

work the magnitude of applied force times the distance through which the force acts; can be thought of as the process by which one form of energy is transformed to another

zone of saturation zone of sediments beneath the surface in which water has collected in all available spaces

Credits

Photographs

Section Openers **Section 1:** © Kindra Clineff; **Section 2:** © Visuals Unlimited/Carolina Biological Supply; **Section 3:** © NASA/Peter Arnold, Inc.; **Section 4:** © Galen Rowell/Mountain Light Photography.

Chapter 1 **Chapter Opener:** © Kindra Clineff; **Figures 1.1, 1.2, 1.13:** Keith Jennings: News Bureau Arizona State.

Chapter 2 **Chapter Opener:** © Walt Anderson/Tom Stack & Associates; **Figure 2.8:** Smithsonian Institute; **Figure 2.13:** Smithsonian Institute; **Box 2.1:** General Motors; **Figure 2.21:** Frank Reed-Arizona State University Media Relations Office.

Chapter 3 **Chapter Opener:** © Kindra Clineff; **Figure 3.1:** Beech Aircraft Corporation; **Figure 3.2:** Smithsonian Institute; **Figure 3.9:** Frank Reed-Arizona State University Media Relations Office; **Figure 3.11:** NASA.

Chapter 4 **Chapter Opener:** © Mark Newman/Tom Stack & Associates; **Figures 4.15A, 4.15B, 4.17:** Arizona Public Service Company.

Chapter 5 **Chapter Opener:** © JMP/Visuals Unlimited; **Figure 5.1:** Bethlehem Steel Corp.; **Figure 5.2:** News Bureau, Arizona State University; photo by John C. Wheatly; **Figure 5.7:** Honeywell, Inc.; **Figure 5.15:** Manville Company.

Chapter 6 **Chapter Opener:** © Kindra Clineff; **Figure 6.1:** © James L. Shaffer; **Figure 6.24:** © Blair Seitz/Photo Researchers, Inc.

Chapter 7 **Chapter Opener:** © Keith Kent/Peter Arnold, Inc.; **Figure 7.14:** Arizona Public Service Company.

Chapter 8 **Chapter Opener:** © Visuals Unlimited; **Figure 8.8:** 1987 Libbey-Owens-Ford Co.

Chapter 9 **Chapter Opener:** © Fermilab/Peter Arnold, Inc.; **Figure 9.1:** News Bureau, Arizona State University; **Figure 9.10:** Frank Reed-Arizona State University Media Relations Office.

Chapter 10 **Chapter Opener:** © J. Hawkins/Peter Arnold, Inc.

Chapter 11 **Chapter Opener:** © John Cunningham/Visuals Unlimited; **Figure 11.1:** Stewart Lindsay-Physics Department, Arizona State University.

Chapter 12 **Chapter Opener:** © Cabisco/Visuals Unlimited; **Figure 12.1:** Science/John Deere/Visuals Unlimited; **Figure 12.11:** Keith Jennings-News Bureau, Arizona State University.

Chapter 13 **Chapter Opener:** © Galen Rowell/Mountain Light Photography.

Chapter 14 **Chapter Opener:** © Cabisco/Visuals Unlimited; **Figures 14.1, 14.10:** American Petroleum Institute.

Chapter 15 **Chapter Opener:** © Visuals Unlimited; **Figure 15.1:** Arizona Public Service Company; **Figure 15.2A & B:** © Fundamental Photographs; **Figures 15.10, 15.17A, 15.17B, 15.19, 15.20:** Arizona Public Service Company.

Chapter 16 **Chapter Opener:** © Dennis Di Cicco/Peter Arnold, Inc.; **Figure 16.1A:** National Optical Astronomy Observatories; **Figure 16.2:** © J. Fuller/Visuals Unlimited; **Figures 16.23, 16.24:** National Optical Astronomy Observatories; **Figure 16.26:** Lick Observatory.

Chapter 17 **Chapter Opener:** © Clyde H. Smith/Peter Arnold, Inc.; **Figures 17.1, 17.13, 17.14, 17.15, 17.16, 17.18A, 17.18B, 17.19B, 17.20:** NASA; **Figure 17.21:** Jet Propulsion Lab; **Figure 17.25:** Lick Observatory; **Figure 17.27:** Center for Meteorite Studies, Arizona State.

Chapter 18 **Chapter Opener:** © Visuals Unlimited; **Figure 18.1:** NASA; **Figure 18.28:** Lick Observatory.

Chapter 19 **Chapter Opener:** © Doug Sokell/Tom Stack & Associates.

Chapter 20 **Chapter Opener:** © Ann Duncan/Tom Stack & Associates; **Figure 20.1:** U.S. Geological Survey.

Chapter 21 **Chapter Opener:** © Galen Rowell / Mountain Light Photography; **Figure 21.1:** Photo by A. Post, USGS Photo Library, Denver, CO.; **Figure 21.4A:** Robert W. Northrop, Photographer/Illustrator; **Figure 21.14D:** Courtesy of National Geophysical Data Center, David Erico, University of Colorado; **Figure 21.18:** NASA.

Chapter 22 **Chapter Opener:** © Galen Rowell/Mountain Light Photography; **Figure 22.1:** Photo by W. R. Hansen, USGS Photo Library, Denver, CO.; **Figure 22.2:** National Park Service Photo by Wm. Belnap, Jr.; **Figure 22.12:** Visuals Unlimited/ Weber; **Figure 22.15:** Photo by R.D. Miller, USGS Photo Library, Denver, CO.

Chapter 23 **Chapter Opener:** © Galen Rowell/Moutain Light Photography; **Figure 23.1:** Robert W. Northrop, Photographer/Illustrator.

Chapter 24 **Chapter Opener:** © Galen Rowell/Mountain Light Photography.

Chapter 25 **Chapter Opener:** H. Feurer/The Image Bank; **Figure 25.1:** © Peter Arnold/Peter Arnold, Inc.; **Figure 25.7:** NOAA; **Figure 25.17:** NOAA; **Figure 25.19:** Cray Research, Inc.

Chapter 26 **Chapter Opener:** © S. J. Kraseman/Peter Arnold, Inc.; **Figures 26.1, 26.9, 26.10:** Salt River Project.

Illustrator Credits

John Forester: Figures 20.7, 20.8, 20.9, 20.20, 20.21a,b,c, Box 20.2, 21.9, 21.19, 22.9, 22.14a,b,c, 23.13, 24.9, 24.10, 24.11, 25.2, 26.3, 26.4, 26.6, 26.7, 26.8, 26.18, 26.19, 26.20, 26.22.

McCullough Graphics (Text art): TA11.1, TA11.2, TA11.3, TA11.4, TA11.5, TA11.6, TA11.7, TA11.8, TA11.9, TA11.10, TA11.11, TA11.12, TA11.13, TA11.14, TA11.15, TA11.19, TA11.20, TA11.21, TA11.22, TA13.1, TA13.2, TA13.3, TA13.4, TA14.15, TA14.16, TA14.18, TA14.20, TA14.21, TA15.1, TA15.2, TA AD2, TA AD3, TA AD5, TA AD6.

Rolin Graphics: Figures 1.4, 1.5, 1.6, 1.7, 1.8, 1.9, 1.10, 1.11, 1.12, 1.14, 1.15, 1.16, 1.17, 2.2, 2.3, 2.4, 2.5, 2.6, 2.7, 2.9, 2.11, 2.12, 2.14, 2.15, 2.16, 2.17, 2.18, 2.19a,b, 2.20, TA2.1, 3.3a,b, 3.4, 3.6, 3.7, 3.8, 3.10, 3.12, 3.13, 3.14, 3.15, 4.2, 4.3, 4.4, 4.5, 4.6, 4.7, 4.8, 4.9, 4.10, 4.11, 4.12, 4.16, 4.18, 4.19, 4.20, 4.21, 4.22, 5.3, 5.4, 5.5, 5.6, 5.8, 5.9, 5.10, 5.11, 5.12, 5.13, 5.14, 5.16, 5.17, Box 15.1, Box 15.2, Box 15.3, 5.18, 5.19, 5.20, 5.21, 5.23, 6.2, 6.3, 6.4, 6.5, 6.6, 6.7, 6.8, 6.9, 6.10, 6.11, 6.12, 6.14, 6.15, 6.16, 6.17, 6.18, 6.19, 6.20, 6.21, 6.22, 6.23, 6.25, 6.26, 6.27, 7.2, 7.3, 7.4, 7.5, 7.6, 7.7, 7.8, 7.9, 7.10, 7.11, 7.12, 7.13, 7.15, 7.16, 7.17, 7.21, 7.22, 7.24, 7.25, 7.26, 7.27, 7.28, 7.29, 7.30, 7.31, 7.32, 7.33, 7.34, 7.35, 7.36, 7.37, Box 7.1, Box 7.3, 8.2, 8.3, 8.4, 8.5, 8.6, 8.9, 8.10, 8.11, 8.12, 8.13, 8.14, 8.15, 8.17, 8.18, 8.19, 8.20, 8.21, 8.22, 8.23, 8.24, 8.25, 8.26, 8.27, Box 8.1, 9.3, 9.4, 9.6, 9.7, 9.8, 9.9, 9.11, 9.12, 9.13, 9.14, 9.16, 9.17, 9.18, 9.19, 9.20, 9.21, 10.3, 10.4, 10.6, 10.7, 10.8, 10.9, 10.10, 10.12, 10.13, 10.14, 10.15, 10.16, 10.17, 10.19, 10.20, 10.21, 10.22, 11.2, 11.3, 11.5, 11.6, 11.7, 11.9, 11.10, 11.12, TA11.16&a, TA11.17&a, TA11.18&a, 11.4, Box 11.1, TA11.23, 12.2, 12.5, 12.7,

12.13, 12.15, 12.16, 12.17, 13.4, 13.5, 13.6, 13.7, 13.8, 13.9, 13.10, 13.11, 13.12, 13.13, 13.14, 13.15, 13.16, 13.17, 13.20, 13.22, 14.2, 14.3, 14.4, 14.5, 14.6, TA14.1, TA14.2, TA14.3, TA14.4, TA14.5, TA14.6, TA14.7, TA14.8, TA14.9, 14.8, 14.9, 14.11, 14.12, 14.13, TA14.10, TA14.11, TA14.12, TA14.13, TA14.14, 14.14, 14.16, 14.17, TA14.17, 14.19, 14.20, 14.21, 14.23, 14.24, 14.25, TA14.19, 14.26, Box 14.1, TA14.22, TA14.23, TA14.24, TA14.25, 15.3, 15.4, 15.5, 15.6, 15.7, 15.8, 15.9, 15.11, 15.12, 15.13, 15.14, 15.15,

15.16, 15.18, 15.21, 15.22, 16.1b, 16.3a,b, 16.4, 16.5, 16.6, 16.7, 16.8, 16.9, 16.10, 16.11, 16.12, 16.13, 16.14, 16.15, 16.16, 16.17, 16.18, 16.19, 16.20, 16.21, 16.22, 16.25, 16.27, 16.28, 16.29, 17.2, 17.3, 17.4, 17.5, 17.6, 17.7, 17.8, 17.9, 17.10, 17.11, 17.12, 17.17, 17.22, 17.23, 17.24, 17.26, 18.2, 18.3, 18.4, 18.5, 18.6, 18.7, 18.8, 18.9, 18.10, 18.11, 18.12, 18.13, 18.14, 18.15, 18.16, 18.17, 18.19, 18.20, 18.22, 18.23, 18.24, 18.27, 18.29, 18.30, 18.31, 18.32, 18.33, 18.34, 19.2, 19.3, 19.6, 19.7, 19.8, TA19.1, 19.11, 19.13, 19.14, 19.15,

20.3, 20.4, 20.5, 20.6, 20.13, Box 20.1, 21.7, 21.16, 23.12, 23.14, 24.2, 24.3, 24.4, 24.5, 24.6, 24.7, 24.8, 24.12, 24.13, 24.14, 24.15, 24.17, 25.3, 25.4, 25.5, 25.6, 25.9, 25.10, 25.11, 25.12, 25.14, 25.15, 25.16, 25.18, 25.21, 25.23, 26.2, 26.13, 26.16, 26.17, TA AA1, TA AD1, TA AD4.

Alice Thiede-Cartographics: Figures 7.23, 18.21, 18.25, 18.26, 20.16, 25.8, 25.13, 25.22, 25.24, 25.25, 26.5, 26.12, 26.21, 26.23.

Index

CD-ROM, 166–67
Celestial
 equator, 346
 meridian, 346
 poles, 346
 sphere, 342, 346
Cellophane, 304
Cellulose, 302
Celsius thermometer scale, 84
Cementation, of sediments to form
 rocks, 431
Cenozoic era, 498
Centrifugal force, 53
Centripetal force, 53
Cepheid variables, 352
Chadwick, James, 124, 186
Chain reaction, of radioactive
 nuclei, 327
Change
 chemical, 203
 physical, 203
Chemical
 bonds, 226–34
 energy, 68, 225
 equation, 225
 equations, 225, 248–52, 256–61
 equations, steps in balancing,
 249–50
 equilibrium, 277
 formula, 228–32, 245–46
 reaction, 225, 253–55
 reaction, definition of, 225
 reactions, categories of, 253–56
 symbols, 206–7
 weathering, 475
Chemistry, 202
 inorganic, definition of, 287
 organic, definition of, 287
Chert, 427
Chinook, 511
Chondrites, 388
Chondrules, 388
Chrysotile, 428
Cinder cone volcano, 468
Circular motion, 52–53
Cirque, 484
Cirrus clouds, 517
Clastic sediments, 430
Clausius, Rudolf, 82
Claystone, 430
Cleavage,
 breaking of minerals, 426
 rock, 432
Climate, 535–40
 changes and the ocean, 538
 how described, 539
 and latitude, 536
 local influences, 537
 zones, 539
Closed tube vibrating air column,
 115
Cloud classification scheme, 517
Clouds, how formed, 524–26
Coal, how formed, 73
Coalescence process, of forming
 precipitation, 526
Cold front, 529
Colors, of quarks, 195

Combination chemical reactions,
 253
Comets, 385–87
 model of, as a dirty-snowball,
 385
 spacecraft missions to, table of,
 385
Compact Disc (CD), 166
Compaction, of sediments to form
 rocks, 431
Composite volcano, 468
Compound motion, 38–40
Compounds
 chemical, 204
 covalent, 230–32
 ionic, 228–30
 names of covalent, 237
 names of ionic, 234–36
Computer, and data storage, 145
Concave mirrors, 162
Concentrated solution, 267
Concept, definition of, 2
Condensation (sound), 105
Condensation nuclei, 516
Condensation of water vapor, 95,
 513
Condensation point, 92
Conduction, of heat, 89–90
Conductors, electrical, 127
Conglomerate, sedimentary rock,
 430
Conservation of energy, law of, 71
Conservation of mass, law of, 248
Constants, numerical, 11
Constellations, 343
Constructive sound interference,
 110
Continental
 air mass, 528
 climate, 539
 divide, 546
 drift, 445
 glaciers, 484
 shelf, 556
 slope, 556
Continuous spectrum, 186
Control rods, of nuclear reactor,
 328
Convection, 90–92
 in atmosphere, 510
 in earth's mantle, 449
Conventional electric current, 132
Convergent boundaries, of earth's
 plates, 447–48
Convex mirrors, 162
Coordinate covalent bond, 232
Copernican system (of sun and
 planets), 367–68
Copernicus, Nicolas, 367
Coprolites, 493
Core, of earth's interior, 441–43
Coriolis effect, 398
Correlation, of rock units, 496
Cost of running electrical
 appliances, 137
Coulomb, Charles, 128
Coulomb's law, 128
Coulomb unit of charge, 127

Covalent bond, 227, 230–34
 polar, 233
Covalent compound, 230–34
 formulas and names, 238
 names of, 237
Creep, 480
Crest, of a wave, 106, 552
Critical angle of refracted light
 ray, 163
Critical mass, of fissionable nuclei,
 327
Crude oil, 294
Crust, of earth, 439–40
Crystal classification system, 423
Cumulus clouds, 517
Curie, Madame Marie, 315
Curie, unit of radioactivity at
 source, 324
Cycle, of a vibration, 103
Cyclone, 530

Dalton, John, 182, 206, 256
Data, 7
Data points, 13
Dating, of rocks,
 geomagnetic, 497
 radiometric, 497
 relative, 496
Daylight saving time, 404
De Broglie, Louis, 190
Deceleration, 29, 37
Decibel scale, 112
Decomposition chemical reactions,
 253
Deflation, 485
Degassing, of planets, 373
Delta, 483
Democritus, 82, 182
Density current, in ocean, 555
Density, 9
 of a mineral, 426
Destructive interference (sound),
 110
Dew, 515
Dew point temperature, 515
Diastrophism, 456
Diesel fuel, 295
Differentiation, of solid earth, 437
Diffraction of light, 166–68
Diffuse reflection of light, 159
Dike, 469
Dilute solution, 267
Dinosaurs, theory of extinction,
 499–501
Dipole, 234
Direct current, 134
Disaccharides, 302
Dispersion of light, 164–65
Dissolving, 271–73
Divergent boundaries, of earth's
 plates, 447
Divide, of stream, 546
Dome, 458
Dolomite, 427, 431
Doppler effect, 117
Double covalent bond, 231
Dry adiabatic lapse rate, 524
Dunes, sand, 485

Earth, future of, 500
Earthflow, 481
Earthquake, 461–66
 epicenter, 463
 focus, 463–64
 Richter Scale table, 466
 safety rules, 465
 waves, 463–64
Earth's
 axis, 371, 396
 hot spots, 451
 interior, theory of formation,
 437
 motions, and ice ages, 412
 shape, 395
Echo, 110
Eclipse, 410
Ecliptic, 343
Edison, Thomas, 147
Einstein, Albert, 171, 325
Elastic
 deformation, 456
 limit, 457
 material, 457
 rebound, 462
Electric
 conductors and nonconductors,
 127
 generator, 146–47
 meters, 143
 motor, 146
 potential energy, 130
 power, 135–36
 resistance, 134–35
 transformer, 147–49
 work, 64, 135, 137
Electrical energy, 69
Electrical forces, 125, 129
 attraction and repulsion, 125
 measurement of, 127
Electric charge, 124–26
 conductors and insulators,
 126–27
 and electrical forces, 125
 electron theory of, 124
 fundamental, 127
 and potential energy, 130
 test charge, for measuring field,
 129
 unit for measuring, 127
Electric circuit, 131–32
Electric current, 131–35
 nature of, 133–34
 quantities and units, summary
 table, 138
 resistance to, 134–35
 two kinds of, 134
 two ways to describe, 132
Electric field, 129–30, 134
 electric field lines, 129
 making a map of, 129
Electrolytes, 273
Electromagnet, 142
Electromagnetic induction, 146
Electromagnetic spectrum, 158
 color of starlight, and
 temperature, 350–52
 wavelengths and frequencies of
 colors, 164–65

Ice ages, cause of, 412
Ice-crystal process, of forming
  precipitation, 526
Ice-forming nuclei, 527
Igneous rocks, 429–30
Images
  real, 162
  virtual, 162
Incandescent, 157
Incident light ray, 161
Index fossils, 496
Index of refraction, 163
Indicators, acid-base, 276
Inertia, 34, 46. *See also* First law
  of motion
Infrasonic, 105
Inorganic chemistry, definition of,
  287
In phase waves, 110
Instability, state of atmosphere,
  525
Insulators,
  electrical, 127
  heat, 89
Intensity, of sound wave, 111
Interference
  of light, 168–69
  of sound waves, 110–11
Intermolecular forces, 270
Internal energy, 86
International date line, 404
International System of Units, 5
Intertropical convergence zone,
  512
Intrusive igneous rocks, 429
Inversion, temperature, 508
Ion, 216
  definition of, 125
  polyatomic, 232
Ion exchange chemical reactions,
  255
Ionic bond, 228
Ionic compounds,
  formulas, how to write, 236–37
  names of, 234–36
Ionization, by water, 273
Ionization counter, as used to
  measure radiation, 322
Ionized substances, 273
Ionosphere, 510
Ions, 215–17
Iron meteorites, 388
Irregular galaxies, 357
Island arcs, 448
Isomers, 289–91
Isostasy, 443
Isotope, 208, 316

Jet stream, 512
Joule, unit of
  equivalence to calories, 87
  as a unit of heat, 86
  as a unit of work, 62
Julian calendar, 405
Jupiter, properties of the planet,
  380–81

Kelvin thermometer scale, 85
Kepler, Johannes, 368
Kepler's laws of planetary motion,
  368
Kerosene, 295
Ketones, 298
Kilogram, 6
Kinetic energy, 66
Kinetic molecular theory, 80–83

La Brea tar pit, 492
Laccolith, 469
Lake, 546
Land, Edwin H., 169
Landforms, and landscapes,
  485–87
Landslide, 480
Lanthanide series, 218
Latent heat, 92–94
  of fusion, 93
  of vaporization, 94
Laterites, 479
Latitude, 399
Latitudes, and solar radiation
  received, 535–36
Lava, 427
Lavoisier, Antoine, 205, 277
Law
  definition of scientific, 15
  periodic, 210–11
Law of
  combining volumes, 257
  conservation of energy, 71
  conservation of mass, 248
  conservation of momentum, 52
  electrical interaction
    (Coulomb's law), 128
  electrical resistance (Ohms'
    law), 134
  gravitation (Newton's law of),
    53–54
  reflection, 161
Laws of Motion
  first law, 46
  second law, 47–49
  third law, 50–51
Laws of planetary motion, 368
Leptons, 195
Leucippus, 182
Life, organic compounds of,
  299–304
Life of a star, 353–55
Light
  diffraction, 166–68
  diffuse reflection, 159
  dispersion, 164–65
  interference, 168–69
  photoelectric effect, 170–71
  polarization, 169–70
  ray model, 159
  reflection, 159–62, 170
    from planets, 344
  refraction, 162–64
  scattering, 170
  starlight, 163, 344
Light year, 347
Limestone, 431
Line spectrum, 186

Liquids, 82, 203
Liter, 7
Lithification, 431
Lithosphere of earth, 441
Little Green Men, 355
Loam, 478
Lodestone, 138
Loess, 485
Lone pairs of electrons, 231, 240
Longitude, 400
Longitudinal wave, 104
Longshore current, 554
Loudness of a sound
  table of noise levels, 112
  unit of loudness, 112
Lowell, Percival, 377
Low pressure zone, 530
LPG, 295
Luminosity, of star, 350
Luminous, 157
Lunar
  eclipse, 410
  highlands, 407
Luster, of a mineral, 426

Macromolecule, 300
Magellan spacecraft, 376
Magma, 427
Magnetic
  declination and dip, 139
  domains, 141
  monopole, 139
  north pole, 138
  poles, 138
Magnetic field, 129, 138–41
  current direction and
    orientation of, 141–42
  and current loops, 142
  of the earth, 139, 141, 443–44
  and electric currents, 146–47
  field lines, 138
  naming convention, 138
  source of, 139–41
Magnetic quantum number, 192
Magnetism, 138–41
  and permanent magnets, 141
  and relation to electric current,
    139, 141–42
Magnetite, 138
Magnitude, 27
Main sequence stars, 352
Manipulated variable, 13
Mantle, of earth's interior, 440–41
Marble, 433
Maria, of moon, 409
Marine climate, 539
Mariner 9 spacecraft, 378
Mariner 10 spacecraft, 374–75
Maritime air mass, 528
Mars,
  properties of the planet,
    377–80
  spacecraft mission to, table,
    378
Mars Observer spacecraft, 380
Martian canals, 377

Mass
  defect, of the nucleus, 325
  density, 9
  and inertia, 46
  number, 210, 316
  spectrometer, 208
  spectrum, 208
  unit for, 6
Mass movement, 480–81
Matter, 201
  ancient Greek reasoning about,
    14, 30–31, 181–82
  classification schemes of, 201–4
  phase changes of, 81–83
  phases of, 81–82, 202–3
Maxwell, James, 129
Mean solar day, 401
Mean solar time, 401
Measurement, 3–6
  definition of, 3
  English system of, 5
  of heat, 86–88
  metric system of, 5–7
  systems of, 4
  of temperature, 83–85
Mechanical energy, 68
Mechanical equivalent of heat,
  86–87
Mechanical weathering, 475
Meitner, Lise, 326
Melting point, 92
Men, Little Green (LGM), 355
Mendeleev, Dmitri, 210
Mercalli scale, 465
Mercury, properties of the planet,
  374–75
Meridians, 399
Mesons, 195
Mesosphere
  of atmosphere, 509
  of earth's interior, 441
Mesozoic era, 498
Metallic bond, 227
Metals, properties of, 201
Metamorphic rocks, 431–32
Meteor, 387
Meteorites, 385, 388
  iron, 388
  stony, 388
  strange, 389
Meteoroids, 387
Meteor shower, 387
Meter, 6
Metric
  prefixes, 6–7
  system of measurement, 5–7
  table of prefixes, 6
Meyer, Lothar, 210
Mica, white, 424
Microclimate, 539
Microwave oven, how it heats, 239
Milky Way galaxy, 355–57
Millibar, 507
Millikan, Robert A., 184
Millirem, unit of radiation, 324
Mineral, 421–29

Polyvinyl chloride (PVC), 304
Pond, 546
Porosity, of sediment, 547
Positive ion, 125, 216
Positron, 195
Post meridiem (P.M.), 401
Potential energy, 65
Power, 63–64
Precambrian era, 498
Precession, of earth's axis, 398
Precipitation, 526
Pressure, 96
Priestley, Joseph, 204
Primary loop, of nuclear power
        plant, 328
Prime meridian, 400
Principal quantum number, 192
Principle, definition of scientific, 15
Principle of
        crosscutting relationships, 495
        faunal succession, 496
        original horizonality, 494
        superposition, 494
        uniformity, 456, 495
Problem solving
        approach and procedures, 17
        format, 12
Products, in chemical equation,
        225
Projectiles, 33
Proof, of an alcohol beverage, 297
Propane, 289, 295
Properties, 3. See also metric
        standard units, e.g.,
        meterfundamental
        properties, 6
Properties of
        acids and bases, theory about,
        275–78
        salts, 279
Proportional relationships, 11
Proteins, 300
Protogalaxies, 359
Proton, 124
        discovery of, 186
Protostar, 348
Psychrometer, 514
Ptolemaic system (of sun and
        planets), 367
Pulsars, 355
Pure sound tones, 113
Pure substances, 203
PVC (polyvinyl chloride), 304
Pyrite, 429

Quad, 87
Quanta, 171, 186
Quanta, of energy, 171
Quantitative use of equations,
        258–61
Quantities, 10

Quantum
        angular momentum, 192
        jumps, 188–90
        magnetic, 192
        mechanics, 191
        number, 187
        principal, 192
        spin, 193
Quark, 195
Quartz, 424
Quartzite, 433

Rad, unit of radiation, 323–24
Radiant energy, 69, 92
Radiation, 92
        background, 324
        blackbody, 158
        consequences of exposure to,
        324
        exposure, 324
        measurement of radioactive,
        322–24
Radioactive decay, 315
        constant, 321
        series, 320–22
        types of, 319–20
Radioactive isotopes, from
        uranium fissioning, 327
Radioactivity, 315, 323–24,
        319–25
Radiometric age, of rocks, 497
Rain, acid, 281
RAM, 145
Rare earth elements, 218
Rarefaction, 105
Ratio, 8–9
Rayon, 304
Reactants, in chemical equation,
        225
Real images, 162
Recycling, how to sort plastic
        bottles, 307
Red giant stars, 352
Redox reactions, 252
Reducing agent, 253
Referent, 3
Reflected light ray, 159–62
Reflection of
        light, 159–62
        sound waves, 109
Refracted light, 162–64
        critical angle of, 163
        dispersion, 164–65
        index of, 163
        mirages, 163–64
        shimmering objects, 163–64
        twinkling stars, 164
Refraction, of sound waves, 109
Rejuvenation, of erosion processes,
        487

Relative humidity, 96, 513
        table, 568
Relay, electromagnetic, 144
Rem, unit of radiation, 324
Replacement
        chemical reactions, 253–55
        fossilization, 492
Representative elements, 213
Reservoir, of surface water, 546
Resonance
        and engineering, 112–13
        of sound, 112, 114
Resonant frequency, sounds,
        114–16
Responding variable, 13
Retrograde
        motion, 344
        rotation, 376
Reverberation, 109
Reversal, of earths magnetic field,
        444
Reverse fault, 461
Revolution
        of the earth, 396
        of a planet, 371
Rhyolite, 430
Richter Scale, of earthquake
        effects, 466
Ridges, on ocean floor, 445–46
Rift valley, 446
Rip current, 554
Rock
        cycle, 433
        texture, 429
Rock flour, 484
Rocks, 422, 429–33
        igneous, 429–30
        metamorphic, 431–32
        sedimentary, 430–31
Rockslide, 481
Rotation
        of the earth, 397–98
        of a planet, 371
Rules of thumb for magnetic field
        direction, 141–42
Running water, as erosional agent,
        481–83
Runoff, 545
Rutherford, Ernest, 124, 184–86,
        315

Salinity, 268, 551
Salts
        properties of, 279
        table of common, 280
San Andreas fault, 461
Sand dunes, 485
Sandstone, 430
Santa Ana wind, 511
Saturated
        air, 95, 513
        solution, 269

Saturated
        fats, 303
        hydrocarbon, 292
Saturn, properties of the planet,
        381–83
Schist, 432
Schrodinger, Erwin, 191
Scientific
        investigations, 14
        law, 15
        method, 14
        notation, how to use, 565–66
        principle, 15
        theory, 17
Scintillation counter, as used to
        measure radiation, 322
Sea breeze, 511
Sea-floor spreading, 446
Seamounts, 558
Seas, 549–50
Seatbelts and airbags, 37
Seawater, 540–56
        movement of, 552–56
        nature of, 550–52
        waves on, 552–53
        why salty, 551
Second, as a time unit, 6
Secondary loop, of nuclear power
        plant, 329
Second law of motion, 47
Sedimentary rocks, 430–31
Sediments, 430
        chemical, 431
        clastic, 430
        size and rocks formed from,
        table, 431
Seismic wave, 438, 461, 463–65
Seismograph, 438, 463
Semiarid climate, 539
Semiconductors, 127, 216
Serpentine, 427, 428
Shadow zones
        P-wave, 441
        S-wave, 441
Shale, 430
Shield volcano, 468
Shell, of atom, 213
Shergotties, meteorites, 389
Shock wave, 117
Sial, 440
Sidereal
        day, 401
        month, 406
        year, 405
Significant figures, 563–64
Silicates, 424
Sill, 469
Siltstone, 430
Sima, 440
Simple harmonic motion, 102
Single covalent bond, 231

Table of Atomic Weights

Element	Symbol	Atomic Number	Atomic Weight	Element	Symbol	Atomic Number	Atomic Weight
Actinium	Ac	89	227.0278	Mercury	Hg	80	200.59
Aluminum	Al	13	26.98154	Molybdenum	Mo	42	95.94
Americium	Am	95	(243)	Neodymium	Nd	60	144.24
Antimony	Sb	51	121.75	Neon	Ne	10	20.179
Argon	Ar	18	39.948	Neptunium	Np	93	237.0482
Arsenic	As	33	74.9216	Nickel	Ni	28	58.69
Astatine	At	85	(210)	Niobium	Nb	41	92.9064
Barium	Ba	56	137.33	Nitrogen	N	7	14.0067
Berkelium	Bk	97	(247)	Nobelium	No	102	(259)
Beryllium	Be	4	9.01218	Osmium	Os	76	190.2
Bismuth	Bi	83	208.9804	Oxygen	O	8	15.9994
Boron	B	5	10.811	Palladium	Pd	46	106.42
Bromine	Br	35	79.704	Phosphorus	P	15	30.97376
Cadmium	Cd	48	112.41	Platinum	Pt	78	195.08
Calcium	Ca	20	40.078	Plutonium	Pu	94	(244)
Californium	Cf	98	(251)	Polonium	Po	84	(209)
Carbon	C	6	12.011	Potassium	K	19	39.0983
Cerium	Ce	58	140.12	Praseodymium	Pr	59	140.9077
Cesium	Cs	55	132.9054	Promethium	Pm	61	(145)
Chlorine	Cl	17	35.453	Protactinium	Pa	91	231.0359
Chromium	Cr	24	51.9961	Radium	Ra	88	226.0254
Cobalt	Co	27	58.9332	Radon	Rn	86	(222)
Copper	Cu	29	63.546	Rhenium	Re	75	186.207
Curium	Cm	96	(247)	Rhodium	Rh	45	102.9055
Dysprosium	Dy	66	162.50	Rubidium	Rb	37	85.4678
Einsteinium	Es	99	(252)	Ruthenium	Ru	44	101.07
Erbium	Er	68	167.26	Samarium	Sm	62	150.36
Europium	Eu	63	151.96	Scandium	Sc	21	44.95591
Fermium	Fm	100	(257)	Selenium	Se	34	78.96
Fluorine	F	9	18.998403	Silicon	Si	14	28.0855
Francium	Fr	87	(223)	Silver	Ag	47	107.8682
Gadolinium	Gd	64	157.25	Sodium	Na	11	22.98977
Gallium	Ga	31	69.723	Strontium	Sr	38	87.62
Germainum	Ge	32	72.59	Sulfur	S	16	32.066
Gold	Au	79	196.9665	Tantalum	Ta	73	180.9479
Hafnium	Hf	72	178.49	Technetium	Tc	43	(98)
Helium	He	2	4.002602	Tellurium	Te	52	127.60
Holmium	Ho	67	164.9304	Terbium	Tb	65	158.9254
Hydrogen	H	1	1.00794	Thallium	Tl	81	204.383
Indium	In	49	114.82	Thorium	Th	90	232.0381
Iodine	I	53	126.9045	Thulium	Tm	69	168.9342
Iridium	Ir	77	192.22	Tin	Sn	50	118.710
Iron	Fe	26	55.847	Titanium	Ti	22	47.88
Krypton	Kr	36	83.80	Tungsten	W	74	183.85
Lanthanum	La	57	138.9055	Uranium	U	92	238.0289
Lawrencium	Lr	103	(260)	Vanadium	V	23	50.9415
Lead	Pb	82	207.2	Xenon	Xe	54	131.29
Lithium	Li	3	6.941	Ytterbium	Yb	70	173.04
Lutetium	Lu	71	174.967	Yttrium	Y	39	88.9059
Magnesium	Mg	12	24.305	Zinc	Zn	30	65.39
Manganese	Mn	25	54.9380	Zirconium	Zr	40	91.224
Mendelevium	Md	101	(258)				